Lectures in Mathematics
ETH Zürich
Department of Mathematics
Research Institute of Mathematics

Managing Editor:
Michael Struwe

Philippe G. LeFloch

Hyperbolic Systems of Conservation Laws

The Theory of Classical and Nonclassical Shock Waves

Springer Basel AG

Author's address:

CNRS Director of Research
Centre de Mathématiques Appliquées
& Centre National de la Recherche Scientifique
Ecole Polytechnique
91128 Palaiseau, France
E-mail: lefloch@cmap.polytechnique.fr

2000 Mathematical Subject Classification 35L65, 35L40, 76L05, 74J40, 74N20

A CIP catalogue record for this book is available from the
Library of Congress, Washington D.C., USA

Deutsche Bibliothek Cataloging-in-Publication Data
LeFloch, Philippe G.:
Hyperbolic systems of conservation laws : the theory of classical and
nonclassical shock waves / Philippe G. LeFloch. - Basel ; Boston ; Berlin :
Birkhäuser, 2002
   (Lectures in mathematics : ETH Zürich)

   ISBN 978-3-7643-6687-2   ISBN 978-3-0348-8150-0 (eBook)
   DOI 10.1007/978-3-0348-8150-0

© 2002 Springer Basel AG
Originally published by Birkhäuser Verlag, Basel-Boston-Berlin in 2002
Printed on acid-free paper produced from chlorine-free pulp. TCF ∞

ISBN 978-3-7643-6687-2

9 8 7 6 5 4 3 2 1   www.birkhauser-science.com

To my wife Claire

To Olivier, Aline, Bruno

Contents

Part 2.   SYSTEMS OF CONSERVATION LAWS

Preface

This set of lecture notes was written for a *Nachdiplom-Vorlesungen* course given at the Forschungsinstitut für Mathematik (FIM), ETH Zürich, during the Fall Semester 2000. I would like to thank the faculty of the Mathematics Department, and especially Rolf Jeltsch and Michael Struwe, for giving me such a great opportunity to deliver the lectures in a very stimulating environment. Part of this material was also taught earlier as an advanced graduate course at the Ecole Polytechnique (Palaiseau) during the years 1995–99, at the Instituto Superior Tecnico (Lisbon) in the Spring 1998, and at the University of Wisconsin (Madison) in the Fall 1998. This project started in the Summer 1995 when I gave a series of lectures at the Tata Institute of Fundamental Research (Bangalore).

One main objective in this course is to provide a self-contained presentation of the **well-posedness theory** for nonlinear hyperbolic systems of first-order partial differential equations in divergence form, also called **hyperbolic systems of conservation laws.** Such equations arise in many areas of continuum physics when fundamental balance laws are formulated (for the mass, momentum, total energy... of a fluid or solid material) and small-scale mechanisms can be neglected (which are induced by viscosity, capillarity, heat conduction, Hall effect...). Solutions to hyperbolic conservation laws exhibit singularities (shock waves), which appear in finite time even from smooth initial data. As is now well-established from pioneering works by Dafermos, Kruzkov, Lax, Liu, Oleinik, and Volpert, weak (distributional) solutions are not unique unless some **entropy condition** is imposed, in order to retain some information about the effect of "small-scales".

Relying on results obtained these last five years with several collaborators, I provide in these notes a complete account of the existence, uniqueness, and continuous dependence theory for the Cauchy problem associated with strictly hyperbolic systems with genuinely nonlinear characteristic fields. The mathematical theory of shock waves originates in Lax's foundational work. The existence theory goes back to Glimm's pioneering work, followed by major contributions by DiPerna, Liu, and others. The **uniqueness** of entropy solutions with bounded variation was established in 1997 in Bressan and LeFloch [2]. Three proofs of the **continuous dependence** property were announced in 1998 and three preprints distributed shortly thereafter; see [3,4,9]. The proof I gave in [4] was motivated by an earlier work ([6] and, in collaboration with Xin, [7]) on linear adjoint problems for nonlinear hyperbolic systems.

In this monograph I also discuss the developing theory of **nonclassical shock waves** for strictly hyperbolic systems whose characteristic fields are not genuinely nonlinear. Nonclassical shocks are fundamental in nonlinear elastodynamics and phase transition dynamics when capillarity effects are the main driving force behind their propagation. While classical shock waves are compressive, independent of small-scale regularization mechanisms, and can be characterized by an **entropy inequality**, nonclassical shocks are **undercompressive** and very sensitive to diffusive and dispersive mechanisms. Their unique selection requires a **kinetic relation**, as I call it following a terminology from material science (for hyperbolic-elliptic problems).

This book is intended to contribute and establish a *unified framework* encompassing both what I call here **classical** and **nonclassical entropy solutions.**

No familiarity with hyperbolic conservation laws is a priori assumed in this course. The well-posedness theory for classical entropy solutions of genuinely nonlinear systems is entirely covered by Chapter I (Sections 1 and 2), Chapter II (Sections 1 and 2), Chapter III (Section 1), Chapter IV (Sections 1 and 2), Chapter V (Sections 1 and 2), Chapter VI (Sections 1 and 2), Chapter VII, Chapter IX (Sections 1 and 2), and Chapter X. The other sections contain more advanced material and provide an introduction to the theory of nonclassical shock waves.

First, I want to say how grateful I am to Peter D. Lax for inviting me to New York University as a Courant Instructor during the years 1990–92 and for introducing me to many exciting mathematical people and ideas. I am particularly indebted to Constantine M. Dafermos for his warm interest to my research and his constant and very helpful encouragement over the last ten years. I also owe Robert V. Kohn for introducing me to the concept of kinetic relations in material science and encouraging me to read the preprint of the paper [1] and to write [6]. I am very grateful to Tai-Ping Liu for many discussions and his constant encouragement; his work [8] on the entropy condition and general characteristic fields was very influential on my research.

It is also a pleasure to acknowledge fruitful discussions with collaborators and colleagues during the preparation of this course, in particular from R. Abeyaratne, F. Asakura, P. Baiti, N. Bedjaoui, J. Knowles, B. Piccoli, M. Shearer, and M. Slemrod. I am particularly thankful to T. Iguchi and A. Mondoloni, who visited me as post-doc students at the Ecole Polytechnique and carefully checked the whole draft of these notes. Many thanks also to P. Goatin, M. Savelieva, and M. Thanh who pointed out misprints in several chapters.

Special thanks to Olivier (for taming my computer), Aline (for correcting my English), and Bruno (for completing my proofs). Last, but not least, this book would not exist without the daily support and encouragement from my wife Claire.

This work was partially supported by the Centre National de la Recherche Scientifique (CNRS) and the National Science Foundation (NSF).

Philippe G. LeFloch

[1] Abeyaratne A. and Knowles J.K., Kinetic relations and the propagation of phase boundaries in solids, Arch. Rational Mech. Anal. 114 (1991), 119–154.

[2] Bressan A. and LeFloch P.G., Uniqueness of entropy solutions for systems of conservation laws, Arch. Rational Mech. Anal. 140 (1997), 301–331.

[3] Bressan A., Liu T.-P., and Yang T., L^1 stability estimate for $n \times n$ conservation laws, Arch. Rational Mech. Anal. 149 (1999), 1–22.

[4] Hu J. and LeFloch P.G., L^1 continuous dependence for systems of conservation laws, Arch. Rational Mech. Anal. 151 (2000), 45–93.

[5] LeFloch P.G., An existence and uniqueness result for two non-strictly hyperbolic systems, in "Nonlinear evolution equations that change type", ed. B.L. Keyfitz and M. Shearer, IMA Vol. Math. Appl., Vol. 27, Springer Verlag, 1990, pp. 126–138.

[6] LeFloch P.G., Propagating phase boundaries: Formulation of the problem and existence via the Glimm scheme, Arch. Rational Mech. Anal. 123 (1993), 153–197.

[7] LeFloch P.G. and Xin Z.-P., Uniqueness via the adjoint problems for systems of conservation laws, Comm. Pure Appl. Math. 46 (1993), 1499–1533.

[8] Liu T.-P., Admissible solutions of hyperbolic conservation laws, Mem. Amer. Math. Soc. 30, 1981.

[9] Liu T.-P. and Yang T., Well-posedness theory for hyperbolic conservation laws, Comm. Pure Appl. Math. 52 (1999), 1553–1580.

CHAPTER I

FUNDAMENTAL CONCEPTS
AND EXAMPLES

In this first chapter, we present some basic definitions and concepts which will be of constant use in this course. We also discuss the main difficulties of the theory and briefly indicate the main results to be established in the forthcoming chapters.

1. Hyperbolicity, genuine nonlinearity, and entropies

We are interested in **systems of N conservation laws** in one-space dimension:

$$\partial_t u + \partial_x f(u) = 0, \quad u(x,t) \in \mathcal{U}, \quad x \in \mathbb{R},\, t > 0, \tag{1.1}$$

where $\mathcal{U}$ is an open and convex subset of $\mathbb{R}^N$ and $f : \mathcal{U} \to \mathbb{R}^N$ is a smooth mapping called the **flux-function** associated with (1.1). In the applications x and t correspond to space and time coordinates, respectively. The dependent variable u is called the **conservative variable.** To formulate the Cauchy problem for (1.1) one prescribes an initial condition at $t = 0$:

$$u(x,0) = u_0(x), \quad x \in \mathbb{R}, \tag{1.2}$$

where the function $u_0 : \mathbb{R} \to \mathcal{U}$ is given. In this section, we restrict attention to smooth solutions of (1.1) which are *continuously differentiable*, at least.

Observe that (1.1) is written in divergence (or conservative) form. Hence, by applying Green's formula on some rectangle $(x_1, x_2) \times (t_1, t_2)$ we obtain

$$\int_{x_1}^{x_2} u(x, t_2)\, dx = \int_{x_1}^{x_2} u(x, t_1)\, dx - \int_{t_1}^{t_2} f(u(x_2, t))\, dt + \int_{t_1}^{t_2} f(u(x_1, t))\, dt. \tag{1.3}$$

In models arising in continuum physics (compressible fluid dynamics, nonlinear elastodynamics, phase transition dynamics) the conservation laws (1.1) are in fact *deduced* from the local balance equations (1.3) which represent fundamental physical principles: conservation laws of mass, momentum, energy, etc. (Examples will be presented in Section 4, below.)

When $\lim_{|x| \to +\infty} u(x,t) = 0$ and the flux is normalized so that $f(0) = 0$, we can let $x_1 \to -\infty$ and $x_2 \to +\infty$ in (1.3) and obtain

$$\int_{-\infty}^{+\infty} u(x,t)\, dx = \int_{-\infty}^{+\infty} u_0(x)\, dx, \quad t \geq 0.$$

Hence, the integral of the solution on the whole space (that is, the total mass, momentum, energy, etc. in the applications) is independent of time.

DEFINITION 1.1. (Hyperbolic systems.) We say that (1.1) is a **first-order, hyperbolic system** of partial differential equations if the Jacobian matrix $A(u) := Df(u)$ admits N real eigenvalues

$$\lambda_1(u) \leq \lambda_2(u) \leq \ldots \leq \lambda_N(u), \quad u \in \mathcal{U},$$

together with a basis of right-eigenvectors $\{r_j(u)\}_{1 \leq j \leq N}$. The eigenvalues are also called the wave speeds or **characteristic speeds** associated with (1.1). The system is said to be strictly hyperbolic if its eigenvalues are distinct:

$$\lambda_1(u) < \lambda_2(u) < \ldots < \lambda_N(u), \quad u \in \mathcal{U}.$$

□

In other words, we have $Df(u)\, r_j(u) = \lambda_j(u)\, r_j(u)$. The pair (λ_j, r_j) is referred to as the j-**characteristic field**. It is assumed that $u \mapsto \lambda_j(u), r_j(u)$ are smooth mappings which for strictly hyperbolic systems follows from the regularity assumption already made on f. For strictly hyperbolic systems the eigenvectors are defined up to a multiplicative constant and, denoting by $\{l_j(u)\}_{1 \leq j \leq N}$ a basis of left-eigenvectors, we will often impose the normalization

$$l_i(u)\, r_j(u) \equiv \delta_{ij},$$

where δ_{ij} is the Kronecker symbol. By convention, $r_j(u)$ is a row-vector while $l_i(u)$ is a column-vector. The transpose of a matrix B is denoted by B^T, so the notation $l_i(u)^T$ stands for the associated row-vector. A dot is used to denote the Euclidian scalar product in $\mathbb{R}^N$. When $N = 1$, there is a single eigenvalue $\lambda_1(u) = f'(u)$ and we set $r_1 = l_1 = 1$.

EXAMPLE 1.2. *Linear advection equation.* When $N = 1$ and $f(u) = a\,u$ where the wave speed a is a constant, (1.1) reduces to the **linear advection** equation

$$\partial_t u + \partial_x (a\, u) = \partial_t u + a\, \partial_x u = 0. \tag{1.4}$$

It is well-known that the solution of the Cauchy problem (1.2) and (1.4) admits the following *explicit formula*:

$$u(x,t) = u_0(x - a\,t), \quad x \in \mathbb{R}, \, t \geq 0. \tag{1.5}$$

□

EXAMPLE 1.3. *Inviscid Burgers-Hopf's equation.* When $N = 1$ and $f(u) = u^2/2$ in (1.1) we arrive at the (inviscid) **Burgers equation**

$$\partial_t u + \partial_x (u^2/2) = \partial_t u + u\, \partial_x u = 0. \tag{1.6}$$

This is an important model for nonlinear wave propagation, originally derived by Burgers for the dynamics of (viscous and turbulent) fluids. Observe that, in (1.6), the wave speed $f'(u)$ truly depends on u.

Comparing (1.6) with (1.4) we are tempted to extend the formula (1.5) found for constant speeds and, for the solution of the nonlinear equation (1.6), to write down the now implicit formula

$$u(x,t) = u_0(x - u(x,t)\,t), \quad x \in \mathbb{R}, \, t \geq 0. \tag{1.7}$$

When the function u_0 is of class C^1 and $\partial_x u_0$ is uniformly bounded on $I\!R$ we have

$$\partial_v\big(v - u_0(x - v\,t)\big) = 1 + t\big(\partial_x u_0\big)(x - v\,t) > 0, \qquad (1.8)$$

for all sufficiently small t. Therefore, by the implicit function theorem, (1.7) defines a unique function u. Interestingly enough, an elementary calculation shows that this function is the (unique) solution of (1.2) and (1.6) (for small times, at least). □

When $N = 1$, Example 1.2 is the prototype of a *linear* equation (f' being a constant) and Example 1.3 is the prototype of a *genuinely nonlinear* equation (f' being strictly monotone). These two examples exhibit distinguished behaviors which are observed in systems, as well. Turning our attention to systems, we now introduce some notions of linearity and nonlinearity for each j-wave family. We will confirm later on that the key quantity here is the rate of change in the wave speed λ_j along the direction of the eigenvector r_j.

DEFINITION 1.4. For each $j = 1, \ldots, N$ we say that the j-characteristic field of (1.1) is **genuinely nonlinear** when

$$\nabla\lambda_j(u) \cdot r_j(u) \neq 0, \quad u \in \mathcal{U}$$

and **linearly degenerate** when

$$\nabla\lambda_j(u) \cdot r_j(u) \equiv 0, \quad u \in \mathcal{U}.$$

□

For genuinely nonlinear fields we will often impose the normalization

$$\nabla\lambda_j(u) \cdot r_j(u) \equiv 1, \quad u \in \mathcal{U},$$

for the general theory. But different normalizations are often more convenient when dealing with specific examples. In view of Definition 1.4, when $N = 1$ the equation (1.1) is genuinely nonlinear if and only if $f''(u) \neq 0$ for all u. It is linearly degenerate if and only if $f''(u) = 0$ for all u.

Definition 1.4 will often be used in connection with the integral curves associated with the system (1.1). By definition, an **integral curve** of the vector-field r_j is a solution $s \mapsto v(s)$ of the ordinary differential equation

$$v'(s) = r_j(v(s)). \qquad (1.9)$$

Relying on (1.9) we see that the j-characteristic field is genuinely nonlinear if

$$\lambda_j \text{ is strictly monotone along the integral curves } s \mapsto v(s),$$

and is linearly degenerate if

$$\lambda_j \text{ remains constant along the integral curves } s \mapsto v(s).$$

This observation sheds further light on Definition 1.4. The above two properties are natural extensions of similar properties already noticed for scalar equations in Examples 1.2 and 1.3.

It is important to keep in mind that, quite often in the applications, the examples of interest fail to be globally genuinely nonlinear. (See again Section 4.) The following scalar equation will serve to exhibit basic features of such models.

EXAMPLE 1.5. *Conservation law with cubic flux.* The equation

$$\partial_t u + \partial_x u^3 = 0, \tag{1.10}$$

is genuinely nonlinear in $\mathcal{U} := \{u > 0\}$ and in $\mathcal{U} := \{u < 0\}$ but fails to be so in any neighborhood of $u = 0$. $\qquad\square$

When the function f is linear the explicit formula derived in Example 1.2 for the linear advection equation extends easily to the system (1.1), as follows.

EXAMPLE 1.6. *Linear hyperbolic systems.* When $f(u) = A u$, where A is a constant matrix with real eigenvalues and a complete basis of eigenvectors, the system (1.1) is hyperbolic and has N linearly degenerate characteristic fields. By setting $\beta_j(x,t) := l_j u(x,t)$ $(1 \le j \le N)$ the **characteristic decomposition**

$$u(x,t) = \sum_{j=1}^{N} \beta_j(x,t)\, r_j$$

transforms (1.1) into N decoupled linear advection equations (Example 1.1) for the **characteristic variables** β_j:

$$\partial_t \beta_j + \lambda_j\, \partial_x \beta_j = 0, \quad 1 \le j \le N.$$

In view of (1.4) and (1.5) we see immediately that the solution of the corresponding Cauchy problem (1.1) and (1.2) is given by the explicit formula

$$u(x,t) = \sum_{j=1}^{N} \beta_j^0(x - \lambda_j\, t)\, r_j, \quad x \in I\!R,\, t \ge 0, \tag{1.11}$$

where the coefficients $\beta_j^0 := l_j u_0$, $1 \le j \le N$, are determined from the data u_0. $\quad\square$

Now, turning our attention to the nonlinear system (1.1) and for a given continuously differentiable solution $u(x,t)$ we attempt to repeat the calculation in Example 1.6. The variable u cannot be used directly for nonlinear equations, but we can decompose (1.1) into N scalar nonlinear equations for the **characteristic variables** defined now by $\alpha_j := l_j(u)\, \partial_x u$, i.e.,

$$\partial_x u(x,t) = \sum_{j=1}^{N} \alpha_j(x,t)\, r_j(u(x,t)). \tag{1.12}$$

Indeed, differentiating (1.1) with respect to x and using (1.12) we obtain

$$\partial_t \sum_{j=1}^{N} \alpha_j\, r_j(u) + \partial_x \sum_{j=1}^{N} \alpha_j\, \lambda_j(u)\, r_j(u) = 0.$$

Multiplying the latter by each left-eigenvector $l_i(u)$ and using the normalization $l_i(u)\, r_j(u) \equiv \delta_{ij}$, we arrive at

$$\partial_t \alpha_i + \partial_x\big(\lambda_i(u)\, \alpha_i\big) = \sum_{1 \le j < k \le N} G_{ijk}(u)\, \alpha_j\, \alpha_k, \quad 1 \le i \le N, \tag{1.13}$$

where the right-hand side depends on the **interaction coefficients**

$$G_{ijk}(u) := \big(\lambda_j(u) - \lambda_k(u)\big)\, l_i(u)\, \big[r_j(u), r_k(u)\big], \quad 1 \le i,j,k \le N, \tag{1.14}$$

the **Poisson bracket** being defined by

$$[r_j(u), r_k(u)] := Dr_k(u)r_j(u) - Dr_j(u)r_k(u).$$

In the special case that all of the coefficients $G_{ijk}(u)$ vanish, (1.13) provides N decoupled equations for the characteristic variables α_i. This is the case in Example 1.6 where $[r_j, r_k] \equiv 0$ for all j, k, and, trivially, in the case of scalar conservation laws. However, in most examples of interest with $N > 1$ no such decoupling arises and one of the main difficulties in extending to systems the arguments of proof known for scalar equations is to cope with the quadratic terms $G_{ijk}(u) \alpha_j \alpha_k$ in (1.13).

In the rest of this section we discuss a fundamental notion of the mathematical theory which will be essential to investigate the properties of weak solutions of systems of conservation laws (in Section 3, below).

DEFINITION 1.7. (Mathematical entropies.) A smooth function $(U, F) : \mathcal{U} \to I\!\!R^2$ is called an **entropy pair** if any continuously differentiable solution of (1.1) satisfies the additional conservation law

$$\partial_t U(u) + \partial_x F(u) = 0.$$

The functions U and F are called **entropy** and **entropy-flux**, respectively.   □

Attempting to pre-multiply (1.1) by $\nabla U(u)^T$ it becomes clear that (U, F) is an entropy pair if and only if

$$\nabla F(u)^T = \nabla U(u)^T Df(u), \quad u \in \mathcal{U}.$$

By differentiation with respect to u we obtain equivalently

$$D^2 F(u) = D^2 U(u) Df(u) + \nabla U(u)^T D^2 f(u).$$

Since $D^2 F(u)$ and $\nabla U(u)^T D^2 f(u)$ are symmetric matrices (which is obvious for the first one and can be checked for the second one by writing the matrix product component by component), we see that the matrix $D^2 U(u) Df(u)$ must be symmetric. (For the converse, one relies on the fact that the set of definition $\mathcal{U}$ is convex and, therefore, connected.) This discussion leads us to a useful criterion for the existence of an entropy, summarized as follows.

THEOREM 1.8. (Characterization of the mathematical entropies.) *A smooth function U is an entropy if and only if*

$$D^2 U(u) Df(u) \quad \text{is a symmetric } N \times N \text{ matrix,} \tag{1.15}$$

which is equivalent to a linear system of $N(N-1)/2$ second-order partial differential equations.   □

For each $j = 1, \dots, N$ a (trivial) entropy pair is defined by

$$U(u) = u_j, \quad F(u) = f_j(u), \quad u \in \mathcal{U}, \tag{1.16}$$

where $u = (u_1, u_2, \dots, u_N)^T$ and $f(u) = (f_1(u), \dots, f_N(u))^T$. However, mathematical entropies of interest should be truly nonlinear in the conservative variable u. A central role will be played by entropies that are **strictly convex**, in the sense that $D^2 U(u)$ is a positive definite symmetric matrix,

$$D^2 U(u) > 0, \quad u \in \mathcal{U},$$

which implies

$$U(u) - U(v) - \nabla U(v) \cdot (u - v) > 0, \quad u \neq v \text{ in } \mathcal{U}.$$

Definition 1.7 is illustrated now with some examples.

EXAMPLE 1.9. *Scalar conservation laws.* When $N = 1$ and $\mathcal{U} = I\!R$ (1.15) imposes no restriction on U, so that any (strictly convex) function $U : \mathcal{U} \to I\!R$ is a (strictly convex) mathematical entropy. The entropy flux F is given by $F'(u) = U'(u) f'(u)$, i.e.,

$$F(u) := F(a) + \int_a^u U'(v) f'(v) \, dv$$

with $a \in \mathcal{U}$ fixed and $F(a)$ chosen arbitrarily.   □

EXAMPLE 1.10. *Decoupled scalar equations.* Consider a system (1.1) of the form

$$u = (u_1, u_2, \ldots, u_N)^T, \quad f(u) = (f_1(u_1), f_2(u_2), \ldots, f_N(u_N))^T.$$

(For example, the linear systems in Example 1.6 have this form if they are written in the characteristic variables.) Such a system is always hyperbolic, with $\lambda_j(u) = f'_j(u_j)$, and the basis $\{r_j(u)\}_{1 \leq j \leq N}$ can be chosen to be the canonical basis of $I\!R^N$. The system is non-strictly hyperbolic, unless for some permutation σ of $\{1, 2, \ldots, N\}$ we have

$$f'_{\sigma(1)}(u_{\sigma(1)}) < f'_{\sigma(2)}(u_{\sigma(2)}) < \ldots < f'_{\sigma(N)}(u_{\sigma(N)}), \quad u \in \mathcal{U}.$$

The j-characteristic field is genuinely nonlinear (respectively linearly degenerate) if and only if $f''_j(u_j) \neq 0$ for all $u \in \mathcal{U}$ (resp. $f''_j(u_j) \equiv 0$ for all $u \in \mathcal{U}$). A class of mathematical entropies is described by the general formula $U(u) := \sum_{j=1}^N U_j(u_j)$ where the functions U_j are arbitrary. All of the entropies have this form if, for instance, all the fields are genuinely nonlinear. The interaction coefficients (1.14) vanish identically.   □

EXAMPLE 1.11. *Symmetric systems.* Consider next

$$\partial_t u + \partial_x f(u) = 0, \quad Df(u) \text{ symmetric}, \tag{1.17}$$

which is hyperbolic but need not be strictly hyperbolic. Since the Jacobian Df is symmetric it coincides with the Hessian matrix of some scalar-valued mapping $\psi : \mathcal{U} \to I\!R$. (Recall here that $u \in \mathcal{U}$ where $\mathcal{U}$ is convex and therefore connected.) Thus $f = \nabla \psi$ and a straightforward calculation shows that

$$U(u) = \frac{|u|^2}{2}, \quad F(u) = \nabla \psi(u) \cdot u - \psi(u) \tag{1.18}$$

is a strictly convex entropy pair of (1.17).   □

More generally, given any system of two conservation laws (i.e., $N = 2$) the condition (1.15) reduces to a single linear hyperbolic partial differential equation of the second order. (See the typical equation (4.9) in Section 4, below.) Based on standard existence theorem for such equations, one can prove that any strictly hyperbolic system of two conservation laws admits a *large family* of non-trivial, entropy pairs. (See the bibliographical notes.)

For systems with three equations at least, (1.15) is generally over-determined so that an arbitrary system of N conservation laws need not admit a non-trivial mathematical entropy. The notion in Definition 1.7 plays an important role however, as every system *arising in continuum physics* and derived from physical conservation principles always admits *one mathematical entropy pair* at least, which often (but not always) is *strictly convex* in the conservative variable. Therefore, in this course, we will restrict attention mostly to strictly hyperbolic systems of conservation laws endowed with a strictly convex mathematical entropy pair.

To close this section we observe that:

THEOREM 1.12. (Symmetrization of systems of conservation laws.) *Any system endowed with a strictly convex entropy pair* (U, F) *may be put in the* **symmetric form**

$$\partial_t g(\hat{u}) + \partial_x h(\hat{u}) = 0,$$
$$Dg(\hat{u}), \; Dh(\hat{u}) \; symmetric, \quad Dg(\hat{u}) \; positive \; definite. \tag{1.19}$$

Conversely, any system of the form (1.19) *can be written in the general form* (1.1) *and admits a strictly convex mathematical entropy pair.*

We prove Theorem 1.12 as follows. On one hand consider the so-called **entropy variable**

$$u \mapsto \hat{u} := \nabla U(u), \tag{1.20}$$

which is a one-to-one change of variable since U is strictly convex. Let us rewrite the conservative variable and the flux in terms of the entropy variable:

$$u = g(\hat{u}), \quad f(u) = h(\hat{u}).$$

It is easily checked that

$$D_{\hat{u}} g(\hat{u}) = \left(D_u^2 U(u) \right)^{-1}, \quad D_{\hat{u}} h(\hat{u}) = D_u f(u) \left(D_u^2 U(u) \right)^{-1}, \tag{1.21}$$

in which the first matrix is clearly symmetric and the second matrix is symmetric thanks to (1.15). This proves (1.19).

On the other hand, given a system of the form (1.19), since $Dg(\hat{u})$ and $Dh(\hat{u})$ are symmetric matrices there exist two scalar-valued functions ϕ and ψ such that

$$g(\hat{u}) = \nabla \phi(\hat{u}), \quad h(\hat{u}) = \nabla \psi(\hat{u}). \tag{1.22}$$

We claim that the **Legendre transform** (G, H) of (ϕ, ψ), defined as usual by

$$G(\hat{u}) := \nabla \phi(\hat{u}) \cdot \hat{u} - \phi(\hat{u}), \quad H(\hat{u}) := \nabla \psi(\hat{u}) \cdot \hat{u} - \psi(\hat{u}), \tag{1.23}$$

is an entropy pair for the system (1.19). Indeed, this follows from

$$\partial_t G(\hat{u}) = \hat{u} \cdot \partial_t \nabla \phi(\hat{u}), \quad \partial_x H(\hat{u}) = \hat{u} \cdot \partial_x \nabla \psi(\hat{u}),$$

and thus, with (1.19),

$$\partial_t G(\hat{u}) + \partial_x H(\hat{u}) = 0.$$

The function $G(\hat{u})$ is strictly convex *in the variable* u since $D_u^2 G(\hat{u}) = Dg(\hat{u})$ is positive definite. Note also that $G(\hat{u}) = U(u)$ and $H(\hat{u}) = F(u)$. This proves the converse statement in Theorem 1.12. $\qquad \square$

2. Shock formation and weak solutions

The existence and the uniqueness of (locally defined in time) smooth solutions of strictly hyperbolic systems of conservation laws follow from standard compactness arguments in Sobolev spaces. Generally speaking, smooth solutions $u = u(x,t)$ of nonlinear hyperbolic equations eventually loose their regularity at some finite critical time, at which the derivative $\partial_x u$ tends to infinity. This breakdown of smooth solutions motivates the introduction of the concept of weak solutions, which allows us to deal with discontinuous solutions of (1.1) such as shock waves.

First of all, in order to clarify the blow-up mechanism, we study the typical case of Burgers equation, introduced in Example 1.3, and we discuss three different approaches demonstrating the non-existence of smooth solutions. Let $u = u(x,t)$ be a continuously differentiable solution of (1.6) satisfying the initial condition (1.2) for some smooth function u_0. Suppose that this solution is defined for small times t, at least.

APPROACH BASED ON THE IMPLICIT FUNCTION THEOREM. Following the discussion in Example 1.3, observe that the implicit function theorem fails to apply to (1.7) when t is too large. More precisely, it is clear that the condition (1.8) always fails if t is sufficiently large, except when

$$u_0 \text{ is a } non\text{-}decreasing \text{ function.} \qquad (2.1)$$

When (2.1) is satisfied, the transformation $v \mapsto v - u_0(x - v\,t)$ remains one-to-one for all times and (1.7) provides the unique solution of (1.6) and (1.2), globally defined in time.

GEOMETRIC APPROACH. Given $y_0 \in I\!R$, the **characteristic curve** $t \mapsto y(t)$ issuing from y_0 is defined (locally in time, at least) by

$$\begin{aligned} y'(t) &= u\big(y(t),t\big), \quad t \geq 0, \\ y(0) &= y_0. \end{aligned} \qquad (2.2)$$

The point y_0 is referred to as the **foot of the characteristic**. Setting

$$v(t) := u\big(y(t),t\big)$$

and using (1.6) and (2.2) one obtains

$$v'(t) = 0.$$

So, the solution is actually *constant* along the characteristic which, therefore, must be a *straight line*. It is geometrically clear that two of these characteristic lines will eventually intersect at some latter time, *except* if the u_0 satisfies (2.1) and the characteristics spread away from each other and never cross.

APPROACH BASED ON THE DERIVATIVE $\partial_x u$. Finally, we show the connection with the well-known blow-up phenomena arising in solutions of ordinary differential equations. Given a smooth solution u, consider its space derivative $\partial_x u$ along a characteristic $t \mapsto y(t)$, that is, set

$$w(t) := (\partial_x u)\big(y(t),t\big).$$

Using that $\partial_{xt} u + u \partial_{xx} u = -(\partial_x u)^2$ we obtain **Riccati equation**

$$w'(t) = -w(t)^2. \tag{2.3}$$

Observe that the right-hand side of this ordinary differential equation is quadratic and that

$$w(t) = \frac{w(0)}{1 + t \, w(0)}.$$

So, w tends to infinity in finite time, *except* if $w(0) \geq 0$ which is, once more, the condition (2.1).

REMARK 2.1. General theorems on blow-up of smooth solutions with small amplitude are known for strictly hyperbolic systems. The proofs rely on the decomposition (1.12)–(1.14) and, in essence, extend to systems the third approach above presented on Burgers equation. Let us just sketch this strategy for a system with genuinely nonlinear fields. Given $y_0 \in \mathbb{R}$, the **i-characteristic curve** issuing from the point y_0 at the time $t = 0$ is, by definition, the solution of the ordinary differential equation

$$\begin{aligned} y'(t) &= \lambda_i(u(y(t), t)), \quad t \geq 0, \\ y(0) &= y_0. \end{aligned} \tag{2.4}$$

Using the notation (1.12) and setting

$$w(t) = \alpha_i(y(t), t),$$

we deduce from (1.13) the **generalized Riccati equation**

$$w'(t) = a(t) \, w(t)^2 + b(t) \, w(t) + c(t) \tag{2.5}$$

with

$$a(t) := -\nabla \lambda_i(u(t)) \cdot r_i(u(t)), \quad b(t) := -\sum_{\substack{k=1 \\ k \neq i}}^{N} \alpha_k(t) \, r_k(u(t)) \cdot \nabla \lambda_i(u(t)),$$

$$c(t) := \sum_{1 \leq j < k \leq N} G_{ijk}(u(t)) \, \alpha_j(t) \, \alpha_k(t).$$

Here, $u(t) := u(y(t), t)$ and $\alpha_j(t) := \alpha_j(y(t), t)$. For genuinely nonlinear fields, after normalization we have $a(t) \equiv -1$, the first term in the right-hand side of (2.5) coincides the right-hand side of (2.3). A rigorous proof of the breakdown for systems requires careful estimates for the remaining terms (particularly, $\alpha_j \alpha_k$ with j, k not both equal to i) in order to establish that one of the i-characteristic components α_i blows-up. (See the bibliographical notes for a reference.) □

From the discussion above we conclude that the class of solutions must be enlarged and should include solutions that are not continuously differentiable and, in fact, are discontinuous. We consider solutions in the space $L^\infty(\mathbb{R} \times \mathbb{R}_+, \mathcal{U})$ of bounded Lebesgue measurable functions $u : \mathbb{R} \times \mathbb{R}_+ \to \mathcal{U}$.

DEFINITION 2.2. (Concept of weak solution.) Given some initial data $u_0 \in L^\infty(I\!R, \mathcal{U})$ we shall say that $u \in L^\infty(I\!R \times I\!R_+, \mathcal{U})$ is a **weak solution** to the Cauchy problem (1.1) and (1.2) if

$$\int_0^{+\infty} \int_{I\!R} \left(u \, \partial_t \theta + f(u) \, \partial_x \theta\right) dx dt + \int_{I\!R} \theta(0) \, u_0 \, dx = 0 \qquad (2.6)$$

for every function $\theta \in C_c^\infty(I\!R \times [0, +\infty))$ (the vector space of real-valued, compactly supported, and infinitely differentiable functions).

Of course, if u is a continuously differentiable solution of (1.1) in the usual sense, then by Green's formula it is also a weak solution. The interest of the definition (2.6) is that it allows u to be a discontinuous function. To construct weak solutions explicitly we will often apply the following criterion.

THEOREM 2.3. (Rankine-Hugoniot jump relations.) *Consider a piecewise smooth function $u : I\!R \times I\!R_+ \to \mathcal{U}$ of the form*

$$u(x, t) = \begin{cases} u_-(x, t), & x < \varphi(t), \\ u_+(x, t), & x > \varphi(t), \end{cases} \qquad (2.7)$$

where, setting $\Omega_\pm := \{x \gtrless \varphi(t)\}$, the functions $u_\pm : \overline{\Omega}_\pm \to \mathcal{U}$ and $\varphi : I\!R_+ \to I\!R$ are continuously differentiable. Then, u is a weak solution of (1.1) if and only if it is a solution in the usual sense in both regions where it is smooth and, furthermore, the following **Rankine-Hugoniot relation** *holds along the curve φ:*

$$-\varphi'(t) \left(u_+(t) - u_-(t)\right) + f(u_+(t)) - f(u_-(t)) = 0, \qquad (2.8)$$

where

$$u_-(t) := \lim_{\substack{\varepsilon \to 0 \\ \varepsilon > 0}} u_-(\varphi(t) - \varepsilon, t), \quad u_+(t) := \lim_{\substack{\varepsilon \to 0 \\ \varepsilon > 0}} u_+(\varphi(t) + \varepsilon, t).$$

PROOF. Given any function θ in $C_c^\infty(I\!R \times (0, +\infty))$ let us rewrite (2.6) in the form

$$\sum_\pm \iint_{\Omega_\pm} \left(u_\pm(x, t) \, \partial_t \theta(x, t) + f(u_\pm(x, t)) \, \partial_x \theta(x, t)\right) dx dt = 0.$$

Applying Green's formula in each region of smoothness $\Omega_\pm$ we obtain

$$\sum_\pm \pm \int_0^{+\infty} \left(-\varphi'(t) \, u_\pm(t) + f(u_\pm(t))\right) \theta(\varphi(t), t) \, dt = 0,$$

which gives (2.8) since θ is arbitrary. $\square$

When u_- and u_+ are constants and φ is linear, say $\varphi(t) = \lambda t$,

$$u(x, t) = \begin{cases} u_-, & x < \lambda t, \\ u_+, & x > \lambda t, \end{cases} \qquad (2.9)$$

Theorem 2.3 implies that (2.9) is a weak solution of (1.1) if and only if the vectors $u_\pm$ and the scalar λ satisfy the Rankine-Hugoniot relation

$$-\lambda \left(u_+ - u_-\right) + f(u_+) - f(u_-) = 0. \qquad (2.10)$$

When $u_- \neq u_+$ the function in (2.9) is called the **shock wave** connecting u_- to u_+ and λ the corresponding **shock speed**.

We will be particularly interested in the **Riemann problem** which is a special Cauchy problem (1.1) and (1.2) corresponding to piecewise constant initial data, i.e.,

$$u(x,0) = u_0(x) = \begin{cases} u_l, & x < 0, \\ u_r, & x > 0, \end{cases} \qquad (2.11)$$

where u_l, $u_r \in \mathcal{U}$ are constants. This problem is central in the theory as it exhibits many important features encountered with general solutions of (1.1) as well. The Riemann solutions will also serve to construct approximation schemes to generate solutions of the general Cauchy problem. At this juncture observe that (2.9) and (2.10) already provide us with a large class of solutions for the Riemann problem (1.1) and (2.11). As we will see, the shock waves do not suffice to solve the Riemann problem and we will also introduce later on another class of solutions, the **rarefaction waves**, which are Lipschitz continuous solutions of (1.1) generated by the integral curves (1.9). We refer to Chapters II and VI below for the explicit construction of the solution of the Riemann problem, for scalar equations and systems respectively, under various assumptions on the flux of (1.1).

When attempting to solve the Riemann problem one essential difficulty of the theory arises immediately. Weak solutions are *not uniquely* determined by their initial data. To illustrate this point we exhibit two typical initial data for which several weak solutions may be found.

EXAMPLE 2.4. *Non-uniqueness for Burgers equation (increasing data).* Observe that, in view of (2.10), a shock wave connecting u_- to u_+ ($u_- \neq u_+$) and propagating at the speed λ satisfies the Rankine-Hugoniot relation for Burgers equation (1.6) if and only if

$$\lambda = \frac{u_- + u_+}{2}. \qquad (2.12)$$

Hence, the Cauchy problem (1.2) and (1.6) with the initial condition

$$u(x,0) = u_0(x) := \begin{cases} -1, & x < 0, \\ 1, & x > 0, \end{cases}$$

admits the (steady) solution

$$u(x,t) = u_0(x) \quad \text{for all } (x,t). \qquad (2.13)$$

It also admits another solution,

$$u(x,t) = \begin{cases} -1, & x < -t, \\ x/t, & -t < x < t, \\ 1, & x > t, \end{cases} \qquad (2.14)$$

which is a *continuous* function of (x,t) in the region $\{t > 0\}$.   □

EXAMPLE 2.5. *Non-uniqueness for Burgers equation (decreasing data).* A Riemann problem may even admit infinitely many solutions. Consider, for instance, the initial condition

$$u_0(x) := \begin{cases} 1, & x < 0, \\ -1, & x > 0. \end{cases}$$

The corresponding Cauchy problem admits the trivial solution (2.13) again, as well as the following *one-parameter family* of solutions:

$$u(x,t) = \begin{cases} 1, & x < -\lambda t, \\ -v, & -\lambda t < x < 0, \\ v, & 0 < x < \lambda t, \\ -1, & x > \lambda t, \end{cases} \qquad (2.15)$$

where $v > 1$ is arbitrary and $\lambda := (v-1)/2$. Here, the initial jump is split into three propagating jumps. For the same Cauchy problem, the formula (2.15) can be easily generalized and solutions having an arbitrary large number of jumps could be also constructed.   □

EXAMPLE 2.6. *Non-existence of weak solutions conserving both u and u^2.* Consider a shock wave connecting u_- to u_+ at the speed λ given by (2.12). We claim that the (additional) conservation law

$$\partial_t \left(\frac{u^2}{2}\right) + \partial_x \left(\frac{u^3}{3}\right) = 0,$$

satisfied by smooth solutions of Burgers equation, cannot be satisfied by *weak* solutions. Otherwise, according to Theorem 2.3 (where u and $f(u)$ should be replaced with $u^2/2$ and $u^3/3$, respectively) we would have

$$\lambda = \frac{2}{3} \frac{u_+^3 - u_-^3}{u_+^2 - u_-^2} = \frac{2}{3} \frac{u_+^2 + u_- u_+ + u_-^2}{u_+ + u_-},$$

contradicting (2.12) if $u_- \neq u_+$.   □

To conclude this section, let us mention that one of the main objectives in this course will be to establish the existence of weak solutions (in a suitable sense to be discussed) to the Cauchy problem (1.1) and (1.2), and to prove uniqueness and continuous dependence results. In particular, we will derive the L^1 **continuous dependence estimate** for any two "solutions" u and v (C being a fixed positive constant)

$$\|u(t_2) - v(t_2)\|_{L^1(\mathbb{R})} \leq C \, \|u(t_1) - v(t_1)\|_{L^1(\mathbb{R})}, \quad 0 \leq t_1 \leq t_2. \qquad (2.16)$$

- The existence of solutions is established in Chapters IV, VII, and VIII, below. See Theorems IV-1.1, IV-2.1, and IV-3.2 and Theorem VIII-1.7 for scalar equations and Theorems VII-2.1 and VIII-3.1 for systems.
- The continuous dependence of solutions is the subject of Chapters V and IX. See Theorems V-2.2, V-3.1, V-3.2, and V-4.2 for scalar equations and Theorems IX-2.3, IX-3.2, and IX-4.1 for systems.
- The uniqueness is established in Chapter X. See Theorems X-3.2, X-4.1, and X-4.3.

3. Singular limits and the entropy inequality

Examples 2.4 and 2.5 show that weak solutions of the Cauchy problem (1.1) and (1.2) are generally non-unique. To single out the solution of interest, we restrict attention to weak solutions realizable as limits ($\varepsilon \to 0$) of smooth solutions of an augmented system

$$\partial_t u^\varepsilon + \partial_x f(u^\varepsilon) = R^\varepsilon, \quad u^\varepsilon = u^\varepsilon(x,t), \tag{3.1}$$

where $\varepsilon > 0$ represents a small-scale parameter corresponding, in the applications, to the viscosity, capillarity, etc. of the physical medium under consideration. The right-hand side of (3.1) may contain a singular regularization R^ε depending upon $u^\varepsilon, \varepsilon\, u^\varepsilon_x, \varepsilon^2\, u^\varepsilon_{xx}, \dots$ and (in a sense clarified by Definition 3.1 below) vanishing when $\varepsilon \to 0$.

We always assume that the system of conservation laws (1.1) is endowed with a strictly convex entropy pair (U, F) and that the singular limit

$$u = \lim_{\varepsilon \to 0} u^\varepsilon \tag{3.2}$$

exists in a sufficiently strong sense. Precisely, there exists a constant $C > 0$ independent of ε such that ($\mathcal{U}$ contains the closed ball with center 0 and radius C and)

$$\|u^\varepsilon\|_{L^\infty(\mathbb{R}\times\mathbb{R}_+)} \leq C \tag{3.3}$$

and the convergence (3.2) holds almost everywhere in (x, t). To arrive to a well-posed Cauchy problem for the hyperbolic system (1.1) we attempt to derive some conditions satisfied by the limit u, which are expected to characterize it among all of the weak solutions of (1.1).

In the present section, under some natural conditions on the smoothing term R^ε, we derive the so-called *entropy inequality* associated with the entropy pair (U, F) (that is, (3.8) below). The entropy inequality plays a fundamental role in the mathematical theory for (1.1). As will be further discussed in Section 5, it does not always completely characterize the limit of (3.1), however.

First of all, we wish that the limit u be a weak solution of (1.1). For each function $\theta \in C_c^\infty(\mathbb{R} \times (0, +\infty))$, relying on (3.1)–(3.3) we find

$$\iint_{\mathbb{R}\times\mathbb{R}_+} \left(u\,\partial_t\theta + f(u)\,\partial_x\theta\right) dx dt = \lim_{\varepsilon \to 0} \iint_{\mathbb{R}\times\mathbb{R}_+} \left(u^\varepsilon\,\partial_t\theta + f(u^\varepsilon)\,\partial_x\theta\right) dx dt$$

$$= \lim_{\varepsilon \to 0} \iint_{\mathbb{R}\times\mathbb{R}_+} R^\varepsilon\,\theta\, dx dt.$$

Therefore, we arrive at the following condition on R^ε which is necessary (and sufficient) for the limit u to be weak solution of (1.1).

DEFINITION 3.1. (Conservative regularization.) The right-hand side R^ε of (3.1) is said to be **conservative** (in the limit $\varepsilon \to 0$) if

$$\lim_{\varepsilon \to 0} \iint_{\mathbb{R}\times\mathbb{R}_+} R^\varepsilon\,\theta\, dx dt = 0, \quad \theta \in C_c^\infty(\mathbb{R} \times (0, +\infty)). \tag{3.4}$$

□

Next, we take advantage of the existence of an entropy pair (U, F). Multiplying (3.1) by $\nabla U(u^\varepsilon)$ we observe that, according to Definition 1.7, the left-hand side admits a *conservative form*, namely

$$\partial_t U(u^\varepsilon) + \partial_x F(u^\varepsilon) = \nabla U(u^\varepsilon) \cdot R^\varepsilon. \tag{3.5}$$

In view of (3.2) and (3.3) the left-hand side of (3.5) converges in the weak sense: For all $\theta \in C_c^\infty\big(\mathbb{R} \times (0, +\infty)\big)$ we have

$$\iint_{\mathbb{R} \times \mathbb{R}_+} \big(U(u^\varepsilon)\, \partial_t \theta + F(u^\varepsilon)\, \partial_x \theta\big)\, dx dt \to \iint_{\mathbb{R} \times \mathbb{R}_+} \big(U(u)\, \partial_t \theta + F(u)\, \partial_x \theta\big)\, dx dt.$$

To deal with the right-hand side of (3.5), we introduce the following definition.

DEFINITION 3.2. (Entropy dissipative regularization.) The right-hand side R^ε of (3.1) is said to be **entropy dissipative** for the entropy U (in the limit $\varepsilon \to 0$) if

$$\limsup_{\varepsilon \to 0} \iint_{\mathbb{R} \times \mathbb{R}_+} \nabla U(u^\varepsilon) \cdot R^\varepsilon\, \theta\, dx dt \le 0, \quad \theta \in C_c^\infty\big(\mathbb{R} \times (0, +\infty)\big), \quad \theta \ge 0. \quad (3.6)$$

$\square$

We summarize our conclusions as follows.

THEOREM 3.3. (Derivation of the entropy inequality.) *Let u^ε be a family of approximate solutions given by (3.1). Suppose that u^ε remains bounded in the L^∞ norm as $\varepsilon \to 0$ and converges almost everywhere towards a limit u; see (3.2) and (3.3). Suppose also that the right-hand side R^ε of (3.1) is conservative (see (3.4)) and entropy dissipative (see (3.6)) for some entropy pair (U, F) of (1.1). Then, u is a weak solution of (1.1) and satisfies the inequality*

$$\iint_{\mathbb{R} \times \mathbb{R}_+} \big(U(u)\, \partial_t \theta + F(u)\, \partial_x \theta\big)\, dx dt \ge 0, \quad \theta \in C_c^\infty\big(\mathbb{R} \times (0, +\infty)\big), \quad \theta \ge 0. \quad (3.7)$$

$\square$

By definition, (3.7) means that in the weak sense

$$\partial_t U(u) + \partial_x F(u) \le 0, \quad (3.8)$$

which is called the **entropy inequality** associated with the pair (U, F). In the following we shall say that a weak solution satisfying (3.8) is an **entropy solution.**

In the rest of this section we check the assumptions (3.4) and (3.6) for two classes of regularizations (3.1). The uniform bound (3.3) is assumed from now on. Consider first the **nonlinear diffusion model**

$$\partial_t u^\varepsilon + \partial_x f(u^\varepsilon) = \varepsilon\, \big(B(u^\varepsilon)\, \partial_x u^\varepsilon\big)_x, \quad (3.9)$$

where the **diffusion matrix** B satisfies

$$v \cdot D^2 U(u)\, B(u) v \ge \kappa\, |B(u)v|^2, \quad v \in \mathbb{R}^N \quad (3.10)$$

for all u under consideration and for some fixed constant $\kappa > 0$.

THEOREM 3.4. (Zero diffusion limit.) *Consider a system of conservation laws (1.1) endowed with a strictly convex entropy pair (U, F). Let u^ε be a sequence of smooth solutions of the model (3.9) satisfying the uniform bound (3.3), tending to a constant state u_* at $x \to \pm\infty$, and such that the derivatives u_x^ε decay to zero at infinity. Suppose also that the initial data satisfy the uniform L^2 bound*

$$\int_{\mathbb{R}} |u^\varepsilon(0) - u_*|^2\, dx \le C, \quad (3.11)$$

where the constant $C > 0$ is independent of ε. Then, the right-hand side of (3.9) is conservative (see (3.4)) and entropy conservative (see (3.6)).

Combining Theorem 3.4 with Theorem 3.3 we conclude that the solution of (3.9) can only converge to a weak solution of (1.1) satisfying the entropy inequality (3.8).

PROOF. Let us first treat the case $B = I$. Then, (3.10) means that U is uniformly convex. It is easy to derive (3.4) since, here,

$$\left| \iint_{I\!\!R \times I\!\!R_+} R^\varepsilon \, \theta \, dx dt \right| \le \iint_{I\!\!R \times I\!\!R_+} \varepsilon \, |u^\varepsilon| \, |\theta_{xx}| \, dx dt$$

$$\le \varepsilon \, \|u^\varepsilon\|_{L^\infty} \, \|\theta_{xx}\|_{L^1} \le C \varepsilon \to 0$$

for all $\theta \in C_c^\infty(I\!\!R \times I\!\!R_+)$.

On the other hand, for the general regularization (3.9) and for general matrices the identity (3.5) takes the form

$$\partial_t U(u^\varepsilon) + \partial_x F(u^\varepsilon) = \varepsilon \left(\nabla U(u^\varepsilon) \cdot B(u^\varepsilon) u_x^\varepsilon \right)_x - \varepsilon \, u_x^\varepsilon \cdot D^2 U(u^\varepsilon) B(u^\varepsilon) u_x^\varepsilon. \quad (3.12)$$

When $B = I$ we find

$$\partial_t U(u^\varepsilon) + \partial_x F(u^\varepsilon) = \varepsilon \, U(u^\varepsilon)_{xx} - \varepsilon \, u_x^\varepsilon \cdot D^2 U(u^\varepsilon) \, u_x^\varepsilon.$$

To derive (3.6) we observe that by integration by parts

$$\iint_{I\!\!R \times I\!\!R_+} \nabla U(u^\varepsilon) \cdot R^\varepsilon \, \theta \, dx dt$$

$$\le \varepsilon \iint_{I\!\!R \times I\!\!R_+} |U(u^\varepsilon)| \, |\theta_{xx}| \, dx dt - \varepsilon \iint_{I\!\!R \times I\!\!R_+} u_x^\varepsilon \cdot D^2 U(u^\varepsilon) \, u_x^\varepsilon \, \theta \, dx dt$$

for all $\theta \in C_c^\infty$ with $\theta \ge 0$. Using the uniform bound (3.3), the first term of the right-hand side tends to zero with ε. The second term is non-positive since $\theta \ge 0$ and the entropy is convex thanks to (3.10). We have thus established that, when $B(u) = I$, the right-hand side of (3.9) is conservative and entropy dissipative.

To deal with the general diffusion matrix, we need to obtain first an a priori bound on the entropy dissipation. This step is based on the uniform convexity assumption (3.10).

By assumption, u^ε decays to some constant state u_* at $x = \pm\infty$ and that u_x^ε decays to zero sufficiently fast. Normalize the entropy flux by $F(u_*) = 0$. Since U is strictly convex and the range of the solutions is bounded a priori, we can always replace the entropy $u \mapsto U(u)$ with $U(u) - U(u_*) - \nabla U(u_*) \cdot (u - u_*)$. The latter is still an entropy, associated with the entropy flux $F(u) - F(u_*) - \nabla U(u_*) \cdot \left(f(u) - f(u_*) \right)$. Moreover it is not difficult to see that U is non-negative. To simplify the notation and without loss of generality, we assume that $u_* = 0 \in \mathcal{U}$, $U(u_*) = 0$, and $\nabla U(u_*) = 0$. Integrating (3.12) over the real line and a finite time interval $[0, T]$ we obtain

$$\int_{I\!\!R} U(u^\varepsilon(T)) \, dx + \varepsilon \int_0^T \int_{I\!\!R} u_x^\varepsilon \cdot D^2 U(u^\varepsilon) B(u^\varepsilon) u_x^\varepsilon \, dx dt = \int_{I\!\!R} U(u^\varepsilon(0)) \, dx \le C,$$

thanks to the bound (3.11) on the initial data $u^\varepsilon(0)$. Using (3.10) and letting $T \to +\infty$ we conclude that the **entropy dissipation** is uniformly bounded:

$$\varepsilon \iint_{I\!\!R \times I\!\!R_+} |B(u^\varepsilon) u_x^\varepsilon|^2 \, dx dt \le C. \quad (3.13)$$

Relying on the uniform energy bound (3.13) we now prove (3.4) and (3.6). Relying on (3.13) we find that for each test-function θ

$$\left| \iint_{I\!R \times I\!R_+} R^\varepsilon \, \theta \, dxdt \right| = \left| \varepsilon \iint_{I\!R \times I\!R_+} \left(B(u^\varepsilon) \, u_x^\varepsilon \right)_x \theta \, dxdt \right|$$

$$\leq \varepsilon \iint_{I\!R \times I\!R_+} |B(u^\varepsilon) \, u_x^\varepsilon| \, |\theta_x| \, dxdt$$

$$\leq C \, \varepsilon \, \|\theta_x\|_{L^2(I\!R \times I\!R_+)} \left(\iint_{I\!R \times I\!R_+} |B(u^\varepsilon) \, u_x^\varepsilon|^2 \, dxdt \right)^{1/2}$$

$$\leq C' \, \varepsilon^{1/2} \to 0,$$

which establishes (3.4).

In view of (3.10) the second term in the right-hand side of (3.12) remains non-positive. On the other hand, the first term in (3.12) tends to zero since following the same lines as above

$$\left| \varepsilon \int_0^{+\infty} \int_{I\!R} \left(\nabla U(u^\varepsilon) B(u^\varepsilon) \, u_x^\varepsilon \right)_x \theta \, dxdt \right| \leq C \, \varepsilon \iint_{I\!R \times I\!R_+} |B(u^\varepsilon) \, u_x^\varepsilon| \, |\theta_x| \, dxdt$$

$$\leq C' \, \varepsilon^{1/2} \to 0.$$

Thus (3.6) holds, which completes the proof of Theorem 3.4. □

Next, we discuss another general regularization of interest, based on the entropy variable $\hat{u} = \nabla U(u)$ introduced in the end of Section 1. Recall that $u \mapsto \hat{u}$ is a change of variable when U is strictly convex. Consider the **nonlinear diffusion-dispersion model**

$$\partial_t u^\varepsilon + \partial_x f(u^\varepsilon) = \varepsilon \, \hat{u}_{xx}^\varepsilon + \delta \, \hat{u}_{xxx}^\varepsilon$$
$$= \varepsilon \, \nabla U(u^\varepsilon)_{xx} + \delta \, \nabla U(u^\varepsilon)_{xxx}, \qquad (3.14)$$

where $\varepsilon > 0$ and $\delta = \delta(\varepsilon) \in I\!R$ are called the *diffusion* and the *dispersion* parameters. Diffusive and dispersive terms play an important role in continuum physics, as illustrated by Examples 4.5 and 4.6 below. Understanding the effect of such terms on discontinuous solutions of (1.1) will be one of our main objectives in this course.

THEOREM 3.5. (Zero diffusion-dispersion limit.) *Consider a system of conservation laws (1.1) endowed with a strictly convex entropy pair (U, F). Let u^ε be a sequence of smooth solutions of the diffusive-dispersive model (3.14) satisfying the uniform bound (3.3), tending to a constant u_* at $x \to \pm\infty$, and such that u_x^ε and u_{xx}^ε decay to zero at infinity. Suppose also that the initial data satisfy the uniform bound (3.11). Then, the right-hand side of (3.14) is conservative (see (3.4)) and entropy dissipative (see (3.6)) in the limit $\varepsilon, \delta \to 0$ with $\delta/\varepsilon \to 0$.*

Again, combining Theorem 3.5 with Theorem 3.3 we conclude that solutions of (3.14) can only converge to a weak solution of (1.1) satisfying the entropy inequality (3.8). Note that (3.8) is derived here only for the entropy U upon which the regularization (3.14) is based.

PROOF. Using the uniform bound (3.3), for all $\theta \in C_c^\infty(I\!R \times I\!R_+)$ we obtain

$$\left| \int_0^{+\infty} \int_{I\!R} R^\varepsilon \, \theta \, dx dt \right| \leq \int_0^{+\infty} \int_{I\!R} \left(\varepsilon \left| \hat{u}^\varepsilon \right| |\theta_{xx}| + \delta \left| \hat{u}^\varepsilon \right| |\theta_{xxx}| \right) dx dt$$

$$\leq C \, \varepsilon \, \|\theta_{xx}\|_{L^1(I\!R \times I\!R_+)} + C \, \delta \, \|\theta_{xxx}\|_{L^1(I\!R \times I\!R_+)} \to 0$$

when $\varepsilon, \delta \to 0$, which leads us to (3.4). Multiplying (3.14) by the entropy variable $\hat{u}$, we find

$$\partial_t U(u^\varepsilon) + \partial_x F(u^\varepsilon) = \varepsilon \, \hat{u}^\varepsilon \cdot \hat{u}_{xx}^\varepsilon + \delta \, \hat{u}^\varepsilon \cdot \hat{u}_{xxx}^\varepsilon$$

$$= \frac{\varepsilon}{2} \left(|\hat{u}^\varepsilon|^2 \right)_{xx} - \varepsilon \, |\hat{u}_x^\varepsilon|^2 + \frac{\delta}{2} \left(\left(|\hat{u}^\varepsilon|^2 \right)_{xx} - 3 \, |\hat{u}_x^\varepsilon|^2 \right)_x. \tag{3.15}$$

All but one (non-positive) term of the right-hand side of (3.15) have a conservative form. After normalization we can always assume that $u_* = 0 \in \mathcal{U}$ and, after normalization, $U(u) \geq 0$, $U(0) = 0$, and $F(0) = 0$. Integrating (3.15) over the whole real line and over a finite time interval $[0, T]$, we obtain

$$\int_{I\!R} U(u^\varepsilon(T)) \, dx + \varepsilon \int_0^T \int_{I\!R} |\hat{u}_x^\varepsilon|^2 \, dx dt = \int_{I\!R} U(u^\varepsilon(0)) \, dx.$$

Provided that the initial data satisfy (3.11) we conclude that

$$\varepsilon \iint_{I\!R \times I\!R_+} |\hat{u}_x^\varepsilon|^2 \, dx dt \leq C. \tag{3.16}$$

To check (3.6) we rely on the identity (3.15) and the uniform bound (3.16), as follows. Taking the favorable sign of one entropy dissipation term into account we obtain for all non-negative $\theta \in C_c^\infty(I\!R \times I\!R_+)$:

$$\iint_{I\!R \times I\!R_+} \nabla U(u^\varepsilon) \cdot R^\varepsilon \, \theta \, dx dt$$

$$\leq \frac{\varepsilon}{2} \iint_{I\!R \times I\!R_+} |\hat{u}^\varepsilon|^2 \, |\theta_{xx}| \, dx dt + \frac{\delta}{2} \iint_{I\!R \times I\!R_+} \left(|\hat{u}^\varepsilon|^2 \, |\theta_{xxx}| + |\hat{u}_x^\varepsilon|^2 \, |\theta_x| \right) dx dt$$

$$\leq C \, \varepsilon \, \|\theta_{xx}\|_{L^1(I\!R \times I\!R_+)} + C \, \delta \, \|\theta_{xxx}\|_{L^1(I\!R \times I\!R_+)} + C \, \delta \, \|\theta_x\|_{L^\infty} \iint_{I\!R \times I\!R_+} |\hat{u}_x^\varepsilon|^2 \, dx dt$$

$$\leq C' \left(\varepsilon + \delta + \delta/\varepsilon \right).$$

As $\varepsilon, \delta \to 0$ with $\delta/\varepsilon \to 0$ we conclude that (3.6) holds, which completes the proof of Theorem 3.5. $\qquad \square$

4. Examples of diffusive-dispersive models

Systems of conservation laws arise in continuum physics a variety of applications. We introduce here several important examples that will be of particular interest in this course.

EXAMPLE 4.1. *Burgers equation.* The simplest example of interest is given by the (inviscid) **Burgers equation**

$$\partial_t u + \partial_x \frac{u^2}{2} = \varepsilon \, u_{xx}. \tag{4.1}$$

Given *any* convex function U let F be a corresponding entropy flux (Example 1.9). Multiplying (4.1) by $U'(u)$ we obtain the entropy balance

$$\partial_t U(u) + \partial_x F(u) = \varepsilon\, U(u)_{xx} - \varepsilon\, U''(u)\, u_x^2,$$

which, formally as $\varepsilon \to 0$, leads to *infinitely many* entropy inequalities

$$\partial_t U(u) + \partial_x F(u) \le 0.$$

□

EXAMPLE 4.2. *Conservation law with cubic flux revisited.* The equation in Example 1.5 may be augmented with diffusive and dispersive terms, as follows

$$\partial_t u + \partial_x u^3 = \varepsilon\, u_{xx} + \delta\, u_{xxx}. \tag{4.2}$$

where $\varepsilon > 0$ and $\delta \in \mathbb{R}$. Using the quadratic entropy $U(u) = u^2$ we obtain

$$\partial_t u^2 + \partial_x\Big(\frac{3\,u^4}{2}\Big) = \varepsilon\,\big(u^2\big)_{xx} - 2\varepsilon\,u_x^2 + \delta\big(2\,u u_{xx} - u_x^2\big)_x$$
$$= \varepsilon\,\big(u^2\big)_{xx} - 2\varepsilon\,u_x^2 + \delta\big((u^2)_{xx} - 3\,u_x^2\big)_x$$

which, in the limit $\varepsilon, \delta \to 0$, yields the *single* entropy inequality

$$\partial_t u^2 + \partial_x \frac{3\,u^4}{2} \le 0. \tag{4.3}$$

We will see later on (Theorem III-2.4 in Chapter III) that for solutions generated by (4.2) the entropy inequality (3.8) does not hold for arbitrary entropies !   □

EXAMPLE 4.3. *Diffusive-dispersive conservation laws.* Consider next the model

$$\partial_t u + \partial_x f(u) = \varepsilon\,\big(b(u)\, u_x\big)_x + \delta\,\big(c_1(u)\,(c_2(u)\, u_x)_x\big)_x, \tag{4.4}$$

where $b(u) > 0$ is a diffusion coefficient and $c_1(u), c_2(u) > 0$ are dispersion coefficients. Let (U_*, F_*) a (strictly convex) entropy pair satisfying

$$U_*''(u) = \frac{c_2(u)}{c_1(u)}, \quad u \in \mathbb{R}.$$

(U_* is unique up to a linear function of u.) Interestingly, the last term in the right-hand side of (4.4) takes a simpler form in this entropy variable $\hat{u} = U_*'(u)$, indeed

$$\delta\,\big(c_1(u)\,(c_2(u)\, u_x)_x\big)_x = \delta\,\big(c_1(u)\,(c_1(u)\, \hat{u}_x)_x\big)_x.$$

Any solution of (4.4) satisfies

$$\partial_t U_*(u) + \partial_x F_*(u) = \varepsilon\,\big(b(u)\, U_*'(u)\, u_x\big)_x - \varepsilon\, b(u)\, U_*''(u)\, |u_x|^2$$
$$+ \delta\,\big(c_1(u)\hat{u}\,(c_1(u)\, \hat{u}_x)_x - |c_2(u)\, u_x|^2/2\big)_x.$$

In the right-hand side above, the contribution due to the diffusion decomposes into a conservative term and a non-positive (dissipative) one. The dispersive term is entirely conservative. In the formal limit $\varepsilon, \delta \to 0$ any limiting function satisfies the *single* entropy inequality

$$\partial_t U_*(u) + \partial_x F_*(u) \le 0.$$

□

EXAMPLE 4.4. *Nonlinear elastodynamics.* The longitudinal deformations of an elastic body with negligible cross-section can be described by the conservation law of total momentum and by the so-called continuity equation, i.e.,

$$\partial_t v - \partial_x \sigma(w) = 0,$$
$$\partial_t w - \partial_x v = 0,$$

(4.5)

respectively. The unknowns v and $w > -1$ represent the **velocity** and **deformation gradient** respectively, while the **stress** $w \mapsto \sigma(w)$ is a given constitutive function depending on the material under consideration. The constrain $w > -1$ arises as follows. Denote by $x \mapsto y(x, t)$ the **Lagrangian variable**, i.e., $y(x, t)$ represents the location (at time t) of the material particle located initially at the point x. The functions v and w are defined from the Lagrangian variable by $v = \partial_t y$, $w = \partial_x (y - x)$. On the other hand, the mapping $y(.,.)$ is constrained by the principle of **impenetrability of matter**, that is, $\partial_x y > 0$ or $w > -1$. The theoretical limit $w \to -1$ corresponds to an infinite compression of the material.

Set $u = \begin{pmatrix} v \\ w \end{pmatrix}$, $f(u) = -\begin{pmatrix} \sigma(w) \\ v \end{pmatrix}$ and $\mathcal{U} = I\!R \times (-1, +\infty)$, and define the **sound speed** as $c(w) = \sqrt{\sigma'(w)}$. For typical elastic materials, we have

$$\sigma'(w) > 0 \quad \text{for all } w > -1$$

(4.6)

so that the system (4.5) is then *strictly hyperbolic* and admits two distinct wave speeds, $\lambda_2 = -\lambda_1 = c(w)$. Left- and right-eigenvectors are chosen to be

$$l_1 = \Big(1, c(w)\Big), \quad r_1 = \begin{pmatrix} c(w) \\ 1 \end{pmatrix}, \quad l_2 = \Big(1, -c(w)\Big), \quad r_2 = \begin{pmatrix} c(w) \\ -1 \end{pmatrix}.$$

Moreover, in view of the relation $\nabla \lambda_j \cdot r_j = -c'(w)$ we see that the two characteristic fields of (4.5) are *genuinely nonlinear* if and only if

$$\sigma''(w) \neq 0, \quad w > -1.$$

(4.7)

However, many materials encountered in applications do not satisfy (4.7) but rather

$$\sigma''(w) \gtrless 0, \quad w \gtrless 0.$$

(4.8)

Using the characterization (1.15) in Theorem 1.8, one easily checks that the entropies $U(v, w)$ of (4.5) satisfy the following second-order, linear hyperbolic equation with non-constant coefficients,

$$U_{ww} - \sigma'(w) U_{vv} = 0.$$

(4.9)

One mathematical entropy pair of particular interest is provided by the **total energy**

$$U(v, w) = \frac{v^2}{2} + \int_0^w \sigma(s) \, ds, \quad F(v, w) = -\sigma(w) \, v,$$

(4.10)

which is *strictly convex* under the assumption (4.6). The change of variables $\hat{v} := v$, $\hat{w} := \sigma(w)$ clearly put (4.6) in a symmetric form,

$$\partial_t \hat{v} - \partial_x \hat{w} = 0, \quad \partial_t \sigma^{-1}(\hat{w}) - \partial_x \hat{v} = 0,$$

in agreement with (1.19) and (1.20). Finally, the functions ϕ and ψ introduced in (1.22) are found to be

$$\phi(v, \sigma(w)) = \frac{v^2}{2} + w\,\sigma(w) - \int_0^w \sigma(s)\,ds, \quad \psi(v, \sigma(w)) = -\sigma(w)\,v. \qquad (4.11)$$

Furthermore, the interaction terms (1.13) and (1.14) are determined from the basic formula

$$[r_1, r_2] = Dr_2 \cdot r_1 - Dr_1 \cdot r_2 = \begin{pmatrix} 2\,c' \\ 0 \end{pmatrix}.$$

For instance, $G_{112} = -4\,c\,c' = -2\,\sigma''$, which vanishes only at the points where the genuine nonlinearity condition (4.7) fails. Away from such points, the two equations in (4.5) are truly coupled. $\qquad\square$

EXAMPLE 4.5. *Phase transitions dynamics.* For a model of phase transitions in solid materials, consider the two conservation laws of elastodynamics (4.5) in which, now, σ is taken to be a *non-monotone* stress-strain function. For instance, for the modeling of a two-phase material one assumes that

$$\begin{aligned} \sigma'(w) > 0, \quad & w \in (-1, w^m) \cup (w^M, +\infty), \\ \sigma'(w) < 0, \quad & w \in (w^m, w^M) \end{aligned} \qquad (4.12)$$

for some constants $w^m < w^M$. In the so-called **unstable phase** (w^m, w^M) the system admits two complex conjugate eigenvalues. All of the solutions of interest from the standpoint of the hyperbolic theory lie *outside* the unstable region. The system is hyperbolic for all $u = \begin{pmatrix} v \\ w \end{pmatrix}$ in the non-connected set $\mathcal{U} := (\mathbb{R} \times (-1, w^m)) \cup (\mathbb{R} \times (w^M, +\infty))$, and most of the algebraic properties described in Example 4.4 remain valid. One important difference concerns the mathematical entropies: the total mechanical energy (4.10) is *convex* in each hyperbolic region but (any extension) is not globally convex in (the convex closure of) $\mathcal{U}$. Hence, the entropy variable (see (1.20)) no longer defines a change of variable. The conservative variable of (4.5) cannot be expressed in the entropy variable $\hat{v} = v$, $\hat{w} = \sigma(w)$, since $w \mapsto \sigma(w)$ fails to be globally invertible. However, we observe that the entropy variable can still be used to express the flux $f(u)$ of (4.5) under the assumption (4.12). $\qquad\square$

EXAMPLE 4.6. *Nonlinear elastodynamics and phase transitions I.* High-order effects such as viscosity and capillarity induce diffusion and dispersion effects which, for instance, have the form

$$\begin{aligned} \partial_t v - \partial_x \sigma(w) &= \varepsilon\,v_{xx} - \delta\,w_{xxx}, \\ \partial_t w - \partial_x v &= 0, \end{aligned} \qquad (4.13)$$

where the stress-strain function σ satisfies the assumptions in Example 4.4 (hyperbolic) or in Example 4.5 (hyperbolic-elliptic). In the right-hand side of (4.13), ε represents the viscosity of the material and δ its capillarity. Observe that

$$\begin{aligned} \partial_t &\left(\frac{v^2}{2} + \int_0^w \sigma(s)\,ds + \frac{\delta}{2}\,w_x^2 \right) - \partial_x\big(v\,\sigma(w)\big) \\ &= \varepsilon\,(v\,v_x)_x - \varepsilon\,v_x^2 + \delta\,(v_x\,w_x - v\,w_{xx})_x. \end{aligned}$$

Clearly the entropy inequality

$$\partial_t \left(\frac{v^2}{2} + \int_0^w \sigma(s)\, ds \right) - \partial_x \big(v\, \sigma(w) \big) \le 0 \qquad (4.14)$$

is formally recovered in the limit $\varepsilon, \delta \to 0$. As for Example 4.2 one can check that, for limiting solutions generated by (4.13), the entropy inequality (4.14) does not hold for arbitrary entropies. $\qquad\square$

EXAMPLE 4.7. *Nonlinear elastodynamics and phase transitions II.* Recall a notation introduced in Example 4.4: $(x, t) \mapsto y(x, t)$ is the Lagrangian variable and the velocity and deformation gradient are determined by $v = \partial_t y$, $w = \partial_x (y - x)$. We assume that an **internal energy** function of the form

$$e = e(w, w_x) = \bar{e}(y_x, y_{xx})$$

is prescribed. The general equations of elastodynamics are then derived (formally) from the postulate that the **action**

$$J(y) = \int_0^T \int_\Omega \left(e(w, w_x) - \frac{v^2}{2} \right) dx\,dt = \int_0^T \int_\Omega \left(\bar{e}(y_x, y_{xx}) - \frac{y_t^2}{2} \right) dx\,dt \qquad (4.15)$$

should be extremal among all "admissible" y. Here $\Omega \subset I\!R$ is the (bounded) interval initially occupied by the fluid and $[0, T]$ is some given time interval.

Let $g : \Omega \times [0, T] \to I\!R$ be a smooth function with compact support. Replacing in (4.15) y with $y + g$ and keeping the first-order terms in g only, we obtain

$$
\begin{aligned}
J(y + g) &= \int_0^T \int_\Omega \left(\bar{e}(y_x + g_x, y_{xx} + g_{xx}) - \frac{1}{2}\left(y_t + g_t \right)^2 \right) dx\,dt \\
&= J(y) + \int_0^T \int_\Omega \left(\frac{\partial \bar{e}}{\partial y_x}(y_x, y_{xx})\, g_x + \frac{\partial \bar{e}}{\partial y_{xx}}(y_x, y_{xx})\, g_{xx} - y_t\, g_t \right) dx\,dt \\
&\quad + O(|g|^2)
\end{aligned}
$$

and, after integration by parts,

$$
\begin{aligned}
J(y + g) =\, &J(y) + \int_0^T \int_\Omega \left(\left(-\frac{\partial \bar{e}}{\partial y_x}(y_x, y_{xx}) \right)_x + \left(\frac{\partial \bar{e}}{\partial y_{xx}}(y_x, y_{xx}) \right)_{xx} + y_{tt} \right) g\, dx\,dt \\
&+ O(|g|^2).
\end{aligned}
$$

Since the solution y should minimize the action J and that g is arbitrary, this formally yields

$$y_{tt} + \left(-\frac{\partial \bar{e}}{\partial y_x}(y_x, y_{xx}) + \left(\frac{\partial \bar{e}}{\partial y_{xx}}(y_x, y_{xx}) \right)_x \right)_x = 0. \qquad (4.16)$$

Returning to the unknown functions v and w and defining the **total stress** as

$$\Sigma(w, w_x, w_{xx}) = \frac{\partial e}{\partial w}(w, w_x) - \left(\frac{\partial e}{\partial w_x}(w, w_x) \right)_x, \qquad (4.17)$$

(4.16) becomes

$$
\begin{aligned}
\partial_t v - \partial_x \Sigma(w, w_x, w_{xx}) &= 0, \\
\partial_t w - \partial_x v &= 0.
\end{aligned}
$$

Finally, if a nonlinear viscosity $\mu(w)$ is also taken into account, we arrive at a general model including viscosity and capillarity effects:

$$\partial_t v - \partial_x \Sigma(w, w_x, w_{xx}) = \big(\mu(w)\, v_x\big)_x,$$
$$\partial_t w - \partial_x v = 0. \tag{4.18}$$

The total energy

$$E(w, v, w_x) := e(w, w_x) + v^2/2$$

again plays the role of a mathematical entropy. We find

$$\partial_t E(w, v, w_x) - \partial_x \big(\Sigma(w, w_x, w_{xx})\, v\big) = \left(v_x\, \frac{\partial e}{\partial w_x}(w, w_x)\right)_x + \big(\mu(w)\, v\, v_x\big)_x - \mu(w)\, v_x^2.$$

Once more, the entropy inequality similar to (4.14) could be obtained.

Finally, let us discuss the properties of the internal energy function e. A standard choice in the literature is for e to be quadratic in w_x. (Linear term should not appear because of the natural invariance of the energy via the transformation $x \mapsto -x$.) Setting, for some positive capillarity coefficient $\lambda(w)$,

$$e(w, w_x) = \varepsilon(w) + \lambda(w)\, \frac{w_x^2}{2}, \tag{4.19}$$

the total stress decomposes as follows:

$$\Sigma(w, w_x, w_{xx}) = \sigma(w) + \lambda'(w)\, \frac{w_x^2}{2} - (\lambda(w)\, w_x)_x, \quad \sigma(w) = \varepsilon'(w). \tag{4.20}$$

The equations in (4.18) take the form

$$\partial_t v - \partial_x \sigma(w) = \left(\lambda'(w)\, \frac{w_x^2}{2} - (\lambda(w)\, w_x)_x\right)_x + \big(\mu(w)\, v_x\big)_x,$$
$$\partial_t w - \partial_x v = 0. \tag{4.21}$$

In this case we have

$$\left(\varepsilon(w) + \frac{v^2}{2} + \lambda(w)\, \frac{w_x^2}{2}\right)_t - (\sigma(w)\, v)_x$$
$$= \big(\mu(w)\, v\, v_x\big)_x - \mu(w)\, v_x^2 + \left(v\, \frac{\lambda'(w)}{2}\, w_x^2 - v\, (\lambda(w)\, w_x)_x + v_x\, \lambda(w)\, w_x\right)_x.$$

Under the simplifying assumption that the viscosity and capillarity are both constants, we can recover Example 4.6 above.   □

5. Kinetic relations and traveling waves

We return to the general discussion initiated in Section 3 and we outline an important standpoint adopted in this course for the study of (1.1). The weak solutions of interest are primary those generated by an augmented model of the general form (3.1). When small physical parameters accounting for the viscosity, heat conduction, or capillarity of the material are negligible with respect to the scale of hyperbolic features, it is desirable to replace (3.1) with the hyperbolic system of conservation laws (1.1). Since the solutions of the Cauchy problem associated with (1.1) are not unique, one must determine suitable admissibility conditions which would pick up the solutions of (1.1) realizable as limits of solutions of (3.1) by incorporating some large-scale effects contained in (3.1) without resolving the small-scales in details. As we will see, *different* regularizations may select *different* weak solutions !

The *entropy inequality* (3.8) was derived for two large classes of regularizations (Section 3) as well as for several specific examples (Section 4). Generally speaking, when more than one mathematical entropy is available ($N \leq 2$), a *single* entropy inequality only is satisfied by the solutions of (1.1). See, for instance, the important Examples 4.2 and 4.6 above. This feature will motivate us to determine first weak solutions of (1.1) satisfying the *single* entropy inequality (3.8). (See Sections II-3 and VI-3, below.)

For systems admitting *genuinely nonlinear* or *linearly degenerate* characteristic fields only, the entropy inequality (3.8) turns out to be sufficiently discriminating to select a *unique* weak solution to the Cauchy problem (1.1) and (1.2). In particular, for such systems, weak solutions are *independent* of the precise regularization mechanism in the right-hand side of (3.1). Such solutions will be called **classical entropy solutions** and a corresponding uniqueness result will be rigorously established in Chapter X (see Theorem X-4.3).

On the other hand, for systems admitting *general characteristic fields* that fail to be globally genuinely nonlinear or linearly degenerate, the entropy inequality (3.8) is *not* sufficiently discriminating. Under the realistic assumptions imposed in the applications, many models arising in continuum physics fail to be globally genuinely nonlinear. For such systems, we will see that weak solutions are *strongly sensitive* to the small-scales that have been neglected at the hyperbolic level of physical modeling, (1.1), but are taken into account in an augmented model, (3.1). In Chapters II and VI we will introduce the corresponding notion of **nonclassical entropy solutions** based on a refined version of the entropy inequality, *more discriminating* than (3.8) and referred to as the **kinetic relation**.

At this juncture, let us describe the qualitative behavior of the solutions of the nonlinear diffusion-dispersion model (3.14), which includes linear diffusive and dispersive terms with "strengths" $\varepsilon > 0$ and δ, respectively. By solving the corresponding Cauchy problem numerically, several markedly different behaviors can be observed, as illustrated in Figure I-1:

- When $|\delta| << \varepsilon^2$, the effect of the dispersion turns out to be negligible. The limiting solutions coincide with the ones generated by the zero-diffusion limit corresponding to $\delta = 0$ and $\varepsilon \to 0$.

- When $|\delta| >> \varepsilon^2$, the dispersion dominates and wild oscillations with high frequencies arise as $\delta \to 0$. The solutions converge in a weak sense only and the conservation laws (1.1) do not truly describe the singular limit in this case.

- The intermediate regime

$$\delta = \gamma \varepsilon^2, \quad \varepsilon \to 0 \text{ and } \gamma \text{ fixed}, \tag{5.1}$$

when diffusion and dispersion are *kept in balance*, is of particular interest in the present course. There is a subtle competition between the parameters ε and δ. The diffusion ε has a regularizing effect on the propagating discontinuities while the dispersion δ generates wild oscillations. It turns out that, in the limit (5.1), the solutions of (3.14) converge (from the numerical standpoint, at least)

to a limit which satisfies the conservation laws (1.1) and the entropy inequality
(3.8). Mild oscillations and spikes are visible near jump discontinuities, only.
Interestingly enough, the limiting solution strongly depends on the parameter
γ. That is, for the same initial data *different* values of γ lead to *different*
shock wave solutions !

$u(x,t)$

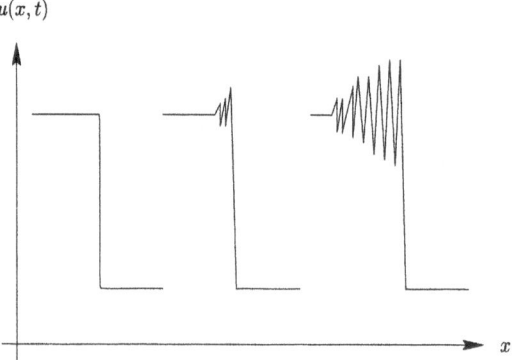

Figure I-1 : Numerical solution for
$|\delta| << \varepsilon^2$, $\delta = \gamma \varepsilon^2$, and $|\delta| >> \varepsilon^2$, respectively.

From this discussion we conclude that no "universal" admissibility criterion can
be postulated for nonlinear hyperbolic systems. Instead, some additional information
should be sought and an *admissibility condition* should be formulated for each problem
(or rather each class of problems) of interest. Before closing this section let us
introduce a few more properties and definitions. First of all, for the entropy inequality
(3.8) we have the obvious analogue of the Rankine-Hugoniot relation derived earlier
in Section 2.

THEOREM 5.1. (Jump relation for the entropy inequality.) *We use the same notation
as in Theorem 2.3. The piecewise smooth function (2.7) satisfies the entropy inequality
(3.8) if and only if*

$$-\varphi'(t)\left(U(u_+(t)) - U(u_-(t))\right) + F(u_+(t)) - F(u_-(t)) \leq 0. \qquad (5.2)$$

*In particular, given a shock wave (2.9) connecting two constant states u_- and u_+ and
associated with the speed λ, the entropy inequality reads*

$$-\lambda\left(U(u_+) - U(u_-)\right) + F(u_+) - F(u_-) \leq 0. \qquad (5.3)$$

$\square$

When dealing with nonclassical solutions, the Rankine-Hugoniot relations (2.10)
and the entropy inequality (5.3) will be supplemented with the following additional
jump condition:

DEFINITION 5.2. (Kinetic relation.) A **kinetic relation** for the shock wave (2.9) is an additional jump relation of the general form

$$\Phi(u_+, u_-) = 0, \tag{5.4}$$

where Φ is a Lipschitz continuous function of its arguments. In particular, a **kinetic relation associated with the entropy** U is the following strengthened version of the entropy inequality (5.3)

$$-\lambda \left(U(u_+) - U(u_-) \right) + F(u_+) - F(u_-) = \phi(u_-, u_+), \tag{5.5}$$

where ϕ is a Lipschitz continuous function of its arguments.

Two remarks are in order:
- Not all propagating waves within a nonclassical solution will require a kinetic relation, but only the so-called *undercompressive* shock waves.
- Suitable assumptions will be imposed on the kinetic functions Φ and ϕ in (5.4) and (5.5), respectively. For instance, an obvious requirement is that the right-hand side of (5.5) be non-positive, that is, $\phi \leq 0$, so that (5.5) implies (5.3).

The role of the kinetic relation in selecting weak solutions to systems of conservation laws will be discussed in this course. We will show that a kinetic relation is necessary and sufficient to set the Riemann problem and the Cauchy problem for (1.1):
- We will establish that the Riemann problem has a unique nonclassical solution characterized by a kinetic relation (Theorems II-4.1 and II-5.4).
- We will also investigate the existence (Theorems IV-3.2, VIII-1.7, and VIII-3.2) and uniqueness (Theorem X-4.1) of nonclassical solutions to the Cauchy problem.

To complete the above analysis, we must determine the kinetic function from a given diffusion-dispersion model like (3.1). The kinetic relation is introduced first in the following "abstract" way. Let us decompose the product $\nabla U(u^\varepsilon) \cdot R^\varepsilon$ arising in (3.6) in the form

$$\nabla U(u^\varepsilon) \cdot R^\varepsilon = Q^\varepsilon + \mu^\varepsilon, \tag{5.6}$$

where $Q^\varepsilon \rightharpoonup 0$ in the sense of distributions and μ^ε is a uniformly bounded sequence of *non-positive L^1 functions*. We refer to μ^ε as the **entropy dissipation measure** for the given model (3.1) and for the given entropy U. (This decomposition was established for the examples (3.9) and (3.14) in the proofs of Theorems 3.4 and 3.5.) After extracting a subsequence if necessary, these measures converge in the weak-star sense to a non-positive bounded measure (Theorem A.1 in the appendix):

$$\mu_U(u) := \lim_{\varepsilon \to 0} \mu^\varepsilon \leq 0. \tag{5.7}$$

The limiting measure $\mu_U(u)$ depends upon the pointwise limit $u := \lim_{\varepsilon \to 0} u^\varepsilon$, but cannot be uniquely determined from it. For regularization-independent shock waves the sole *sign* of the entropy dissipation measure $\mu_U(u)$ suffices and one simply writes down the entropy inequality (3.8). However, for regularization-sensitive shock waves, the *values* taken by the measure $\mu_U(u)$ play a crucial role in selecting weak solutions. The corresponding **kinetic relation** takes the form

$$\partial_t U(u) + \partial_x F(u) = \mu_U(u) \leq 0, \tag{5.8}$$

where $\mu_U(u)$ is a non-positive, locally bounded measure depending on the solution u under consideration. Clearly, the measure $\mu_U(u)$ cannot be prescribed arbitrarily and, in particular, must vanish on the set of continuity points of u.

DEFINITION 5.3. (Traveling waves.) Consider a propagating jump discontinuity connecting two states u_- and u_+ at some speed λ. A function $u^\varepsilon(x,t) = w(y)$ with $y := (x - \lambda t)/\varepsilon$ is called a **traveling wave** of (3.1) connecting u_- to u_+ at the speed λ if it is a smooth solution of (3.1) satisfying

$$w(-\infty) = u_-, \quad w(+\infty) = u_+, \tag{5.9}$$

and

$$\lim_{|y| \to +\infty} w'(y) = \lim_{|y| \to +\infty} w''(y) = \ldots = 0. \tag{5.10}$$

For instance consider, in (3.1), conservative regularizations of the form

$$R^\varepsilon = \left(S(u^\varepsilon, \varepsilon\, u_x^\varepsilon, \varepsilon^2\, u_{xx}^\varepsilon, \ldots) \right)_x$$

with the natural condition $S(u, 0, \ldots) = 0$ for all u. Then the traveling waves w of (3.1) are given by the ordinary differential equation

$$-\lambda\, w' + f(w)' = S(w, w', w'', \ldots)'$$

or, equivalently, after integration over intervals $(-\infty, y]$ by

$$S(w, w', w'', \ldots) = f(w) - f(u_-) - \lambda\,(w - u_-). \tag{5.11}$$

It is straightforward (but fundamental) to check that:

THEOREM 5.4. *If w is a traveling wave solution, then the pointwise limit*

$$u(x,t) := \lim_{\varepsilon \to 0} w\left(\frac{x - \lambda t}{\varepsilon}\right) = \begin{cases} u_-, & x < \lambda t, \\ u_+, & x > \lambda t, \end{cases} \tag{5.12}$$

is a weak solution of (1.1) satisfying the entropy inequality (3.8). In particular, the Rankine-Hugoniot relation (2.10) follows from (5.11) by letting $y \to +\infty$.

Moreover, the solution u satisfies the kinetic relation (5.8) where the dissipation measure is given by

$$\mu_U(u) = M\,\delta_{x - \lambda t},$$

$$M := -\int_{I\!R} S(w(y), w'(y), w''(y), \ldots) \cdot D^2 U(w)\, w'(y)\, dy,$$

where $\delta_{x-\lambda t}$ denotes the Dirac measure concentrated on the line $x - \lambda t = 0$. $\quad\square$

We will see in Chapter III that traveling wave solutions determine the kinetic relation:

- For the scalar equation with cubic flux (Example 4.2) the kinetic relation can be determined explicitly; see Theorem III-2.3.
- For the more general model in Example 4.3 a careful analysis of the existence of traveling wave solutions leads to many interesting properties of the associated kinetic function (monotonicity, asymptotic behavior); see Theorem III-3.3. This analysis allows one to identify the terms Φ, ϕ, and $\mu_U(u)$ in (5.4), (5.5), and (5.7), respectively.
- Systems of equations such as those in Examples 4.6 and 4.7 can be covered by the same approach; see Remark III-5.4 and the bibliographical notes.

PART 1

SCALAR CONSERVATION LAWS

CHAPTER II

THE RIEMANN PROBLEM

In this chapter, we study the Riemann problem for scalar conservation laws. In Section 1 we discuss several formulations of the entropy condition. Then, in Section 2 we construct the *classical* entropy solution satisfying, by definition, all of the entropy inequalities; see Theorems 2.1 to 2.4. Next in Section 3, imposing only that solutions satisfy a single entropy inequality, we show that *undercompressive* shock waves are also admissible and we determine a *one-parameter* family of solutions to the Riemann problem; see Theorem 3.5. Finally in Sections 4 and 5, we construct *nonclassical* entropy solutions which, by definition, satisfy a single entropy inequality together with a *kinetic relation;* see Theorem 4.1 for *concave-convex* flux-functions and Theorem 5.4 for *convex-concave* flux-functions.

1. Entropy conditions

Consider the Riemann problem for the scalar conservation law

$$\partial_t u + \partial_x f(u) = 0, \quad u = u(x,t) \in I\!\!R, \tag{1.1}$$

where $f : I\!\!R \to I\!\!R$ is a smooth mapping. That is, we restrict attention to the initial data

$$u(x,0) = \begin{cases} u_l, & x < 0, \\ u_r, & x > 0, \end{cases} \tag{1.2}$$

where u_l and u_r are constants. Following the discussion in Sections I-3 to I-5 we seek for a weak solution of (1.1) and (1.2) satisfying some form of the entropy condition. As was pointed out in Theorem I-3.4, solutions determined by the zero diffusion limit satisfy

$$\partial_t U(u) + \partial_x F(u) \le 0 \quad \text{for } all \text{ convex entropy pairs } (U, F), \tag{1.3}$$

while for more general regularizations (Examples I-4.2 and I-4.3)

$$\partial_t U(u) + \partial_x F(u) \le 0 \quad \text{for a } single \text{ strictly convex pair } (U, F). \tag{1.4}$$

Recall that, in (1.3) and (1.4), U is a convex function and $F(u) = \int^u U'(v)\, f'(v)\, dv$.

First, we establish an equivalent formulation of (1.3), which is easier to work with.

THEOREM 1.1. (Oleinik entropy inequalities.) *A shock wave solution of* (1.1) *having the form*

$$u(x,t) = \begin{cases} u_-, & x < \lambda t, \\ u_+, & x > \lambda t, \end{cases} \tag{1.5}$$

for some constants u_-, u_+, and λ with $u_- \neq u_+$, satisfies the infinite set of entropy inequalities (1.3) *if and only if* **Oleinik entropy inequalities**

$$\frac{f(v) - f(u_-)}{v - u_-} \geq \frac{f(u_+) - f(u_-)}{u_+ - u_-} \quad \textit{for all } v \textit{ between } u_- \textit{ and } u_+ \qquad (1.6)$$

are satisfied. Moreover, (1.3) *and* (1.6) *imply* **Lax shock inequalities**

$$f'(u_-) \geq \lambda \geq f'(u_+). \qquad (1.7)$$

According to the Rankine-Hugoniot relation (derived in Theorem I-2.3), the speed λ in (1.5) is determined uniquely from the states u_- and u_+:

$$\begin{aligned} \lambda = \overline{a}(u_-, u_+) : &= \frac{f(u_+) - f(u_-)}{u_+ - u_-} \\ &= \int_0^1 a(u_- + s(u_+ - u_-))\, ds, \end{aligned} \qquad (1.8)$$

where

$$a(u) = f'(u), \quad u \in \mathbb{R}.$$

The (geometric) condition (1.6) simply means that the graph of f is below (above, respectively) the line connecting u_- to u_+ when $u_+ < u_-$ ($u_+ > u_-$, resp.). The condition (1.7) shows that the characteristic lines impinge on the discontinuity from both sides. The shock wave is said to be **compressive** and will be referred to as a **classical shock.**

THEOREM 1.2. (Lax shock inequality.) *When the function f is convex all of the conditions* (1.3), (1.4), (1.6), (1.7), *and* **Lax shock inequality**

$$u_- \geq u_+ \qquad (1.9)$$

are equivalent.

PROOFS OF THEOREMS 1.1 AND 1.2. For the function in (1.5) the inequalities in (1.3) are equivalent to (see Theorem I-5.1)

$$E(u_-, u_+) := -\overline{a}(u_-, u_+)\left(U(u_+) - U(u_-)\right) + F(u_+) - F(u_-) \leq 0,$$

that is,

$$\begin{aligned} E(u_-, u_+) &= \int_{u_-}^{u_+} U'(v)\left(-\lambda + f'(v)\right) dv \\ &= -\int_{u_-}^{u_+} U''(v)\left(-\lambda(v - u_-) + f(v) - f(u_-)\right) dv \\ &= -\int_{u_-}^{u_+} U''(v)(v - u_-)\left(\frac{f(v) - f(u_-)}{v - u_-} - \frac{f(u_+) - f(u_-)}{u_+ - u_-}\right) dv \\ &\leq 0, \end{aligned} \qquad (1.10)$$

where (1.8) was used to cancel the boundary terms in the integration by parts formula. Since U'' is arbitrary (1.10) and (1.6) are equivalent.

On the other hand, it is geometrically obvious that (1.6) is also equivalent to

$$\frac{f(v) - f(u_+)}{v - u_+} \leq \frac{f(u_+) - f(u_-)}{u_+ - u_-} \quad \text{for all } v \text{ between } u_- \text{ and } u_+. \qquad (1.11)$$

To derive the two inequalities in (1.7) we simply let $v \to u_-$ in (1.6) and $v \to u_+$ in (1.11).

The particular case of a convex flux is straightforward from (1.6). We just observe that the single entropy inequality (1.4) provides a sufficient condition. Indeed, the integrand in (1.10) has a constant sign when f is convex and that this sign is favorable if and only if (1.9) holds.   □

2. Classical Riemann solver

The shock waves (1.5) and (1.8) form a special class of solutions for the Riemann problem (1.1) and (1.2). Given a left-hand state u_- let us define the **(classical) shock set** $\mathcal{S}(u_-)$ as the set of all right-hand states attainable by shock waves satisfying (1.3). When the flux is convex, in view of the characterization derived in Theorem 1.2 we find

$$\mathcal{S}(u_-) = (-\infty, u_-]. \tag{2.1}$$

In fact, by Theorem 1.2, a single entropy inequality is sufficient to characterize the solution.

Next, we search for *smooth solutions* of (1.1) that are centered and of the self-similar form

$$u(x,t) = w(\xi), \quad \xi := \frac{x}{t}. \tag{2.2}$$

Necessarily we have

$$-\xi\, w_\xi + f(w)_\xi = 0,$$

thus (assuming that $w_\xi(\xi) \neq 0$)

$$f'(w(\xi)) = \xi \quad \text{for all } \xi \text{ under consideration.}$$

By differentiation we find

$$f''(w(\xi))\, w_\xi(\xi) = 1 \quad \text{for all } \xi \text{ under consideration.}$$

Therefore, w is well-defined and strictly monotone except if $f''(w(\xi))$ vanishes at some point ξ, i.e., if the genuine nonlinearity fails at some value $w(\xi)$. In the latter case, $w_\xi(\xi)$ becomes infinite at some finite value ξ.

Given u_- and u_+, suppose that the function $f'(u)$ is *increasing* when u varies *from u_- to u_+*. Then, the inverse function of f', say g, is well-defined on the interval $[f'(u_-), f'(u_+)]$, and the formula

$$u(x,t) = \begin{cases} u_-, & x < t\, f'(u_-), \\ g(x/t), & t\, f'(u_-) < x < t\, f'(u_+), \\ u_+, & x > t\, f'(u_+) \end{cases} \tag{2.3}$$

defines a smooth and monotone solution of the conservation law (1.1). This solution is called a (centered) **rarefaction wave** connecting u_- to u_+. By definition, the **rarefaction set** $\mathcal{R}(u_-)$ is made of all right-hand states attainable through a rarefaction wave. Using the condition that $x \mapsto f'(u(x,t))$ be increasing in the **rarefaction fan** (that is, in the interval $x \in [t\, f'(u_-), t\, f'(u_+)]$), when the flux is convex we find

$$\mathcal{R}(u_-) = [u_-, +\infty). \tag{2.4}$$

We are now ready to combine together elementary waves and construct a self-similar solution of the Riemann problem. For clarity in the presentation, we call **P** the **class of piecewise smooth functions** $u = u(x/t)$ which are made of finitely many constant states separated with shock waves or rarefaction fans. In the present chapter, all the existence and uniqueness results will be stated in this class. Of course, much more general existence and uniqueness results will be established later in this course.

THEOREM 2.1. (Riemann problem – Convex flux.) *Suppose that the flux f is convex and fix some Riemann data u_l and u_r. Then, the Riemann problem (1.1) and (1.2) admits a* unique *classical entropy solution (in the class* **P** *), composed of shock waves satisfying (1.3) and rarefaction waves, given as follows:*
  (a) *If $u_r \geq u_l$, the solution u is rarefaction wave connecting continuously and monotonically u_l to u_r.*
  (b) *If $u_r < u_l$, the solution is a shock wave connecting u_l to u_r.*
In both cases the solution is always monotone.

The construction when f is concave is completely analogous.

PROOF. Observe that, obviously, for a wave pattern to be realizable in the physical space, one needs the wave speed to be a *monotone increasing* function of the self-similar variable x/t. It is clear that the function described in the theorem is an admissible weak solution of the Riemann problem. On the other hand, the following two claims are immediate:
  • A shock connecting a state u_- to a state $u_+ < u_-$ cannot be followed by another shock or by a rarefaction.
  • A rarefaction cannot be followed by a shock. (But a rarefaction can always be continued by attaching to it another rarefaction.)
Hence, a Riemann solution contains exactly one wave and, therefore, the solution given in the theorem is the only possible combination of shocks and rarefactions. This establishes the uniqueness of the solution in the class **P**.   □

Next, we consider flux-functions having a *single inflection point,* normalized to be $u = 0$. Suppose first that f is a **concave-convex function**, in the sense that (see Figure II-1)

$$u f''(u) > 0 \quad (u \neq 0), \quad f'''(0) \neq 0,$$
$$\lim_{|u| \to +\infty} f'(u) = +\infty. \tag{2.5}$$

The prototype of interest is the cubic flux $f(u) = u^3 + a u$ with $a \in \mathbb{R}$. With some minor modification the following discussion can be extended to functions f' having finite limits at infinity, and functions having several inflection points. Consider the graph of the function f in the (u, f)-plane. For any $u \neq 0$ there exists a unique line that passes through the point with coordinates $(u, f(u))$ and is tangent to the graph at a point $\left(\varphi^\natural(u), f(\varphi^\natural(u))\right)$ with $\varphi^\natural(u) \neq u$. In other words we set (Figure II-1)

$$f'\left(\varphi^\natural(u)\right) = \frac{f(u) - f\left(\varphi^\natural(u)\right)}{u - \varphi^\natural(u)}. \tag{2.6}$$

Note that $u \varphi^\natural(u) < 0$ and define $\varphi^\natural(0) = 0$. Thanks to (2.5) the map $\varphi^\natural : \mathbb{R} \to \mathbb{R}$ is monotone decreasing and onto, and so is invertible. Denote by $\varphi^{-\natural} : \mathbb{R} \to \mathbb{R}$ its inverse function. Obviously, $\varphi^\natural \circ \varphi^{-\natural} = \varphi^{-\natural} \circ \varphi^\natural = id$. By the implicit function theorem, the functions $\varphi^\natural$ and $\varphi^{-\natural}$ are smooth. (This is clear away from $u = 0$, while the discussion of the regularity at $u = 0$ is postponed to Remark 4.4 below.) Moreover, we have $\varphi^{\natural'}(0) = -1/2$ and $\varphi^{-\natural'}(0) = -2$.

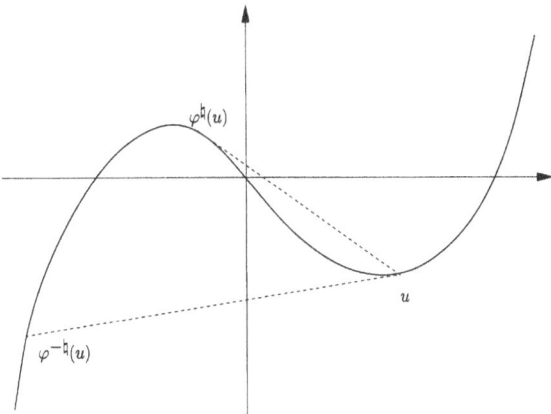

Figure II-1 : Concave-convex flux-function

When the flux is a concave-convex function and all of the entropy inequalities (1.4) are enforced, Oleinik entropy inequalities (1.6) in Theorem 1.1 imply

$$S(u_-) = \begin{cases} [\varphi^\natural(u_-), u_-], & u_- \geq 0, \\ [u_-, \varphi^\natural(u_-)], & u_- \leq 0. \end{cases} \tag{2.7}$$

In passing we point out that, for functions having one inflection point (or none), Lax shock inequalities (1.7) and Oleinik entropy inequalities (1.6) are equivalent. On the other hand, following the general discussion given before the statement of Theorem 2.1, the rarefaction set is easily found to be

$$\mathcal{R}(u_-) = \begin{cases} [u_-, +\infty), & u_- > 0, \\ (-\infty, +\infty), & u_- = 0, \\ (-\infty, u_-], & u_- < 0. \end{cases} \tag{2.8}$$

THEOREM 2.2. (Riemann problem – Concave-convex flux.) *Suppose that the function f is concave-convex (see (2.5)) and fix some Riemann data u_l and u_r. Then, the Riemann problem (1.1) and (1.2) admits a unique classical entropy solution (in the class **P**), made of shock waves satisfying (1.3) and rarefaction waves, given as follows when, for definiteness, $u_l \geq 0$:*

(a) *If $u_r \geq u_l$, the solution u is a rarefaction wave connecting continuously and monotonically u_l to u_r.*

(b) *If $u_r \in [\varphi^\natural(u_l), u_l)$, the solution is a single (classical) shock wave.*

(c) *If $u_r < \varphi^\natural(u_l)$, the solution is composed of a classical shock connecting to $\varphi^\natural(u_l)$ followed by a rarefaction connecting to u_r. The shock is a right-contact* wave, *that is, the shock speed coincides with the right-hand characteristic speed:*

$$\frac{f(u_l) - f(\varphi^{\natural}(u_l))}{u_l - \varphi^{\natural}(u_l)} = f'(\varphi^{\natural}(u_l)), \tag{2.9}$$

that is, the rarefaction is "attached" to the shock.

It is obvious that, in Theorem 2.2 (as well as in Theorem 2.3 below) the Riemann solution is monotone and, when it contains two waves, the intermediate state (specifically here $\varphi^{\natural}(u_l)$) depends continuously upon the data u_l and u_r and converges to u_l or to u_r when passing from one case to another. These important properties of *classical* solutions will no longer hold with *nonclassical* solutions. (See the weaker statement after Theorem 4.1, below.)

PROOF. Observe that in Case (c) above, after a right-contact wave one can add a rarefaction fan, precisely because the left-hand of the rarefaction fan travels with a speed faster than or equal to (in fact, equal to) the shock speed; see (2.9). In view of (2.7) and (2.8), the function described in the theorem is an admissible weak solution of the Riemann problem. To establish that this is the *unique* solution made of elementary waves, we make the following observations:

- After a shock connecting u_- to u_+, no other wave can be added except when $u_+ = \varphi^{\natural}(u_-)$. (The shock is then a right-contact and can be followed with a rarefaction preserving the monotonicity of the solution.)

- After a rarefaction connecting u_- to u_+, no other wave can be added except another rarefaction.

We conclude that a Riemann solution is monotone and contains at most two elementary waves. This establishes the desired uniqueness result. □

When the flux is a **convex-concave function**, in the sense that

$$u\, f''(u) < 0 \quad (u \neq 0), \quad f'''(0) \neq 0,$$
$$\lim_{|u| \to +\infty} f'(u) = -\infty \tag{2.10}$$

and all of the entropies are enforced, we obtain

$$\mathcal{S}(u_-) = \begin{cases} \left(-\infty, \varphi^{-\natural}(u_-)\right] \cup [u_-, +\infty), & u_- \geq 0, \\ \left(-\infty, u_-\right] \cup \left[\varphi^{-\natural}(u_-), +\infty\right), & u_- \leq 0, \end{cases} \tag{2.11}$$

and

$$\mathcal{R}(u_-) = \begin{cases} [0, u_-], & u_- > 0, \\ \{0\}, & u_- = 0, \\ [u_-, 0], & u_- < 0. \end{cases} \tag{2.12}$$

We state without proof:

THEOREM 2.3. (Riemann problem – Convex-concave flux.) *Suppose that the function f is* convex-concave *(see (2.10)) and fix some Riemann data u_l and u_r. Then, the Riemann problem (1.1) and (1.2) admits a* unique *classical entropy solution in the class* **P**, *made of shock waves satisfying all of the entropy inequalities (1.3) and rarefaction waves which, assuming $u_l \geq 0$, is given as follows:*

(a) *If $u_r \geq u_l$, the solution u is a (classical) shock wave connecting u_l to u_r.*

(b) *If $u_r \in [0, u_l)$, the solution is a rarefaction wave connecting monotonically u_l to u_r.*

(c) *If $\varphi^{-\natural}(u_l) < u_r < 0$, the solution is composed of a rarefaction connecting u_l to $\varphi^\natural(u_r)$, followed by a shock wave connecting to u_r. The shock is a left-contact, that is, the shock speed coincides with the left-hand characteristic speed:*

$$\frac{f(u_r) - f\left(\varphi^\natural(u_r)\right)}{u_r - \varphi^\natural(u_r)} = f'\left(\varphi^\natural(u_r)\right). \tag{2.13}$$

In particular, the rarefaction is attached to the shock.

(d) *If $u_r \in \left(-\infty, \varphi^{-\natural}(u_l)\right)$, the solution is a (classical) shock wave connecting u_l to u_r.*

□

Finally, when the flux-function f admits more than one inflection point but, for clarity, has only *finitely many inflection points*, the Riemann problem (1.1)–(1.3) can also be solved explicitly. The construction is based on the convex hull (when $u_l < u_r$) or the concave hull (when $u_l > u_r$) of the function f in the interval limited by the Riemann data u_l and u_r. Denoting this envelop by $\hat{f}$ and assuming for instance that $u_l > u_r$ we can decompose the interval $[u_r, u_l]$ in the following way: There exist states

$$u_l = u^1 \geq u^2 > \ldots > u^{N-1} > u^N = u_r$$

such that for all relevant values of p

$$\begin{aligned} \hat{f}(u) &= f(u), \quad u \in \left(u^{2p}, u^{2p+1}\right), \\ \hat{f}(u) &< f(u), \quad u \in \left(u^{2p+1}, u^{2p+3}\right). \end{aligned} \tag{2.14}$$

The intervals in which $\hat{f}$ coincides with f correspond to rarefaction fans in the solution of the Riemann problem, while the intervals where $\hat{f}$ is strictly below f correspond to shock waves.

It is not difficult to check from Oleinik entropy inequalities (1.6) that:

THEOREM 2.4. (Riemann problem – General flux.) *Suppose that the function f has finitely many inflection points and fix some Riemann data u_l and u_r. Then, the Riemann problem (1.1) and (1.2) admits a unique classical entropy solution (in the class $\mathbf{P}$), made of shock waves satisfying (1.3) and rarefaction waves which, when $u_l \geq u_r$, is given by*

$$u(x, t) = \begin{cases} u_l, & x < t\,\hat{f}(u_l), \\ \left((\hat{f})'\right)^{-1}\left(\frac{x}{t}\right), & t\,\hat{f}(u_l) < x < t\,\hat{f}(u_r), \\ u_r, & x > t\,\hat{f}(u_r). \end{cases} \tag{2.15}$$

□

Observe that $\hat{f}$ is convex and, therefore, $\left(\hat{f}\right)'$ is non-decreasing and its inverse is well-defined but may be discontinuous.

3. Entropy dissipation function

In the rest of this chapter we solve the Riemann problem (1.1) and (1.2) when the *single entropy inequality* (1.4) is imposed on the solutions. When the flux $f : I\!R \to I\!R$ is convex (or concave) the single inequality (1.4) is equivalent to the infinite list (1.3) and we immediately recover the solution in Theorem 2.1. Consequently, in the rest of this chapter we focus on *non-convex* flux-functions and explain how to construct *nonclassical* entropy solutions of the Riemann problem.

Our general strategy is as follows. First, in the present section we describe the class of Riemann solutions satisfying (1.1), (1.2), and (1.4), and we exhibit a *one-parameter family* of Riemann solutions, obtained by combining shock waves and rarefaction waves. Second, in the following section we explain how a *kinetic relation* may be imposed in order to formulate a well-posed Riemann problem.

In this section and in Section 4 the function f is assumed to be *concave-convex*, in the sense (2.5). We will use the function $\varphi^\natural : I\!R \to I\!R$ defined in (2.6) together with its inverse denoted by $\varphi^{-\natural} : I\!R \to I\!R$. Recall that, by (2.5), both $\varphi^\natural$ and $\varphi^{-\natural}$ are monotone decreasing and onto.

Consider a shock wave of the general form (1.5) connecting two states u_- and u_+, where the speed $\lambda = \bar{a}(u_-, u_+)$ is given by the Rankine-Hugoniot relation (1.8). Recall that the entropy inequality (1.4) holds if and only if the **entropy dissipation**

$$E(u_-, u_+) := -\bar{a}(u_-, u_+)\big(U(u_+) - U(u_-)\big) + F(u_+) - F(u_-) \qquad (3.1)$$

is non-positive. We can prove that $u_+ \mapsto E(u_-, u_+)$ achieves a maximum negative value at $u_+ = \varphi^\natural(u_-)$ and vanishes exactly twice. For definiteness we take $u_- > 0$ in the rest of the discussion. Dealing with the other case is completely similar and can also be deduced from the forthcoming results, based on the skew-symmetry of the function E, i.e., $E(u_-, u_+) = -E(u_+, u_-)$.

THEOREM 3.1. (Entropy dissipation for concave-convex flux.) *For any left-hand state* $u_- > 0$ *the function* $E(u_-, .)$ *is monotone decreasing in* $\big(-\infty, \varphi^\natural(u_-)\big]$ *and monotone increasing in* $\big[\varphi^\natural(u_-), +\infty\big)$. *More precisely, we have*

$$\begin{aligned}
&\partial_{u_+} E(u_-, .) < 0 \text{ in the interval } \big(-\infty, \varphi^\natural(u_-)\big), \\
&\partial_{u_+} E(u_-, .) > 0 \text{ in the intervals } \big(\varphi^\natural(u_-), u_-\big) \cup \big(u_-, +\infty\big), \\
&E(u_-, u_-) = 0, \\
&E(u_-, \varphi^\natural(u_-)) < 0, \quad E(u_-, \varphi^{-\natural}(u_-)) > 0.
\end{aligned} \qquad (3.2)$$

Therefore, for $u > 0$ *there exists some value* $\varphi_0^\flat(u)$ *satisfying*

$$E(u, \varphi_0^\flat(u)) = 0, \quad \varphi_0^\flat(u) \in \big(\varphi^{-\natural}(u), \varphi^\natural(u)\big).$$

The definition of $\varphi_0^\flat(u)$ *for* $u \leq 0$ *is analogous, and the function* $\varphi_0^\flat : I\!R \to I\!R$ *is monotone decreasing with*

$$\partial_u \varphi_0^\flat(u) < 0, \quad u \in I\!R, \qquad (3.3)$$

and

$$\varphi_0^\flat(\varphi_0^\flat(u)) = u, \quad u \in I\!R. \qquad (3.4)$$

We refer to Figure II-2 for a graphical representation of the **zero-entropy dissipation function** $\varphi_0^\flat$. To motivate our notation we stress that $\varphi_0^\flat$ will determine a critical limit for the range of the kinetic functions $\varphi^\flat$ introduced later in Section 4. On the other hand, in Chapter III the function $\varphi_0^\flat$ will also arise from diffusive-dispersive approximations when a diffusion parameter tends to 0.

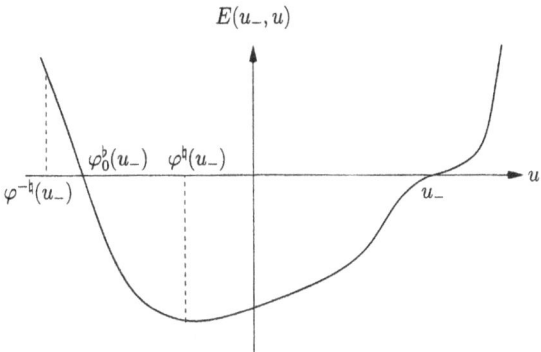

Figure II-2 : Entropy dissipation function.

To the function $\varphi_0^\flat$ we associate its companion function $\varphi_0^\sharp : I\!\!R \to I\!\!R$ defined by (see Figure II-4 below)

$$\frac{f(u) - f(\varphi_0^\sharp(u))}{u - \varphi_0^\sharp(u)} = \frac{f(u) - f(\varphi_0^\flat(u))}{u - \varphi_0^\flat(u)}, \qquad (3.5)$$

so that the points with coordinates

$$\left(\varphi_0^\flat(u), f(\varphi_0^\flat(u))\right), \quad \left(\varphi_0^\sharp(u), f(\varphi_0^\sharp(u))\right), \quad (u, f(u))$$

are aligned. Since $\varphi^{-\natural}(u) < \varphi_0^\flat(u) < \varphi^\natural(u)$ we also have

$$\varphi^\natural(u) < \varphi_0^\sharp(u) < u, \quad u > 0.$$

More generally, when $u_+ \neq u_-, \varphi^\natural(u_-)$ it will be useful to define $\rho(u_-, u_+) \in I\!\!R$ by

$$\frac{f(\rho(u_-, u_+)) - f(u_-)}{\rho(u_-, u_+) - u_-} = \frac{f(u_+) - f(u_-)}{u_+ - u_-}, \quad \rho(u_-, u_+) \neq u_-, u_+, \qquad (3.6)$$

and to extend the mapping ρ by continuity.

PROOF OF THEOREM 3.1. Observe first that some of the properties (3.2) are obvious from the formula

$$E(u_-, u_+) = -\int_{u_-}^{u_+} U''(v)\, (v - u_-) \left(\frac{f(v) - f(u_-)}{v - u_-} - \frac{f(u_+) - f(u_-)}{u_+ - u_-}\right) dv \qquad (3.7)$$

derived earlier in the proof of Theorem 1.1. For instance, when $u_+ \leq \varphi^{-\natural}(u_-)$ or when $u_+ \geq u_-$, the term in the integrand of (3.7) have a *constant* sign and we see that $E(u_-, u_+) > 0$. On the other hand, for values u_+ near u_- the dissipation $E(u_-, u_+)$ is equivalent to $(u_+ - u_-)^3$ (up to a multiplicative constant). Thus, locally near u_-, $E(u_-, u_+) > 0$ for $u_+ > u_-$ and $E(u_-, u_+) < 0$ for $u_+ < u_-$.

To show the first two statements in (3.2), note that the differentials of the functions $E(u_-, u_+)$ and $\bar{a}(u_-, u_+)$ are closely related. Indeed, a calculation based on differentiating (1.8) and (3.1) with respect to u_+ yields

$$\partial_{u_+} E(u_-, u_+) = b(u_-, u_+)\, \partial_{u_+} \bar{a}(u_-, u_+),$$
$$b(u_-, u_+) := U(u_-) - U(u_+) - U'(u_+)\,(u_- - u_+) > 0$$

for $u_- \neq u_+$ and

$$\partial_{u_+} \bar{a}(u_-, u_+) = \frac{f'(u_+) - \bar{a}(u_-, u_+)}{u_+ - u_-}. \tag{3.8}$$

In view of (3.8) and (2.5) it is clear that

$$\partial_{u_+} \bar{a}(u_-, u_+) < 0, \quad u_+ < \varphi^\natural(u_-),$$
$$\partial_{u_+} \bar{a}(u_-, u_+) > 0, \quad u_+ > \varphi^\natural(u_-).$$

This leads us to the conclusions listed in (3.2). Then, in view of (3.2) there exists $\varphi_0^\flat(u_-)$ satisfying $E(u_-, \varphi_0^\flat(u_-)) = 0$.

By definition, for any $u \neq 0$

$$E\big(u, \varphi_0^\flat(u)\big) = 0, \quad u \neq \varphi_0^\flat(u)$$

and, since a similar result as (3.2) holds for negative left-hand side,

$$E\big(\varphi_0^\flat(u), \varphi_0^\flat(\varphi_0^\flat(u))\big) = 0, \quad \varphi_0^\flat(u) \neq \varphi_0^\flat(\varphi_0^\flat(u)).$$

Thus, (3.4) follows from the fact that the entropy dissipation has a single "non-trivial" zero and from the symmetry property

$$E\big(\varphi_0^\flat(u), u\big) = -E\big(u, \varphi_0^\flat(u)\big) = 0.$$

Finally, by the implicit function theorem it is clear that the function $\varphi_0^\flat$ is smooth, at least away from $u = 0$. (The regularity at $u = 0$ is discussed in Remark 4.4 below.) Using again the symmetry property $E(u_-, u_+) = -E(u_+, u_-)$ we have

$$\big(\partial_{u_-} E\big)\big(u, \varphi_0^\flat(u)\big) = -\big(\partial_{u_+} E\big)\big(\varphi_0^\flat(u), u\big). \tag{3.9}$$

Thus, differentiating the identity $E\big(u, \varphi_0^\flat(u)\big) = 0$ we obtain

$$\frac{d\varphi_0^\flat}{du}(u) = -\frac{\big(\partial_{u_-} E\big)\big(u, \varphi_0^\flat(u)\big)}{\big(\partial_{u_+} E\big)\big(u, \varphi_0^\flat(u)\big)} = \frac{\big(\partial_{u_+} E\big)\big(\varphi_0^\flat(u), u\big)}{\big(\partial_{u_+} E\big)\big(u, \varphi_0^\flat(u)\big)},$$

where we used (3.9). For $u > 0$ we have already established (see (3.2)) that

$$\big(\partial_{u_+} E\big)\big(u, \varphi_0^\flat(u)\big) < 0,$$

and since a similar result as (3.2) holds for negative left-hand side and that we have $(\varphi_0^\flat \circ \varphi_0^\flat)(u) = u$, we conclude that

$$\big(\partial_{u_+} E\big)\big(\varphi_0^\flat(u), u\big) > 0$$

and therefore $d\varphi_0^\flat/du < 0$. $\square$

REMARK 3.2. For the choice $U(u) = u^2/2$ the function $\varphi_0^\flat$ is given geometrically by an analogue of Maxwell's **equal area rule**. Namely, rewriting (3.1) in the form

$$E(u_-, u_+) = -\int_{u_-}^{u_+} \left(f(v) - f(u_-) - \frac{f(u_+) - f(u_-)}{u_+ - u_-}(v - u_-) \right) dv,$$

we see that the line connecting $(u_-, f(u_-))$ to $(\varphi_0^\flat(u_-), f(\varphi_0^\flat(u_-)))$ cut the graph of f in two regions with equal areas. This property arises also in the context of elastodynamics (Example I-4.4) and phase transition dynamics (Example I-4.5). □

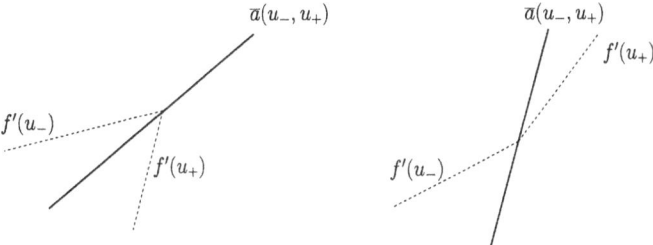

Figure II-3 : Compressive and undercompressive shock waves.

Using the notation in Theorem 3.1 we reach the following conclusion.

LEMMA 3.3. (Single entropy inequality.) *A shock wave of the form* (1.5) *and* (1.8) *satisfies the single entropy inequality* (1.4) *if and only if*

$$u_+ \in \begin{cases} [\varphi_0^\flat(u_-), u_-], & u_- \geq 0, \\ [u_-, \varphi_0^\flat(u_-)], & u_- \leq 0. \end{cases} \tag{3.10}$$

DEFINITION 3.4. Among the propagating discontinuities satisfying (3.10) some satisfy also Oleinik entropy inequalities (1.6) (and therefore Lax shock inequalities (1.7)) and will be called **classical shocks** or **Lax shocks**. They correspond to the intervals

$$u_+ \in \begin{cases} [\varphi^\natural(u_-), u_-], & u_- \geq 0, \\ [u_-, \varphi^\natural(u_-)], & u_- \leq 0. \end{cases} \tag{3.11}$$

On the other hand, the propagating discontinuities satisfying (3.10) but violating Oleinik entropy inequalities, i.e.,

$$u_+ \in \begin{cases} [\varphi_0^\flat(u_-), \varphi^\natural(u_-)), & u_- \geq 0, \\ (\varphi^\natural(u_-), \varphi_0^\flat(u_-)], & u_- \leq 0, \end{cases} \tag{3.12}$$

will be called **nonclassical shocks**.

Observe (see also Figure II-3) that nonclassical shocks are **slow undercompressive** in the sense that characteristics on both sides of the discontinuity pass through the shock:

$$f'(u_\pm) \geq \bar{a}(u_-, u_+). \tag{3.13}$$

This is in strong contrast with Lax shock inequalities which impose that the characteristics impinge on the discontinuity from both sides. Undercompressive waves are a potential source of *non-uniqueness*, as will become clear shortly.

The rarefaction waves associated with the equation (1.1) were already studied in Section 2. For concave-convex fluxes we found (see (2.8)):

$$\mathcal{R}(u_-) = \begin{cases} [u_-, +\infty), & u_- > 0, \\ (-\infty, +\infty), & u_- = 0, \\ (-\infty, u_-], & u_- < 0. \end{cases} \tag{3.14}$$

We are now in the position to solve the Riemann problem (1.1), (1.2), and (1.4).

THEOREM 3.5. (*One-parameter family of Riemann solutions for concave-convex functions.*) *Suppose that the flux f is a concave-convex function (see (2.5)) and fix some Riemann data u_l and u_r. Restricting attention to solutions satisfying the entropy inequality (1.4) for a given strictly convex entropy pair (U, F) and assuming for definiteness that $u_l \geq 0$, the Riemann problem (1.1) and (1.2) admits the following solutions in the class* **P**:

(a) *If $u_r \geq u_l$, the solution is unique and consists of a rarefaction connecting continuously u_l to u_r.*

(b) *If $u_r \in [\varphi_0^\natural(u_l), u_l)$, the solution is unique and consists of a classical shock connecting u_l to u_r.*

(c) *If $u_r \in [\varphi_0^\flat(u_l), \varphi_0^\natural(u_l))$, there exist infinitely many solutions, consisting of a nonclassical shock connecting u_l to some intermediate state u_m followed by*
   – a classical shock if $u_m < \rho(u_l, u_r)$ (the function ρ being defined in (3.6)),
   – or a rarefaction if $u_m \geq u_r$.
   The values $u_r \in [\varphi^\natural(u_l), \varphi_0^\natural(u_l)]$ can also be attained with a single classical shock.

(d) *If $u_r \leq \varphi_0^\flat(u_l)$, there exist infinitely many solutions, consisting of a nonclassical shock connecting u_l to some intermediate state $u_m \in [\varphi_0^\flat(u_l), \varphi^\natural(u_l)]$ followed by a rarefaction connecting continuously to u_r.*

In Cases (c) and (d) above there exists a **one-parameter family of Riemann solutions**. Note that, in Case (c), the solution at time $t > 0$ may contain two shocks and have a total variation which is *larger* than the one of its initial data.

PROOF. The functions described above are clearly solutions of the Riemann problem. The only issue is to see whether they are the *only* admissible solutions. The argument below is based on the two key properties (3.3) and (3.4). We recall that two wave fans can be combined only when the largest speed of the left-hand wave is less than or equal to the smaller speed of the right-hand one.

> Claim 1 : A nonclassical shock connecting u_- to $u_+ \in [\varphi_0^\flat(u_-), \varphi^\natural(u_-))$ can be followed only by a shock connecting to a value $u_2 \in [u_+, \rho(u_-, u_+))$ or else by a rarefaction to $u_2 \leq u_+$.

Indeed, each state $u_2 \in [u_+, \rho(u_-, u_+))$ is associated with a classical shock propagating at the speed $\bar{a}(u_2, u_+)$, which is greater than $\bar{a}(u_-, u_+)$. These states are thus attainable by adding a classical shock after the nonclassical one. On the other hand, a state $u_2 \in (\varphi_0^\natural(u_+), \varphi_0^\flat(u_+)]$ cannot be reached by adding a second shock after the non-classical one since, by the property (3.4), $\varphi_0^\flat(u_+) = u_-$ and therefore any shock connecting u_+ to some state $u_2 > \varphi_0^{\natural}(u_+)$ travels with a smaller speed: $\bar{a}(u_+, u_2) < \bar{a}(u_-, u_+)$. Finally, the states $u_2 < u_+$ cannot be reached since they are associated with rarefactions which travel faster than the nonclassical shock.

> Claim 2 : After a classical shock leaving from a state u_- and reaching u_+, no other wave can be added except when $u_+ = \varphi^{\natural}(u_-)$ and, in that case, a rarefaction only can follow the classical shock.

It is easy to see using the condition on the ordering of waves that a classical shock cannot be added after another classical shock, nor a rarefaction except when $u_+ = \varphi^{\natural}(u_-)$. Consider next a nonclassical shock issuing from u_+ and reaching u_2. Consider for instance the case $u_+ < 0$. For the nonclassical shock to be admissible one needs

$$u_2 \leq \varphi_0^{\flat}(u_+),$$

but the speeds should be ordered,

$$\bar{a}(u_+, u_2) > \bar{a}(u_-, u_+),$$

and therefore $u_2 > u_-$. By combining the condition (3.4),the monotonicity property of $\varphi_0^{\flat}$, and the inequality $u_+ > \varphi_0^{\flat}(u_-)$ we find also

$$u_- = \varphi_0^{\flat}(\varphi_0^{\flat}(u_-)) \geq \varphi_0^{\flat}(u_+) \geq u_2,$$

which is a contradiction.

Claims 1 and 2 prevent us from combining together more than two waves and this completes the proof of Theorem 3.5.   □

4. Nonclassical Riemann solver for concave-convex flux

In view of the results in Section 3 it is necessary to supplement the Riemann problem with an additional selection criterion which we call a "kinetic relation". The approach followed now, in particular the assumptions placed on the kinetic function, will be fully justified a posteriori by the results in Chapter III, devoted to deriving kinetic functions from a traveling wave analysis of diffusive-dispersive models.

Imposing the single entropy inequality (1.4) already severely restricts the class of admissible solutions. One free parameter, only, remains to be determined and the range of nonclassical shocks is constrained by the zero-entropy dissipation function $\varphi_0^{\flat}$ discovered in Theorem 3.1.

Let $\varphi^{\flat} : I\!R \mapsto I\!R$ be a **kinetic function**, i.e., by definition, a monotone decreasing and Lipschitz continuous mapping such that

$$\begin{aligned} \varphi_0^{\flat}(u) < \varphi^{\flat}(u) \leq \varphi^{\natural}(u), \quad u > 0, \\ \varphi^{\natural}(u) \leq \varphi^{\flat}(u) < \varphi_0^{\flat}(u), \quad u < 0, \end{aligned} \tag{4.1}$$

The kinetic function will be applied to select nonclassical shock waves. Observe that (3.4) and (4.1) imply the following **contraction property**:

$$|\varphi^{\flat}(\varphi^{\flat}(u))| < |u|, \quad u \neq 0. \tag{4.2}$$

From $\varphi^{\flat}$ we also define its companion function $\varphi^{\natural} : I\!R \to I\!R$ by

$$\frac{f(u) - f(\varphi^{\natural}(u))}{u - \varphi^{\natural}(u)} = \frac{f(u) - f(\varphi^{\flat}(u))}{u - \varphi^{\flat}(u)}, \quad u \neq 0. \tag{4.3}$$

Note that, by (4.1),

$$\varphi^{\natural}(u) \le \varphi^{\sharp}(u) < \varphi^{\sharp}_0(u), \quad u > 0,$$

$$\varphi^{\sharp}_0(u) < \varphi^{\sharp}(u) \le \varphi^{\natural}(u), \quad u < 0.$$

(See Figure II-4 for an illustration of the respective positions of these functions.)

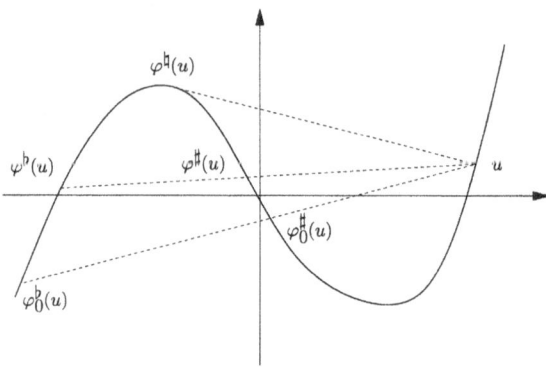

Figure II-4 : Concave-convex flux-function

From Theorem 3.5 we easily reach the following conclusion. (See Figure II-5.)

DEFINITION AND THEOREM 4.1. (Riemann solution for concave-convex flux – First formulation.) *Let $\varphi^{\flat}$ be a given kinetic function. Under the assumptions of Theorem 3.5, a weak solution (in the class* **P***) is called a* **nonclassical entropy solution** *(associated with the kinetic function $\varphi^{\flat}$) if any nonclassical shock connecting two states u_- and u_+ satisfies the* **kinetic relation**

$$u_+ = \varphi^{\flat}(u_-) \quad \text{for all nonclassical shocks,} \tag{4.4}$$

The Riemann problem (1.1), (1.2), and (1.4) admits an (essentially unique) nonclassical entropy solution (in the class **P***), given as follows when $u_l > 0$:*
- (a) *If $u_r \ge u_l$, the solution is a rarefaction connecting u_l to u_r.*
- (b) *If $u_r \in [\varphi^{\sharp}(u_l), u_l)$, the solution is a classical shock.*
- (c) *If $u_r \in [\varphi^{\flat}(u_l), \varphi^{\sharp}(u_l))$, the solution consists of a nonclassical shock connecting u_l to $\varphi^{\flat}(u_l)$ followed by a classical shock.*
- (d) *If $u_r \le \varphi^{\flat}(u_l)$, the solution consists of a nonclassical shock connecting u_l to $\varphi^{\flat}(u_l)$ followed by a rarefaction connecting to u_r.*

In Cases (a), (b), and (d) the solution is monotone, while it is non-monotone in Case (c). The classical Riemann solution (Theorem 2.2) is also admissible as it contains only classical waves (for which (4.4) is irrelevant). □

Observe that the value $\varphi^{\sharp}(u_l)$ determines an important *transition* from a one-wave to a two-wave pattern. The nonclassical Riemann solution fails to depend pointwise continuously upon its initial data, in the following sense. The solution in Case (c) contains the middle state $\varphi^{\flat}(u_l)$ which *does not converge to u_l nor u_r* when the right state converges to $\varphi^\natural(u_l)$. We point out that the Riemann problem with left data $u_l < 0$ is solved in a completely similar manner using the value $\varphi^\flat(u_l) > 0$. For $u_l = 0$ the Riemann problem is simply a rarefaction wave connecting monotonically u_l to u_r. *Different* kinetic functions yield *different* Riemann solver. This reflects the fact that different regularizations of the conservation law, in general, yield different limits. With the trivial choice $\varphi^\flat = \varphi^\natural$ we recover the classical Riemann solution, while with the choice $\varphi^\flat = \varphi_0^\flat$ we select nonclassical shocks with zero entropy dissipation. (See also Chapter III.)

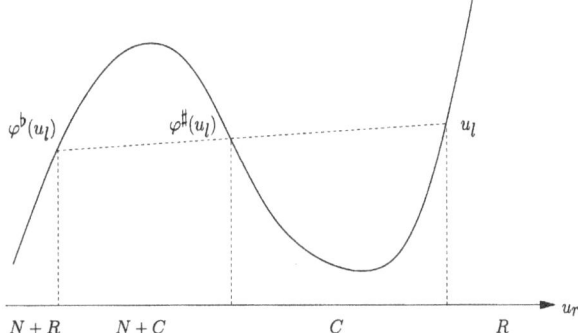

Figure II-5 : The four wave patterns for the Riemann solution.

REMARK 4.2. If, in addition to (4.1), the stronger condition

$$\varphi^\natural(u) \le \varphi^\natural(u) < 0, \quad u > 0,$$
$$0 < \varphi^\natural(u) \le \varphi^\natural(u), \quad u < 0,$$

is assumed on the kinetic function, then the solution of the Riemann problem is classical as long as the left- and right-hand states have the same sign. In particular, this is always the case when $U = u^2/2$ and $f(u) = u^3$ (or, more generally, f is a skew-symmetric function of u) since then $\varphi_0^\natural \equiv 0$.   □

In the rest of this section we propose a *reformulation* of the kinetic relation (4.4), along the following lines:
- Since the entropy dissipation E in (3.1) played a central role in restricting the range of nonclassical shocks (see (3.12)) it is natural to set the kinetic relation in terms of the function E, also.
- Speaking loosely, we regard a nonclassical shock as a "propagating boundary" separating two "phases" of a material. In continuum physics, an analogue of E is called a **driving force** acting on the propagating discontinuity, and one typically postulates a one-to-one, monotonic relation between the driving force and the propagation speed.

- The (second) formulation in Theorem 4.3 below allows us to eliminate the classical Riemann solution still left out in the (first) formulation given in Theorem 4.1.

For a nonclassical shock connecting some states u_- and u_+ at some speed $\lambda = \overline{a}(u_-, u_+)$ we now write the **kinetic relation** in the form

$$E(u_-, u_+) = \begin{cases} \Phi^+(\lambda), & u_+ < u_- \\ \Phi^-(\lambda), & u_+ > u_- \end{cases} \quad \text{for all nonclassical shocks,} \quad (4.5)$$

where, by definition, the **kinetic functions** $\Phi^\pm : [f'(0), +\infty) \to I\!\!R_-$ are Lipschitz continuous mappings satisfying

$$\Phi^\pm(f'(0)) = 0,$$
$$\Phi^\pm \text{ is monotone decreasing,} \quad (4.6)$$
$$\Phi^\pm(\lambda) \geq E^\pm(\lambda).$$

In the latter inequality, the lower bounds $E^\pm$ are the **maximum negative entropy dissipation** function defined by

$$E^\pm(\lambda) := E\big(u, \varphi^\natural(u)\big), \quad \lambda = f'(\varphi^\natural(u)) \quad \text{for } \pm u \leq 0. \quad (4.7)$$

Observe that given $\lambda > f'(0)$ there are exactly one positive root and one negative root u such that $\lambda = f'(u)$. This is why we have to introduce two kinetic functions $\Phi^\pm$ associated with decreasing and increasing jumps, respectively. Note also that $f'(0)$ is a lower bound for all wave speeds. As we will see shortly, (4.5) is equivalent to a relation

$$u_+ = \varphi^\flat(u_-),$$

from which we also define $\varphi^\sharp(u_-)$ as in (4.3).

Finally, in order to exclude the classical entropy solution we impose the following **nucleation criterion.** For every shock connecting u_- to u_+ we have

$$E(u_-, u_+) \geq E(u_-, \varphi^\sharp(u_-)) =: E^\sharp(u_-). \quad (4.8)$$

This condition enforces that a discontinuity having an entropy dissipation larger than the critical threshold $E^\sharp(u_-)$ must "nucleate", that is, gives rise to nonclassical waves.

THEOREM 4.3. (Riemann problem for concave-convex flux – Second formulation.) *Fix some kinetic functions* $\Phi^\pm : [f'(0), +\infty) \to I\!\!R_-$ *(satisfying, in particular, (4.6)). Then, under the assumptions of Theorem 3.5 the kinetic relation (4.5) selects a* unique nonclassical shock *for each left-hand state* u_-. *On the other hand, the nucleation criterion (4.8) excludes the classical solution. As a consequence, the Riemann problem admits a* unique nonclassical entropy solution *(in the class* **P***), described in Theorem 4.1 above.*

PROOF. For $u_- > 0$ fixed we claim that there is a unique nonclassical connection to a state u_+ satisfying the jump relation and the kinetic relation (4.5). Let us write the entropy dissipation as a function of the speed λ:

$$\Psi(\lambda) = E\big(u_-, u_+(\lambda)\big), \quad \lambda = \frac{f\big(u_+(\lambda)\big) - f(u_-)}{u_+(\lambda) - u_-}.$$

Setting

$$\lambda^{\natural} := f'(\varphi^{\natural}(u_-)), \quad \lambda_0 := \frac{f(\varphi_0^{\flat}(u_-)) - f(u_-)}{\varphi_0^{\flat}(u_-) - u_-},$$

from Theorem 3.1 and the assumption (4.6) it follows that

$$\Psi \text{ is monotone increasing for } \lambda \in [\lambda^{\natural}, \lambda_0],$$
$$\Psi(\lambda^{\natural}) = E^+(\lambda^{\natural}) \le \Phi^+(\lambda^{\natural}), \qquad (4.9)$$
$$\Psi(\lambda_0) = 0 \ge \Phi^+(\lambda_0).$$

All of the desired properties are obvious, except the fact that Ψ is increasing. But, note that $u_+ \mapsto E(u_-, u_+)$ is decreasing in the relevant interval $u_+ \in [\varphi_0^{\flat}(u_-), \varphi^{\natural}(u_-)]$. The mapping $\lambda \mapsto u_+(\lambda)$ is also decreasing for $\lambda \in (\lambda^{\natural}, \lambda_0]$ since

$$u_+'(\lambda)\,\lambda + u_+(\lambda) - u_- = f'(u_+(\lambda))\,u_+'(\lambda).$$

Thus

$$u_+'(\lambda) = \frac{u_+(\lambda) - u_-}{f'(u_+(\lambda)) - \lambda} < 0.$$

Finally, in view of (4.9) and since Φ^+ is monotone decreasing the equation

$$\Psi(\lambda) = \Phi^+(\lambda)$$

admits exactly one solution. (See also Figure II-6.) This completes the proof that the nonclassical shock is unique.

We now deal with the nucleation criterion (4.8). By the monotonicity properties of the function E (Theorem 3.1) the condition (4.8) implies that the classical shocks connecting u_- to $u_+ \in [\varphi^{\natural}(u_-), \varphi^{\sharp}(u_-))$ are not admissible. On the other hand, since the speeds of the shock connecting u_- to $\varphi^{\flat}(u_-)$ and the one of the shock connecting u_- to $\varphi^{\sharp}(u_-)$ coincide, we have

$$E(u_-, \varphi^{\sharp}(u_-)) - E(u_-, \varphi^{\flat}(u_-)) = E(\varphi^{\flat}(u_-), \varphi^{\sharp}(u_-)) \le 0.$$

Moreover, the inequality above is a consequence of Theorem 3.1 and the fact that $\varphi^{\sharp}(u_-) < \varphi^{\natural}(\varphi^{\flat}(u_-))$ (which is clear geometrically). So, we have

$$E(u_-, \varphi^{\flat}(u_-)) \ge E^{\natural}(u_-), \qquad (4.10)$$

which means that the nonclassical value satisfies the nucleation criterion. In conclusion, (4.8) excludes the classical solution precisely when the nonclassical solution is available. This completes the proof of Theorem 4.3. □

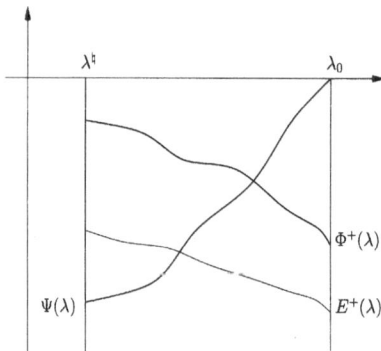

Figure II-6 : Dissipation and kinetic functions versus shock speed.

REMARK 4.4.
- It easily follows from the implicit function theorem that, since the flux function f is smooth, all of the functions $\varphi^{\natural}$, $\varphi^{-\natural}$, $\varphi_0^{\flat}$, and $\varphi_0^{\sharp}$ are smooth away from $u = 0$ at least. To discuss the regularity at $u = 0$ we will rely on the assumption made in (2.5) that the flux is non-degenerate at 0 in the sense that

$$f'''(0) \neq 0.$$

The regularity of the function $\varphi^{\natural}$ at $u = 0$ is obtained by applying the implicit function theorem to the (differentiable) function

$$H(u, \varphi) = \frac{f(u) - f(\varphi) - (u - \varphi)\,f'(\varphi)}{(u - \varphi)^2}$$

$$= \int_0^1 \int_0^1 f''(\varphi + m\,s\,(u - \varphi))\,m\,s\,dsdm,$$

which satisfies $H(0,0) = 0$ and $(\partial H / \partial \varphi)(0,0) = 2\,f'''(0) \neq 0$. A similar argument would establish the regularity of $\varphi^{-\natural}$. The regularity of the function $\varphi_0^{\flat}$ follows also from the implicit function theorem by relying on the (differentiable) function

$$H(u, \varphi) = \frac{F(u) - F(\varphi) - (U(u) - U(\varphi))\,\overline{a}(u, \varphi)}{(u - \varphi)^3},$$

$$= \int_0^1 \int_0^1 \int_0^1 \int_0^1 f''(\varphi + m\,(u - \varphi) + p\,(s - m)\,(u - \varphi))$$
$$U''(\varphi + qs\,(u - \varphi))\,(s - m)\,s\,dsdmdpdq,$$

which satisfies $H(0,0) = 0$ and $(\partial H / \partial \varphi)(0,0) = f'''(0)\,U''(0)/24 \neq 0$.
- As we will see in the applications in Chapter III, it is natural to assume that the kinetic function $\varphi^{\flat}$ is solely Lipschitz continuous. The Lipschitz continuity of the companion function $\varphi^{\sharp}$ follows from a generalization of the implicit

function theorem for Lipschitz continuous mappings. (See the bibliographical notes for a reference.) One should use here the function

$$H(u, \varphi) = \frac{\overline{a}(u, \varphi) - \overline{a}(u, \varphi^\flat(u))}{u - \varphi^\flat(u)}$$

$$= \int_0^1 \int_0^1 f'' \left(\varphi^\flat(u) + s \left(u - \varphi^\flat(u) \right) + m \left(1 - s \right) \left(\varphi - \varphi^\flat(u) \right) \right) m \left(1 - s \right) ds dm,$$

which satisfies $H(0,0) = 0$ and $(\partial H / \partial \varphi)(0,0) = f'''(0)/9 \neq 0$.

$\square$

5. Nonclassical Riemann solver for convex-concave flux

In this last section we restrict attention to flux-functions satisfying the convex-concave property (2.10). Strictly speaking, the case (2.10) could be deduced from the case (2.5), provided the Riemann solution of (1.1) and (1.2) would be described by fixing the right-hand state u_r and using u_l as a parameter. We shall omit most of the proofs in this section since they are similar to the ones in Sections 3 and 4. First of all, the functions $\varphi^\natural$ and $\varphi^{-\natural}$ are defined as in Section 2. Again, we consider a shock wave of the form (1.5) and (1.8) connecting two states u_- and u_+ at the speed $\lambda = \overline{a}(u_-, u_+)$. We study the entropy dissipation $E(u_-, u_+)$ (see (3.1)) by keeping u_- fixed.

THEOREM 5.1. (Entropy dissipation for convex-concave flux.) *Given $u_- > 0$, the function $E(u_-, .)$ is monotone increasing in $\left(-\infty, \varphi^\natural(u_-) \right]$ and monotone decreasing in $\left[\varphi^\natural(u_-), +\infty \right)$. More precisely, we have*

$$\partial_{u_+} E(u_-, .) > 0 \quad \text{in the interval } \left(-\infty, \varphi^\natural(u_-) \right),$$

$$\partial_{u_+} E(u_-, .) < 0 \quad \text{in the intervals } \left(\varphi^\natural(u_-), u_- \right) \cup \left(u_-, +\infty \right),$$

$$E(u_-, u_-) = 0, \quad E \left(u_-, \varphi^\natural(u_-) \right) > 0, \quad E \left(u_-, \varphi^{-\natural}(u_-) \right) < 0.$$

Hence, for each $u > 0$ there exists $\varphi_0^\flat(u) \in \left(\varphi^{-\natural}(u), \varphi^\natural(u) \right)$ such that $E(u, \varphi_0^\flat(u)) = 0$. The definition of $\varphi_0^\flat(u)$ for $u \leq 0$ is analogous and the function $\varphi_0^\flat : \mathbb{R} \to \mathbb{R}$ is monotone decreasing (as are both $\varphi^\natural$ and $\varphi^{-\natural}$) with

$$\partial_u \varphi_0^\flat(u) < 0, \quad u \in \mathbb{R}.$$

$\square$

To the function $\varphi_0^\flat$ we associate the function $\varphi_0^\sharp : \mathbb{R} \to \mathbb{R}$ given by (4.3). It can be checked that

$$\varphi^\natural(u) \leq \varphi_0^\sharp(u) < u, \quad u > 0,$$

$$u < \varphi_0^\sharp(u) \leq \varphi^\natural(u), \quad u < 0.$$

We conclude from Theorem 5.1 that:

LEMMA 5.2. (Single entropy inequality.) *A shock wave of the form (1.5) and (1.8) satisfies the single entropy inequality (1.4) if and only if*

$$u_+ \in \begin{cases} \left(-\infty, \varphi_0^\flat(u_-) \right] \cup \left[u_-, +\infty \right), & u_- \geq 0, \\ \left(-\infty, u_- \right] \cup \left[\varphi_0^\flat(u_-), +\infty \right), & u_- \leq 0. \end{cases}$$

□

With the terminology in Definition 3.4 *classical shocks* or *Lax shocks* correspond here to right states

$$u_+ \in \begin{cases} \left(-\infty, \varphi^{-\natural}(u_-)\right] \cup \left[u_-, +\infty\right), & u_- \geq 0, \\ \left(-\infty, u_-\right] \cup \left[\varphi^{-\natural}(u_-), +\infty\right), & u_- \leq 0, \end{cases} \tag{5.1}$$

while the *nonclassical shocks* correspond to

$$u_+ \in \begin{cases} \left(\varphi^{-\natural}(u_-), \varphi_0^{\flat}(u_-)\right], & u_- \geq 0, \\ \left[\varphi_0^{\flat}(u_-), \varphi^{-\natural}(u_-)\right), & u_- \leq 0. \end{cases} \tag{5.2}$$

Observe that, now, nonclassical shocks are **fast undercompressive:**

$$f'(u_\pm) \leq \bar{a}(u_-, u_+).$$

On the other hand, according to (2.12) the rarefaction waves for a convex-concave flux are

$$\mathcal{R}(u_-) = \begin{cases} [0, u_-], & u_- > 0, \\ \{0\}, & u_- = 0, \\ [u_-, 0], & u_- < 0. \end{cases}$$

The Riemann problem admits a class of solutions, described as follows. Recall that the function ρ was defined earlier (after (3.6)). In addition, we denote by $\varphi_0^{-\flat}$ the *inverse* of the zero-entropy dissipation function.

THEOREM 5.3. (One-parameter family for convex-concave flux.) *Suppose that f is a convex-concave function (see (2.10)) and fix some Riemann data u_l and u_r. Restricting attention to solutions satisfying (1.4) for a given entropy pair (U, F), the Riemann problem (1.1) and (1.2) admits the following solutions (in the class $\mathbf{P}$) when $u_l \geq 0$:*

(a) *If $u_r \geq u_l$, the solution is unique and consists of a classical shock wave connecting u_l to u_r.*

(b) *If $u_r \in [0, u_l)$, the solution is unique and consists of a rarefaction wave connecting monotonically u_l to u_r.*

(c) *If $u_r \in [\varphi^{-\natural}(u_l), 0)$, there are infinitely many solutions, consisting of a rarefaction wave connecting u_l to some intermediate state u_m with $0 \leq u_m \leq \varphi_0^{-\flat}(u_r) \leq u_l$, followed with a classical or nonclassical shock connecting to u_r.*

(d) *If $u_r \in \left(-\infty, \varphi_0^{\flat}(u_l)\right)$, the solution may contain a classical shock connecting u_l to some state $u_m > u_l$, followed with a classical or nonclassical shock connecting to u_r. This happens when there exists u_m satisfying with $\rho(u_m, u_r) < u_l < u_m < \varphi_0^{-\flat}(u_r)$.*

(e) *Finally, if $u_r \in \left(-\infty, \varphi^{-\natural}(u_l)\right)$, there exists a solution connecting u_l to u_r by a classical shock wave.*

□

In Case (d), the solution contains two shocks and has a larger total variation than its initial data. Note that the intervals of right-hand states in Cases (c), (d), and (e) overlap.

The kinetic relation is based on a prescribed *kinetic function* $\varphi^\flat : I\!R \mapsto I\!R$ which, by definition, is a monotone decreasing function such that

$$\varphi^{-\natural}(u) \leq \varphi^\flat(u) < \varphi_0^\flat(u), \quad u > 0,$$
$$\varphi_0^\flat(u) < \varphi^\flat(u) \leq \varphi^{-\natural}(u), \quad u < 0. \tag{5.3}$$

We now have the property

$$|u| < \left|\varphi^\flat\left(\varphi^\flat(u)\right)\right|, \quad u \neq 0. \tag{5.4}$$

To the function $\varphi^\flat$ we associate its companion function $\varphi^\sharp : I\!R \to I\!R$, as was defined in (4.3). Furthermore, relying on the monotonicity property of the kinetic function, it is not hard to see that, to any point $u_l > 0$, we can associate a point $\rho^\flat(u_l) > u_l$ such that the speed of the classical shock connecting u_l to $\rho^\flat(u_l)$ be identical with the speed of the nonclassical shock connecting $\rho^\flat(u_l)$ to $\varphi^\flat \circ \rho^\flat(u_l)$. This latter corresponds to a transition in the Riemann solver described now. In addition, we denote by $\varphi^{-\flat}$ the *inverse* of the kinetic function. (See Figure II-7.)

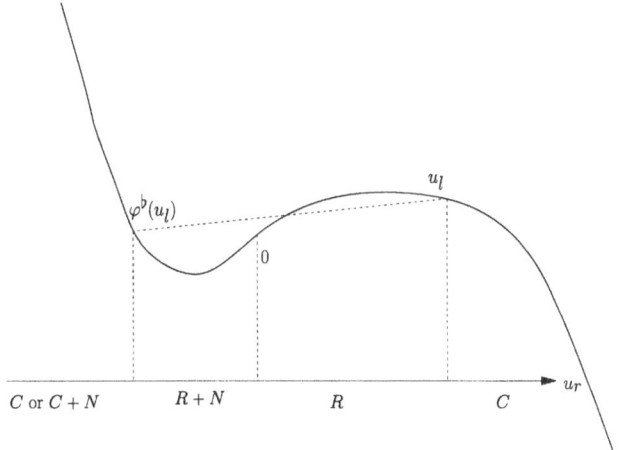

Figure II-7 : The four wave patterns for the Riemann solution.

THEOREM 5.4. (Riemann solution for convex-concave flux.) *Under the assumptions of Theorem 5.3 let us prescribe that any nonclassical shock connecting two states u_- and u_+ satisfies the* kinetic relation

$$u_+ = \varphi^\flat(u_-) \quad \text{for all nonclassical shocks}, \tag{5.5}$$

where $\varphi^\flat$ is a given kinetic function (satisfying (5.3)). Then, the Riemann problem (1.1), (1.2), (1.4), and (5.5) admits an (essentially unique) nonclassical entropy solution (in the class **P***), given as follows when $u_l > 0$:*

(a) *If $u_r \geq u_l$, the solution is unique and consists of a classical shock wave connecting u_l to u_r.*

(b) *If $u_r \in [0, u_l)$, the solution is unique and consists of a rarefaction wave connecting monotonically u_l to u_r.*

(c) *If $u_r \in [\varphi^b(u_l), 0)$, the solution contains a rarefaction wave connecting u_l to $u_m = \varphi^{-b}(u_r)$, followed with a nonclassical shock connecting to u_r.*

(d) *If $u_r \leq \varphi^b(u_l)$, the solution contains:*

   – if $u_l > \rho(\varphi^{-b}(u_r), u_r)$, a classical shock connecting u_l to $u_m = \varphi^{-b}(u_r)$ followed by a nonclassical shock connecting u_m to u_r,

   – if $u_l < \rho(\varphi^{-b}(u_r), u_r)$, a single classical shock connecting u_l to u_r.

In Cases (a), (b), and (c), the Riemann solution is monotone, while it is nonmonotone in Case (d). The solution depends continuously upon its initial data in the L^1 norm. Furthermore, the classical Riemann solution (Theorem 2.3) is also admissible as it contains only classical waves.   □

Note that the condition $u_l > \rho(\varphi^{-b}(u_r), u_r)$ precisely determines that the shock connecting u_l to $u_m = \varphi^{-b}(u_r)$ is slower than the one connecting to u_r. Finally, by following the same lines as in Theorem 4.3 and imposing a nucleation criterion we can exclude the classical Riemann solution and select a unique nonclassical Riemann solution for convex-concave flux.

REMARK 5.5. In Sections 3 to 5, to develop the theory of nonclassical solutions to the Riemann problem we have first set a strictly convex entropy pair (U, F) and determine the corresponding zero-entropy function φ_0^b, which was then used to restrict the range of the kinetic function. This approach is justified by the examples discussed earlier in Chapter I (Examples I-4.2 and I-4.3). However, the theory can be extended to encompass even more general kinetic functions which need not arise from a regularized model. For concave-convex flux-functions (Section 4) it is sufficient to assume, instead of (4.1), that

$$\varphi^{-\natural}(u) < \varphi^b(u) \leq \varphi^{\natural}(u), \quad u > 0,$$
$$\varphi^{\natural}(u) \leq \varphi^b(u) < \varphi^{-\natural}(u), \quad u < 0,$$
(5.6)

and

$$|\varphi^b(\varphi^b(u))| < |u|, \quad u \neq 0.$$
(5.7)

For convex-concave flux-functions (Section 5) it is sufficient to assume, instead of (5.3), that (5.6) holds together with

$$|u| < |\varphi^b(\varphi^b(u))|, \quad u \neq 0.$$
(5.8)

□

CHAPTER III

DIFFUSIVE-DISPERSIVE TRAVELING WAVES

In this chapter we study a large class of *diffusive-dispersive* equations associated with scalar conservation laws. We investigate the existence of *traveling wave* solutions which, as was pointed out earlier (Theorem I–5.4), converge to shock wave solutions of (1.1) as the diffusion and the dispersion tend to zero. The corresponding shock set can be determined and compared with the one obtained in Chapter II by applying entropy inequalities. The present chapter demonstrates the relevance of the construction given in Chapter II. We confirm here that classical shock waves are independent of the small-scale mechanisms, while nonclassical shock waves require the kinetic relation determined by the given diffusive-dispersive operator. In Section 1 we consider the effect of the *diffusion only*; see Theorem 1.2. In Section 2 we determine the kinetic relation explicitly for the conservation law with *cubic* flux and linear diffusion-dispersion terms; see Theorem 2.3. The main result in this chapter for *general* flux-functions are stated in Section 3; see Theorem 3.3. The proofs of the results given in Section 3 are postponed to Sections 4 and 5.

1. Diffusive traveling waves

Consider the scalar conservation law

$$\partial_t u + \partial_x f(u) = 0, \quad u = u(x,t) \in I\!\!R, \tag{1.1}$$

where $f : I\!\!R \to I\!\!R$ is a smooth mapping. In this section we restrict attention to the **nonlinear diffusion model**

$$\partial_t u + \partial_x f(u) = \varepsilon \left(b(u)\, u_x \right)_x, \quad u = u^\varepsilon(x,t) \subset I\!\!R, \tag{1.2}$$

where $\varepsilon > 0$ is a small parameter. The *diffusion function* $b : I\!\!R \to I\!\!R_+$ is assumed to be smooth and bounded below:

$$b(u) \geq \bar{b} > 0, \tag{1.3}$$

so that the equation (1.2) is **uniformly parabolic**. We are going to establish that the shock set associated with the traveling wave solutions of (1.2) coincides with the one described by Oleinik entropy inequalities (see (II–1.6)).

Recall that a **traveling wave** of (1.2) is a solution depending only upon the variable

$$y := \frac{x - \lambda t}{\varepsilon} \tag{1.4}$$

for some constant speed λ. Note that, after rescaling, the corresponding **trajectory** $y \mapsto u(y)$ is independent of the parameter ε. Fixing the left-hand state u_- we search for traveling waves of (1.2) **connecting** u_- **to some state** u_+, that is, solutions $y \mapsto u(y)$ of the ordinary differential equation

$$-\lambda\, u_y + f(u)_y = \big(b(u)\, u_y\big)_y \tag{1.5}$$

satisfying the boundary conditions

$$\lim_{y \to -\infty} u(y) = u_-, \quad \lim_{y \to +\infty} u(y) = u_+, \quad \lim_{|y| \to +\infty} u_y(y) = 0. \tag{1.6}$$

In view of (1.6) the equation (1.5) can be integrated once:

$$b(u(y))\, u_y(y) = -\lambda\, (u(y) - u_-) + f(u(y)) - f(u_-), \quad y \in I\!R. \tag{1.7}$$

The Rankine-Hugoniot condition

$$-\lambda\, (u_+ - u_-) + f(u_+) - f(u_-) = 0 \tag{1.8}$$

follows by letting $y \to +\infty$ in (1.7). The equation (1.7) is an ordinary differential equation (O.D.E) on the real line. The qualitative behavior of the solutions is easily determined, as follows.

THEOREM 1.1. (*Diffusive traveling waves.*) *Consider the scalar conservation law* (1.1) *with general flux-function* f *together with the diffusive model* (1.2). *Fix a left-hand state* u_- *and a right-hand state* $u_+ \neq u_-$. *Then, there exists a traveling wave of* (1.7) *associated with the nonlinear diffusion model* (1.2) *if and only if* u_- *and* u_+ *satisfy* **Oleinik entropy inequalities in the strict sense**, *that is:*

$$\frac{f(v) - f(u_-)}{v - u_-} > \frac{f(u_+) - f(u_-)}{u_+ - u_-} \quad \text{for all } v \text{ lying strictly between } u_- \text{ and } u_+. \tag{1.9}$$

PROOF. All the trajectories of interest are bounded, i.e., cannot escape to infinity. Namely, the shock profile satisfies the equation

$$u' = \frac{u - u_-}{b(u)} \left(\frac{f(u) - f(u_-)}{u - u_-} - \frac{f(u_+) - f(u_-)}{u_+ - u_-} \right). \tag{1.10}$$

It is not difficult to see that the solution exists and connects monotonically u_- to u_+ provided Oleinik entropy inequalities hold and the right-hand side of (1.10) keeps (strictly) a constant sign (except at the end point $y = \pm\infty$ where it vanishes). □

By analogy with the approach followed in Chapter II, for each left-hand state u_- we define the **shock set** associated with the nonlinear diffusion model as

$$S(u_-) := \big\{ u_+ \ / \ \text{there exists a solution of (1.6)-(1.8)} \big\}.$$

Combining Theorem 1.1 with the results obtained earlier in Section II-2 we reach the following conclusion.

THEOREM 1.2. (*Shock set based on diffusive limits.*) *Consider the scalar conservation law* (1.1) *when the flux* f *is convex, concave-convex, or convex-concave. (See Section II-2 for the definitions.) Then, for any* u_-, *the shock set* $S(u_-)$ *associated with the nonlinear diffusion model* (1.2) *and* (1.3) *is independent of the diffusion function* b, *and the closure of* $S(u_-)$ *coincides with the shock set characterized by Oleinik entropy inequalities (or, equivalently, Lax shock inequalities).* □

REMARK 1.3. The conclusions of Theorem 1.2 do not hold for more general flux-functions. This is due to the fact that a strict inequality is required in (1.9) for the existence of the traveling waves. The set based on traveling waves may be strictly smaller than the one based on Oleinik entropy inequalities. $\square$

2. Kinetic functions for the cubic flux

Investigating traveling wave solutions of diffusive-dispersive regularizations of (1.1) is considerably more involved than what was done in Section 1. Besides proving the existence of associated (classical and nonclassical) traveling waves our main objective will be to derive the corresponding kinetic functions for nonclassical shocks, which were discovered in Chapter II.

To explain the main difficulty and ideas it will be useful to treat first, in the present section, the specific **diffusive-dispersive model with cubic flux** (Example I–4.2)

$$\partial_t u + \partial_x u^3 = \varepsilon\, u_{xx} + \delta\, u_{xxx}, \qquad (2.1)$$

which, formally as $\varepsilon, \delta \to 0$, converges to the **conservation law with cubic flux**

$$\partial_t u + \partial_x u^3 = 0. \qquad (2.2)$$

We are interested in the singular limit $\varepsilon \to 0$ in (2.1) when the ratio

$$\alpha = \frac{\varepsilon}{\sqrt{\delta}} \qquad (2.3)$$

is kept constant. We assume also that the dispersion coefficient δ is positive. Later, in Theorem 3.5 below, we will see that all traveling waves are classical when $\delta < 0$ which motivates us to restrict attention to $\delta > 0$.

We search for traveling wave solutions of (2.1) depending on the rescaled variable

$$y := \alpha\, \frac{x - \lambda t}{\varepsilon} = \frac{x - \lambda t}{\sqrt{\delta}}. \qquad (2.4)$$

Proceeding along the same lines as those in Section 1 we find that a traveling wave $y \mapsto u(y)$ should satisfy

$$-\lambda\, u_y + (u^3)_y = \alpha\, u_{yy} + u_{yyy}, \qquad (2.5)$$

together with the boundary conditions

$$\lim_{y \to \pm\infty} u(y) = u_\pm, $$
$$\lim_{y \to \pm\infty} u_y(y) = \lim_{y \to \pm\infty} u_{yy}(y) = 0, \qquad (2.6)$$

where $u_- \neq u_+$ and λ are constants. Integrating (2.5) once we obtain

$$\alpha\, u_y(y) + u_{yy}(y) = -\lambda\,(u(y) - u_-) + u(y)^3 - u_-^3, \quad y \in \mathbb{R}, \qquad (2.7)$$

which also implies

$$\lambda = \frac{u_+^3 - u_-^3}{u_+ - u_-} = u_-^2 + u_-\, u_+ + u_+^2. \qquad (2.8)$$

To describe the family of traveling waves it is convenient to fix the left-hand state (with for definiteness $u_- > 0$) and to use the speed λ as a parameter. Given u_-, there is a range of speeds,

$$\lambda \in (3\, u_-^2/4, 3\, u_-^2),$$

for which the line passing through the point with coordinates (u_-, u_-^3) and with slope λ intersects the graph of the flux $f(u) := u^3$ at three distinct points. For the discussion in this section we restrict attention to this situation, which is most interesting. There exist **three equilibria** at which the right-hand side of (2.7) vanishes. The notation

$$u_2 < u_1 < u_0 := u_-$$

will be used, where u_2 and u_1 are the two distinct roots of the polynomial

$$u^2 + u_0\, u + u_0^2 = \lambda. \tag{2.9}$$

Observe in passing that $u_2 + u_1 + u_0 = 0$.

Consider a trajectory $y \mapsto u(y)$ leaving from u_- at $-\infty$. We want to determine which point, among u_1 or u_2, the trajectory will reach at $+\infty$. Clearly, the trajectory is associated with a **classical shock** if it reaches u_1 and with a **nonclassical shock** if it reaches u_2. (See Section II–3 for the definitions). Accordingly, we will refer to it as a **classical trajectory** or as a **nonclassical trajectory**, respectively.

We reformulate (2.7) as a differential system of two equations,

$$\frac{d}{dy}\begin{pmatrix} u \\ v \end{pmatrix} = K(u, v), \tag{2.10}$$

where

$$K(u, v) = \begin{pmatrix} v \\ -\alpha\, v + g(u, \lambda) - g(u_-, \lambda) \end{pmatrix}, \quad g(u, \lambda) = u^3 - \lambda\, u. \tag{2.11}$$

The function K vanishes precisely at the three equilibria $(u_0, 0)$, $(u_1, 0)$, and $(u_2, 0)$ of (2.10). The eigenvalues of the Jacobian matrix of $K(u, v)$ at any point $(u, 0)$ are $-\alpha/2 \pm \sqrt{\alpha^2/4 + g_u'(u, \lambda)}$. So we set

$$\begin{aligned} \underline{\mu}(u) &= \frac{1}{2}\left(-\alpha - \sqrt{\alpha^2 + 4\,(3\,u^2 - \lambda)}\right), \\ \overline{\mu}(u) &= \frac{1}{2}\left(-\alpha + \sqrt{\alpha^2 + 4\,(3\,u^2 - \lambda)}\right). \end{aligned} \tag{2.12}$$

At this juncture, we recall the following standard definition and result. (See the bibliographical notes for references.)

DEFINITION 2.1. (Nature of equilibrium points.) Consider a differential system of the form (2.10) where K is a smooth mapping. Let $(u_*, v_*) \in I\!\!R^2$ be an equilibrium point, that is, a root of $K(u_*, v_*) = 0$. Denote by $\underline{\mu} = \underline{\mu}(u_*, v_*)$ and $\overline{\mu} = \overline{\mu}(u_*, v_*)$ the two (real or complex) eigenvalues of the Jacobian matrix of K at (u_*, v_*), and suppose that a basis of corresponding eigenvectors $\underline{r}(u_*, v_*)$ and $\overline{r}(u_*, v_*)$ exists. Then, the equilibrium (u_*, v_*) is called

- a **stable point** if $Re(\underline{\mu})$ and $Re(\overline{\mu})$ are both negative,
- a **saddle point** if $Re(\underline{\mu})$ and $Re(\overline{\mu})$ have opposite sign,
- or an **unstable point** if $Re(\underline{\mu})$ and $Re(\overline{\mu})$ are both positive.

Moreover, a stable or unstable point is called a **node** if the eigenvalues are real and a **spiral** if they are complex conjugate.

THEOREM 2.2. (Local behavior of trajectories.) *Consider the differential system (2.10) under the same assumptions as in Definition 2.1. If (u_*, v_*) is a saddle point, there are two trajectories defined on some interval $(-\infty, y_*)$ and two trajectories defined on some interval $(y_*, +\infty)$ and converging to (u_*, v_*) at $-\infty$ and $+\infty$, respectively. The trajectories are tangent to the eigenvectors $\underline{r}(u_*, v_*)$ and $\overline{r}(u_*, v_*)$, respectively.* $\square$

Returning to (2.11) and (2.12) we conclude that, since $g'_u(u, \lambda) = 3u^2 - \lambda$ is positive at both $u = u_2$ and $u = u_0$, we have

$$\underline{\mu}(u_0) < 0 < \overline{\mu}(u_0), \quad \underline{\mu}(u_2) < 0 < \overline{\mu}(u_2).$$

Thus both points u_2 and u_0 are *saddle points*. On the other hand, since we have $g'_u(u_1, \lambda) < 0$, the point u_1 is stable: it is a *node* if $\alpha^2 + 4\,(3\,u_1^2 - \lambda) \geq 0$ or a *spiral* if $\alpha^2 + 4\,(3\,u_1^2 - \lambda) < 0$. In summary, as illustrated by Figure III-1, for the system (2.10)-(2.11)

$$u_2 \text{ and } u_0 \text{ are saddle points and} \tag{2.13}$$
$$u_1 \text{ is a stable point (either a node or a spiral).}$$

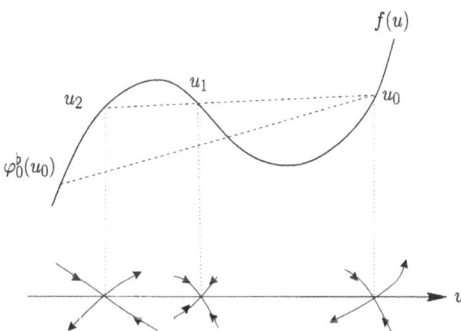

Figure III-1 : Qualitative behavior when $\alpha^2 + 4\,(3\,u_1^2 - \lambda) > 0$.

In the present section we check solely that, in some range of the parameters u_0, λ, and α, there exists a *nonclassical trajectory* connecting the two saddle points u_0 and u_2. **Saddle-saddle connections** are not "generic" and, as we will show, arise only when a special relation (the kinetic relation) holds between u_0, λ, and α or, equivalently, between u_0, u_2, and α; see (2.15) below.

For the cubic model (2.1) an *explicit formula* is now derived for the nonclassical trajectory. Motivated by the fact that the function g in (2.11) is a cubic, we a priori assume that $v = u_y$ is a **parabola in the variable** u. Since v must vanish at the two equilibria we write

$$v(y) = a\,(u(y) - u_2)\,(u(y) - u_0), \quad y \in {I\!\!R}, \tag{2.14}$$

where a is a constant to be determined. Substituting (2.14) into (2.10)-(2.11), we obtain an expression of v_y:

$$
\begin{aligned}
v_y &= -\alpha\,v + u^3 - u_0^3 - \lambda\,(u - u_0) \\
&= -\alpha\,v + (u - u_2)\,(u - u_0)\,(u + u_0 + u_2) \\
&= v\left(-\alpha + \frac{1}{a}\,(u + u_0 + u_2)\right).
\end{aligned}
$$

But, differentiating (2.14) directly we have also

$$
\begin{aligned}
v_y &= a\,u_y\,(2\,u - u_0 - u_2) \\
&= a\,v\,(2\,u - u_0 - u_2).
\end{aligned}
$$

The two expressions of v_y above coincide if we choose

$$
\frac{1}{a} = 2\,a, \quad -\alpha + \frac{1}{a}\,(u_0 + u_2) = -a(u_0 + u_2).
$$

So, $a = 1/\sqrt{2}$ (since clearly we need $v < 0$) and the three parameters u_0, u_2, and α satisfy the **explicit relation**

$$
u_2 = -u_0 + \frac{\sqrt{2}}{3}\,\alpha. \tag{2.15}
$$

Since $u_1 = -u_0 - u_2$ we see that the trajectory (2.14) is the saddle-saddle connection we are looking for, only if $u_2 < u_1$ as expected, that is, only if

$$
u_0 > \frac{2\sqrt{2}}{3}\,\alpha. \tag{2.16}
$$

Now, by integrating (2.14), it is not difficult to arrive at the following **explicit formula for the nonclassical trajectory**:

$$
\begin{aligned}
u(y) &= \frac{u_0 + u_2}{2} - \frac{u_0 - u_2}{2}\,\tanh\left(\frac{u_0 - u_2}{2\sqrt{2}}\,y\right) \\
&= \frac{\alpha}{3\sqrt{2}} - \left(u_- - \frac{\alpha}{3\sqrt{2}}\right)\tanh\left(\left(u_- - \frac{\alpha}{3\sqrt{2}}\right)\frac{y}{\sqrt{2}}\right).
\end{aligned} \tag{2.17}
$$

We conclude that, given any left-hand state $u_0 > 2\sqrt{2}\,\alpha/3$, there exists a saddle-saddle connection connecting u_0 to $-u_0 + \sqrt{2}\,\alpha/3$ which is given by (2.17). Later, in Section 3 and followings, we will prove that the trajectory just found is actually the *only* saddle-saddle trajectory leaving from $u_0 > 2\sqrt{2}\,\alpha/3$ and that no such trajectory exists when u_0 is below that threshold.

Now, denote by $S_\alpha(u_-)$ the set of all right-hand states u_+ attainable through a diffusive-dispersive traveling wave of (2.1) with $\delta > 0$ and $\varepsilon/\sqrt{\delta} = \alpha$ fixed. In the case of the equation (2.1) the results to be established in the following sections can be summarized as follows. (See also Figure III-2.)

THEOREM 2.3. (Kinetic function and shock set for the cubic flux.) *The **kinetic function** associated with the diffusive-dispersive model* (2.1) *is*

$$
\varphi_\alpha^\flat(u_-) = \begin{cases} -u_- - \tilde{\alpha}/2, & u_- \leq -\tilde{\alpha}, \\ -u_-/2, & |u_-| \leq \tilde{\alpha}, \\ -u_- + \tilde{\alpha}/2, & u_- \geq \tilde{\alpha}, \end{cases} \tag{2.18}
$$

with $\tilde{\alpha} := 2\alpha\sqrt{2}/3$, while the corresponding **shock set** is

$$\mathcal{S}_\alpha(u_-) = \begin{cases} (u_-, \tilde{\alpha}/2] \cup \{-u_- - \tilde{\alpha}/2\}, & u_- \leq -\tilde{\alpha}, \\ [-u_-/2, u_-), & -\tilde{\alpha} \leq u_- \leq \tilde{\alpha}, \\ \{-u_- + \tilde{\alpha}/2\} \cup [-\tilde{\alpha}/2, u_-), & u_- \geq \tilde{\alpha}. \end{cases} \qquad (2.19)$$

In agreement with the general theory in Chapter II the kinetic function (2.18) is monotone decreasing and lies between the limiting functions $\varphi^\natural(u) := -u/2$ and $\varphi_0^\flat(u) := -u$. Depending on u_- the shock set can be either an interval or the union of a point and an interval.

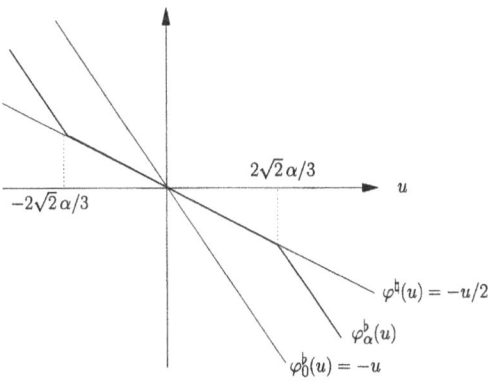

Figure III-2 : Kinetic function for the cubic flux.

Consider next the **entropy dissipation** associated with the nonclassical shock:

$$E(u_-; \alpha, U) := -\left(\varphi_\alpha^\flat(u_-)^2 + \varphi_\alpha^\flat(u_-)\, u_- + u_-^2\right)\left(U(\varphi_\alpha^\flat(u_-)) - U(u_-)\right) + F(\varphi_\alpha^\flat(u_-)) - F(u_-), \qquad (2.20)$$

where (U, F) is any convex entropy pair of the equation (2.2). By multiplying (2.5) by $U'(u(y))$ and integrating over $y \in \mathbb{R}$ we find the equivalent expression

$$\begin{aligned} E(u_-; \alpha, U) &= \int_{\mathbb{R}} U'(u(y))\left(\alpha\, u_{yy}(y) + u_{yyy}(y)\right) dy \\ &= \int_{\mathbb{R}} \left(-\alpha\, U''(u)\, u_y^2 + U'''(u)\, u_y^3/2\right) dy. \end{aligned} \qquad (2.21)$$

So, the sign of the entropy dissipation can also be determined from the explicit form (2.17) of the traveling wave.

THEOREM 2.4. (Entropy inequalities.)

(i) *For the quadratic entropy*

$$U(z) = z^2/2, \quad z \in \mathbb{R},$$

the entropy dissipation $E(u_-; \alpha, U)$ *is non-positive for all real* u_- *and all* $\alpha \geq 0$.

(ii) *For all convex entropy U the entropy dissipation $E(u_-; \alpha, U)$ is non-positive for all $\alpha > 0$ and all $|u_-| \le 2\sqrt{2}\,\alpha/3$.*

(iii) *Consider $|u_-| > 2\sqrt{2}\,\alpha/3$ and any (convex) entropy U whose third derivative is sufficiently small, specifically*

$$\left(|u_-| - \alpha/(3\sqrt{2})\right)^2 |U'''(z)| \le 2\alpha\sqrt{2}\,U''(z), \quad z \in \mathbb{R}. \tag{2.22}$$

Then, the entropy dissipation $E(u_-; \alpha, U)$ is also non-positive.

(iv) *Finally given any $|u_-| > 2\sqrt{2}\,\alpha/3$ there exists infinitely many strictly convex entropies for which $E(u_-; \alpha, U)$ is positive.*

PROOF. When U is quadratic (with $U'' \ge 0$ and $U''' \equiv 0$) we already observed that (i) follows immediately from (2.21). The statement (ii) is also obvious since the function $\varphi^\flat$ reduces to a classical value in the range under consideration. Under the condition (2.22) the *integrand* of (2.21) is non-positive, as follows from the inequality (see (2.14))

$$|u_y| \le \frac{1}{4\sqrt{2}}\,(u_0 - u_2)^2 = \frac{1}{\sqrt{2}}\left(u_- - \alpha/(3\sqrt{2})\right)^2.$$

This implies the statement (iii). Finally, to derive (iv) we use the (Lipschitz continuous) **Kruzkov entropy pairs**

$$U_k(z) := |z - k|, \quad F_k(z) := \mathrm{sgn}(z - k)(z^3 - k^3), \quad z \in \mathbb{R}, \tag{2.23}$$

with the choice $k = -u_-/2$. We obtain

$$E(u_-; \alpha, U_k) = \frac{3}{4}\,|u_-|\left(|u_-| - 2\alpha\sqrt{2}/3\right)^2 > 0.$$

By continuity, $E(u_-; \alpha, U_k)$ is also strictly positive for all k in a small neighborhood of $-u_-/2$. The desired conclusion follows by observing that any smooth convex function can be represented by a weighted sum of Kruzkov entropies. $\qquad\square$

REMARK 2.5. We collect here the explicit expressions of some functions associated with the model (2.1), introduced earlier in Chapter II or to be defined later in this chapter. From now on we restrict attention to the entropy pair

$$U(u) = u^2/2, \quad F(u) = 3\,u^4/4.$$

First of all, recall that for the equation (2.2) the following two functions

$$\varphi^\natural(u) = -\frac{u}{2}, \quad \varphi^\flat_0(u) = -u, \quad u \in \mathbb{R}. \tag{2.24}$$

determine the admissible range of the kinetic functions.

We define the **critical diffusion-dispersion ratio**

$$A(u_0, u_2) = \frac{3}{\sqrt{2}}\,(u_0 + u_2) \tag{2.25}$$

for $u_0 \ge 0$ and $u_2 \in (-u_0, -u_0/2)$ and for $u_0 \le 0$ and $u_2 \in (-u_0/2, -u_0)$. In view of Theorem 2.3 (see also (2.15)), a nonclassical trajectory connecting u_0 to u_2 exists if and only if the parameter $\alpha = \varepsilon/\sqrt{\delta}$ equals $A(u_0, u_2)$. The function A increases monotonically in u_2 from the value 0 to the **threshold diffusion-dispersion ratio** $(u_0 > 0)$

$$A^\natural(u_0) = \frac{3\,u_0}{2\sqrt{2}}. \tag{2.26}$$

For each fixed state $u_0 > 0$ there exists a nonclassical trajectory leaving from u_0 if and only if α is less than $A^{\natural}(u_0)$. On the other hand, for each fixed α there exists a nonclassical trajectory leaving from u_0 if and only if the left-hand state u_0 is greater than $A^{\natural^{-1}}(\alpha)$. The function $A^{\natural}$ is a linear function (for $u_0 > 0$) with range extending therefore from $\underline{A^{\natural}} = 0$ to $\overline{A^{\natural}} = +\infty$. $\qquad\qquad\Box$

REMARK 2.6. It is straightforward to check that if (2.1) is replaced with the more general equation

$$\partial_t u + \partial_x \left(K\, u^3 \right) = \varepsilon\, u_{xx} + \delta\, C\, u_{xxx}, \tag{2.27}$$

where C and K are positive constants, then (2.26) becomes

$$A^{\natural}(u_0) = \frac{3\, u_0}{2\sqrt{2}} \sqrt{K\, C}. \tag{2.28}$$

$\qquad\qquad\Box$

REMARK 2.7. Clearly, there is a one-parameter family of traveling waves connecting the same end states: If $u = u(y)$ is a solution of (2.5) and (2.6), then the translated function $u = u(y + b)$ ($b \in I\!\!R$) satisfies the same conditions. However, one could show that the *trajectory* in the phase plane connecting two given end states is *unique*. $\quad\Box$

3. Kinetic functions for general flux

Consider now the general **diffusive-dispersive conservation law**

$$\partial_t u + \partial_x f(u) = \varepsilon \left(b(u)\, u_x \right)_x + \delta \left(c_1(u)\, (c_2(u)\, u_x)_x \right)_x, \quad u = u^{\varepsilon,\delta}(x,t), \tag{3.1}$$

where the diffusion coefficient $b(u) > 0$ and dispersion coefficients $c_1(u), c_2(u) > 0$ are given smooth functions. Following the discussion in Chapter II we assume that $f : I\!\!R \to I\!\!R$ is a *concave-convex* function, that is,

$$\begin{aligned}
&u\, f''(u) > 0 \quad \text{for all } u \neq 0, \\
&f'''(0) \neq 0, \quad \lim_{|u| \to +\infty} f'(u) = +\infty.
\end{aligned} \tag{3.2}$$

As in Section 2 above we are interested in the singular limit $\varepsilon \to 0$ when $\delta > 0$ and the ratio $\alpha = \varepsilon/\sqrt{\delta}$ is kept constant. The limiting equation associated with (3.1), formally, is the scalar conservation law

$$\partial_t u + \partial_x f(u) = 0, \quad u = u(x,t) \in I\!\!R.$$

Earlier (see Example I–4.3) we also proved that the entropy inequality

$$\partial_t U(u) + \partial_x F(u) \leq 0$$

holds, provided the entropy pair (U, F) is chosen such that

$$U''(u) := \frac{c_2(u)}{c_1(u)}, \quad F'(u) := U'(u)\, f'(u), \quad u \in I\!\!R, \tag{3.3}$$

which we assume in the rest of this chapter. Since $c_1, c_2 > 0$ the function U is strictly convex.

Given two states $u_\pm$ and the corresponding propagation speed

$$\lambda = \bar{a}(u_-, u_+) := \begin{cases} \frac{f(u_+) - f(u_-)}{u_+ - u_-}, & u_+ \neq u_-, \\ f'(u_-), & u_+ = u_-, \end{cases}$$

we search for traveling wave solutions $u = u(y)$ of (3.1) depending on the rescaled variable $y := (x - \lambda t)\,\alpha/\varepsilon$. Following the same lines as those in Sections 1 and 2 we find that the trajectory satisfies

$$c_1(u)\,(c_2(u)\,u_y)_y + \alpha\,b(u)u_y = -\lambda\,(u - u_-) + f(u) - f(u_-), \quad u = u(y), \qquad (3.4)$$

and the boundary conditions

$$\lim_{y \to \pm\infty} u(y) = u_\pm, \quad \lim_{y \to \pm\infty} u_y(y) = 0. \qquad \qquad \cdot$$

Setting now

$$v = c_2(u)\,u_y,$$

we rewrite (3.4) in the general form (2.10) for the unknowns $u = u(y)$ and $v = v(y)$ ($y \in \mathbb{R}$), i.e.,

$$\frac{d}{dy}\begin{pmatrix} u \\ v \end{pmatrix} = K(u, v) \qquad (3.5)$$

with

$$K(u, v) = \begin{pmatrix} \frac{v}{c_2(u)} \\ -\alpha\,\frac{b(u)}{c_1(u)c_2(u)}\,v + \frac{g(u,\lambda) - g(u_-, \lambda)}{c_1(u)} \end{pmatrix}, \quad g(u, \lambda) := f(u) - \lambda\,u, \qquad (3.6)$$

while the boundary conditions take the form

$$\lim_{y \to \pm\infty} u(y) = u_\pm, \quad \lim_{y \to \pm\infty} v(y) = 0. \qquad (3.7)$$

The function K in (3.6) vanishes at the **equilibrium points** $(u, v) \in \mathbb{R}^2$ satisfying

$$g(u, \lambda) = g(u_-, \lambda), \quad v = 0. \qquad (3.8)$$

In view of the assumption (3.2), given a left-hand state u_- and a speed λ there exist at most three equilibria u satisfying (3.8) (including u_- itself). Considering a trajectory leaving from u_- at $-\infty$, we will determine whether this trajectory diverges to infinity or else which equilibria (if there is more than one equilibria) it actually connects to at $+\infty$. Before stating our main result (Theorem 3.3 below) let us derive some fundamental inequalities satisfied by states u_- and u_+ connected by a traveling wave.

Consider the **entropy dissipation**

$$E(u_-, u_+) := -\bar{a}(u_-, u_+)\,\big(U(u_+) - U(u_-)\big) + F(u_+) - F(u_-) \qquad (3.9)$$

or, equivalently, using (3.3) and (3.7)

$$\begin{aligned} E(u_-, u_+) &= \int_{-\infty}^{+\infty} U'(u(y))\,\big(-\lambda\,u_y(y) + f(u(y))_y\big)\,dy \\ &= -\int_{-\infty}^{+\infty} U''(u(y))\,\big(-\lambda\,(u(y) - u_-) + f(u) - f(u_-)\big)\,u_y(y)\,dy \qquad (3.10) \\ &= -\int_{u_-}^{u_+} \big(g(z, \bar{a}(u_-, u_+)) - g(u_-, \bar{a}(u_-, u_+))\big)\,\frac{c_2(z)}{c_1(z)}\,dz. \end{aligned}$$

In view of

$$E(u_-, u_+) = \int_{-\infty}^{+\infty} U'(u) \left(\alpha \left(b(u)\, u_y \right)_y + \left(c_1(u) \left(c_2(u)\, u_y \right)_y \right)_y \right) dy$$

$$= - \int_{-\infty}^{+\infty} \alpha\, U''(u)\, b(u)\, u_y^2 \, dy,$$

we have immediately the following.

LEMMA 3.1. (Entropy inequality.) *If there exists a traveling wave of* (3.4) *connecting* u_- *to* u_+, *then the corresponding entropy dissipation is non-positive,*

$$E(u_-, u_+) \leq E(u_-, u_-) = 0.$$

$\square$

We will use the same notation as in Chapter II. From the graph of the function f we define the functions $\varphi^\natural$ and $\lambda^\natural$ by (see Figure III-3)

$$\lambda^\natural(u) := f'\left(\varphi^\natural(u)\right) = \frac{f(u) - f\left(\varphi^\natural(u)\right)}{u - \varphi^\natural(u)}, \quad u \neq 0.$$

We have $u\,\varphi^\natural(u) < 0$ and by continuity $\varphi^\natural(0) = 0$ and, thanks to (3.2), the map $\varphi^\natural : I\!R \to I\!R$ is decreasing and onto. It is invertible and its inverse function is denoted by $\varphi^{-\natural}$. Observe in passing that, u_- being kept fixed, $\lambda^\natural(u_-)$ is a *lower bound* for all shock speeds λ satisfying the Rankine-Hugoniot relation

$$-\lambda\left(u_+ - u_-\right) + f(u_+) - f(u_-) = 0$$

for some u_+.

The properties of the entropy dissipation (3.9) were already investigated in Chapter II where the **zero-entropy dissipation** function $\varphi_0^\flat$ was introduced. Let us recall that:

LEMMA 3.2. (Entropy dissipation function.) *There exists a decreasing function* $\varphi_0^\flat$: $I\!R \to I\!R$ *such that for all* $u_- > 0$ *(for instance)*

$$E(u_-, u_+) = 0 \text{ and } u_+ \neq u_- \quad \text{if and only if} \quad u_+ = \varphi_0^\flat(u_-),$$

$$E(u_-, u_+) < 0 \quad \text{if and only if} \quad \varphi_0^\flat(u_-) < u_+ < u_-,$$

and

$$\varphi^{-\natural}(u_-) < \varphi_0^\flat(u_-) < \varphi^\natural(u_-).$$

$\square$

In passing, define also the function $\varphi_0^\sharp = \varphi_0^\sharp(u_-)$ and the speed $\lambda_0 = \lambda_0(u_-)$ by

$$\lambda_0(u_-) = \frac{f(u_-) - f(\varphi_0^\flat(u_-))}{u_- - \varphi_0^\flat(u_-)} = \frac{f(u_-) - f(\varphi_0^\sharp(u_-))}{u_- - \varphi_0^\sharp(u_-)}, \quad u_- \neq 0. \qquad (3.11)$$

Combining Lemmas 3.1 and 3.2 together we conclude that, if there exists a traveling wave connecting u_- to u_+, necessarily

$$u_+ \text{ belongs to the interval } [\varphi_0^\flat(u_-), u_-]. \qquad (3.12)$$

In particular, the states $u_+ > u_-$ and $u_+ < \varphi^{-\natural}(u_-)$ *cannot* be reached by a traveling wave and, therefore, it is not restrictive to focus on the case that three equilibria exist.

Next, for each $u_- > 0$ we define the **shock set** generated by the diffusive-dispersive model (3.1) by

$$\mathcal{S}_\alpha(u_-) := \Big\{ u_+ \ / \ \text{there exists a traveling wave of (3.4) connecting } u_- \text{ to } u_+ \Big\}.$$

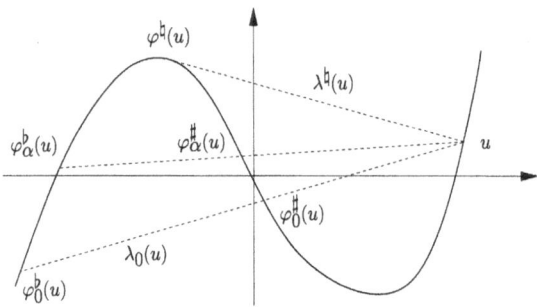

Figure III-3 : Concave-convex flux-function.

THEOREM 3.3. (Kinetic function and shock set for general flux.) *Given a concave-convex flux-function f (see (3.2)), consider the diffusive-dispersive model (3.1) in which the ratio $\alpha = \varepsilon/\sqrt{\delta} > 0$ is fixed. Then, there exists a locally Lipschitz continuous and decreasing* **kinetic function** *$\varphi^\flat_\alpha : \mathbb{R} \to \mathbb{R}$ satisfying*

$$\begin{aligned}
\varphi^\natural(u) \le \varphi^\flat_\alpha(u) < \varphi^\flat_0(u), \quad u < 0, \\
\varphi^\flat_0(u) < \varphi^\flat_\alpha(u) \le \varphi^\natural(u), \quad u > 0,
\end{aligned} \tag{3.13}$$

and such that

$$\mathcal{S}_\alpha(u_-) = \begin{cases} [u_-, \varphi^\sharp_\alpha(u_-)) \cup \{\varphi^\flat_\alpha(u_-)\}, & u_- < 0, \\ \{\varphi^\flat_\alpha(u_-)\} \cup (\varphi^\sharp_\alpha(u_-), u_-], & u_- > 0. \end{cases} \tag{3.14}$$

Here, the function $\varphi^\sharp_\alpha$ is defined from the kinetic function $\varphi^\flat_\alpha$ by

$$\frac{f(u) - f\big(\varphi^\sharp_\alpha(u)\big)}{u - \varphi^\sharp_\alpha(u)} = \frac{f(u) - f\big(\varphi^\flat_\alpha(u)\big)}{u - \varphi^\flat_\alpha(u)}, \quad u \ne 0,$$

with the constraint

$$\begin{aligned}
\varphi^\sharp_0(u) < \varphi^\sharp_\alpha(u) \le \varphi^\natural(u), \quad u < 0, \\
\varphi^\natural(u) \le \varphi^\sharp_\alpha(u) < \varphi^\sharp_0(u), \quad u > 0.
\end{aligned} \tag{3.15}$$

Moreover, there exists a function

$$A^\natural : \mathbb{R} \to [0, +\infty),$$

called the **threshold diffusion-dispersion ratio**, *which is smooth away from $u = 0$, Lipschitz continuous at $u = 0$, increasing in $u > 0$, and decreasing in $u < 0$ with*

$$A^\natural(u) \sim C \, |u| \quad \text{as } u \to 0, \tag{3.16}$$

(where $C > 0$ depends upon f, b, c_1, and c_2 only) and such that

$$\varphi_\alpha^\flat(u) = \varphi^\natural(u) \quad \text{when } \alpha \geq A^\natural(u). \tag{3.17}$$

Additionally we have

$$\varphi_\alpha^\flat(u) \to \varphi_0^\flat(u) \quad \text{as } \alpha \to 0 \quad \text{for each } u \in I\!R. \tag{3.18}$$

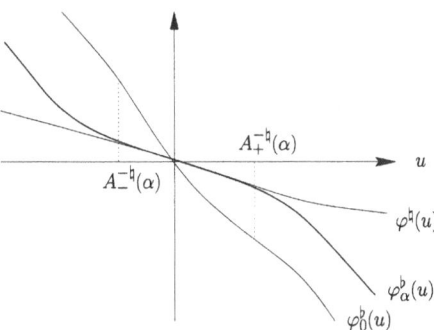

Figure III-4 : Kinetic function for general flux.

The proof of Theorem 3.3 will be the subject of Sections 4 and 5 below. The kinetic function $\varphi_\alpha^\flat : I\!R \to I\!R$ (sketched on Figure III-4) completely characterizes the dynamics of the nonclassical shock waves associated with (3.1). In view of Theorem 3.3 the theory in Chapter II applies. The kinetic function $\varphi_\alpha^\flat$ is decreasing and its range is limited by the functions $\varphi^\natural$ and $\varphi_0^\flat$. Therefore we can solve the Riemann problem, uniquely in the class of nonclassical entropy solutions selected by the kinetic function $\varphi_\alpha^\flat$.

The statements (3.17) and (3.18) provide us with important qualitative properties of the nonclassical shocks:

- The shocks leaving from u_- are always classical if the ratio α is chosen to be sufficiently large or if u_- is sufficiently small.
- The shocks leaving from u_- are always nonclassical if the ratio α is chosen to be sufficiently small.

Furthermore, under a mild assumption on the growth of f at infinity, one could also establish that the shock leaving from u_- are always nonclassical if the state u_- is sufficiently large. (See the bibliographical notes.)

In this rest of this section we introduce some important notation and investigate the limiting case when the diffusion is identically zero ($\alpha = 0$). We always suppose that $u_- > 0$ (for definiteness) and we set

$$u_0 = u_-.$$

The shock speed λ is regarded as a parameter allowing us to describe the set of attainable right-hand states. Precisely, given a speed in the interval

$$\lambda \in \left(\lambda^\natural(u_0), f'(u_0)\right),$$

there exist exactly three distinct solutions denoted by u_0, u_1, and u_2 of the equation (3.8) with

$$u_2 < \varphi^\natural(u_0) < u_1 < u_0. \tag{3.19}$$

(Recall that no trajectory exists when λ is chosen outside the interval $\left[\lambda^\natural(u_0), f'(u_0)\right]$.)

From Lemmas 3.1 and 3.2 (see (3.12)) it follows that a trajectory either is *classical* if u_0 is connected to

$$u_1 \in \left[\varphi^\natural(u_0), u_0\right] \text{ with } \lambda \in \left[\lambda^\natural(u_0), f'(u_0)\right] \tag{3.20}$$

or else is *nonclassical* if u_0 is connected to

$$u_2 \in \left[\varphi_0^\flat(u_0), \varphi^\natural(u_0)\right) \text{ with } \lambda \in \left(\lambda^\natural(u_0), \lambda_0(u_0)\right]. \tag{3.21}$$

For the sake of completeness we cover here both cases of positive and negative dispersions. For the statements in Lemma 3.4 and Theorem 3.5 below *only* we will set $\alpha := \varepsilon/\sqrt{|\delta|}$ and $\eta = \mathrm{sgn}(\delta) = \pm 1$. If (u,v) is an equilibrium point, the eigenvalues of the Jacobian matrix of the function $K(u,v)$ in (3.6) are found to be

$$\mu = \frac{1}{2}\left(-\eta\,\alpha\,\frac{b(u)}{c_1(u)c_2(u)} \pm \sqrt{\alpha^2\,\frac{b(u)^2}{c_1(u)^2 c_2(u)^2} + 4\eta\,\frac{f'(u) - \lambda}{c_1(u)\,c_2(u)}}\right).$$

So, we set

$$
\begin{aligned}
\underline{\mu}(u;\lambda,\alpha) &= \frac{\eta\,\alpha}{2}\frac{b(u)}{c_1(u)c_2(u)}\left(-1 - \eta\sqrt{1 + \frac{4\eta}{\alpha^2}\frac{c_1(u)c_2(u)}{b(u)^2}\left(f'(u) - \lambda\right)}\right), \\
\overline{\mu}(u;\lambda,\alpha) &= \frac{\eta\,\alpha}{2}\frac{b(u)}{c_1(u)c_2(u)}\left(-1 + \eta\sqrt{1 + \frac{4\eta}{\alpha^2}\frac{c_1(u)c_2(u)}{b(u)^2}\left(f'(u) - \lambda\right)}\right).
\end{aligned}
\tag{3.22}
$$

LEMMA 3.4. (Nature of equilibrium points.) *Fix some values u_- and λ and denote by $(u_*, 0)$ any one of the three equilibrium points satisfying (3.8).*
- *If $\eta = +1$ and $f'(u_*) - \lambda < 0$, then $(u_*, 0)$ is a stable point.*
- *If $\eta\left(f'(u_*) - \lambda\right) > 0$, then $(u_*, 0)$ is a saddle point.*
- *If $\eta = -1$ and $f'(u_*) - \lambda > 0$, then $(u_*, 0)$ is an unstable point.*

Furthermore, in the two cases that $\eta\left(f'(u_*) - \lambda\right) < 0$ we have the additional result: When $\alpha^2\,b(u_*)^2 + 4\,\eta\,c_1(u_*)\,c_2(u_*)\left(f'(u_*) - \lambda\right) \geq 0$ the equilibrium is a node, and is a spiral otherwise. □

For negative dispersion coefficient δ, that is, when $\eta = -1$, we see that both u_1 and u_2 are *unstable points* which no trajectory can attain at $+\infty$, while u_1 is a stable point. So, in this case, we obtain immediately:

THEOREM 3.5. (Traveling waves for negative dispersion.) *Consider the diffusive-dispersive model (3.1) where the flux satisfies (3.2). If $\varepsilon > 0$ and $\delta < 0$, then only classical trajectories exist.* □

Some additional analysis (along similar lines) would be necessary to establish the existence of these classical trajectories and conclude that

$$S_\alpha(u_-) = S(u_-) := \begin{cases} [\varphi^\natural(u_-), u_-], & u_- \geq 0 \\ \\ [u_-, \varphi^\natural(u_-)], & u_- \leq 0 \end{cases} \qquad \text{when } \delta < 0,$$

which is the shock set already found in Section 1 when $\delta = 0$.

We return to the case of a positive dispersion which is of main interest here. (From now on $\eta = +1$.) Since $g'_u(u, \lambda)$ is positive at both $u = u_2$ and $u = u_0$, we have

$$\underline{\mu}(u_0) < 0 < \overline{\mu}(u_0), \quad \underline{\mu}(u_2) < 0 < \overline{\mu}(u_2),$$

and both points u_2 and u_0 are *saddle*. On the other hand, since $g'_u(u_1, \lambda) < 0$, the equilibrium u_1 is a stable point which may be a *node* or a *spiral*. These properties are the same as the ones already established for the equation with cubic flux. (See Figure III-1.) The following result is easily checked from the expressions (3.22).

LEMMA 3.6. (Monotonicity properties of eigenvalues.) *In the range of parameters where $\underline{\mu}(u, \lambda, \alpha)$ and $\overline{\mu}(u; \lambda, \alpha)$ remain real-valued, we have*

$$\frac{\partial \underline{\mu}}{\partial \lambda}(u; \lambda, \alpha) > 0, \quad \frac{\partial \underline{\mu}}{\partial \alpha}(u; \lambda, \alpha) < 0 \quad \frac{\partial \overline{\mu}}{\partial \lambda}(u; \lambda, \alpha) < 0,$$

and, under the assumption $f'(u) - \lambda > 0$,

$$\frac{\partial \overline{\mu}}{\partial \alpha}(u; \lambda, \alpha) < 0.$$

□

To the state u_0 and the speed $\lambda \in (\lambda^\natural(u_0), \lambda_0(u_0))$ we associate the following function of the variable u, which will play an important role throughout,

$$G(u; u_0, \lambda) := \int_{u_0}^{u} (y(z, \lambda) - y(u_0, \lambda)) \frac{c_2(z)}{c_1(z)} \, dz.$$

Observe, using (3.10), that the functions G and E are closely related:

$$G(u; u_0, \lambda) = -E(u_0, u) \quad \text{when} \quad \lambda = \overline{a}(u_0, u). \tag{3.23}$$

Note also that the derivative $\partial_u G(u; u_0, \lambda)$ vanishes exactly at the equilibria u_0, u_1, and u_2 satisfying (3.8). Using the function G we rewrite now the main equations (3.5)-(3.6) in the form

$$c_2(u) u_y = v, \tag{3.24a}$$

$$c_2(u) v_y = -\alpha \frac{b(u)}{c_1(u)} v + G'_u(u; u_0, \lambda), \tag{3.24b}$$

which we will often use in the rest of the discussion.

We collect now some fundamental properties of the function G. (See Figure III-5.)

THEOREM 3.7. (Monotonicity properties of the function G.) *Fix some $u_0 > 0$ and $\lambda \in \left(\lambda^\natural(u_0), f'(u_0)\right)$ and consider the associated states u_1 and u_2. Then, the function $u \mapsto \tilde{G}(u) := G(u; u_0, \lambda)$ satisfies the monotonicity properties*

$$\tilde{G}'(u) < 0, \quad u < u_2 \text{ or } u \in (u_1, u_0),$$
$$\tilde{G}'(u) > 0, \quad u \in (u_2, u_1) \text{ or } u > u_0.$$

Moreover, if $\lambda \in \left(\lambda^\natural(u_0), \lambda_0(u_0)\right)$ we have

$$\tilde{G}(u_0) = 0 < \tilde{G}(u_2) < \tilde{G}(u_1), \tag{3.25i}$$

while, if $\lambda = \lambda_0(u_0)$,

$$\tilde{G}(u_0) = \tilde{G}(u_2) = 0 < \tilde{G}(u_1) \tag{3.25ii}$$

and finally, if $\lambda \in \left(\lambda_0(u_0), f'(u_0)\right)$,

$$\tilde{G}(u_2) < 0 = \tilde{G}(u_0) < \tilde{G}(u_1). \tag{3.25iii}$$

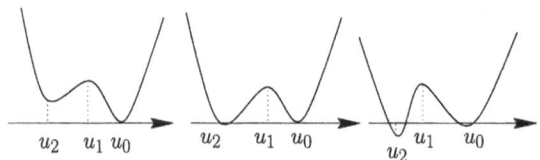

Figure III-5 : The function G
when $\lambda \in (\lambda^\natural(u_0), \lambda_0(u_0))$; $\lambda = \lambda_0(u_0)$; $\lambda = (\lambda_0(u_0), f'(u_0))$.

PROOF. The sign of $\tilde{G}'$ is the same as the sign of the function

$$g(u, \lambda) - g(u_0, \lambda) = (u - u_0) \left(\frac{f(u) - f(u_0)}{u - u_0} - \lambda \right).$$

So, the sign of $\tilde{G}'$ is easy determined geometrically from the graph of the function f. To derive (3.25) note that $\tilde{G}(u_0) = 0$ and (by the monotonicity properties above) $\tilde{G}(u_1) > \tilde{G}(u_0)$. To complete the argument we only need the sign of $\tilde{G}(u_2)$. But by (3.23) we have $\tilde{G}(u_2) = -E(u_0, u_2)$ whose sign is given by Lemma 3.2.   □

We conclude this section with the special case that the diffusion is zero. Note that the shock set below *is not* the obvious limit from (3.14).

THEOREM 3.8. (Dispersive traveling waves.) *Consider the traveling wave equation (3.4) in the limiting case $\alpha = 0$ (not included in Theorem 3.3) under the assumption that the flux f satisfies (3.2). Then, the corresponding* **shock set** *reduces to*

$$\mathcal{S}_0(u_-) = \left\{ \varphi_0^\flat(u_-), u_- \right\}, \quad u_- \in I\!R.$$

PROOF. Suppose that there exists a trajectory connecting a state $u_- > 0$ to a state $u_+ \neq u_-$ for the speed $\lambda = \bar{a}(u_-, u_+)$ and satisfying (see (3.24))

$$c_2(u)\, u_y = v,$$
$$c_1(u)\, v_y = g(u, \lambda) - g(u_-, \lambda). \tag{3.26}$$

Multiplying the second equation in (3.26) by $v/c_1(u) = c_2(u)\, u_y/c_1(u)$, we find

$$\frac{1}{2}\left(v^2\right)_y = \left(g(u, \lambda) - g(u_-, \lambda)\right) \frac{c_2(u)}{c_1(u)}\, u_y$$

and, after integration over some interval $(-\infty, y]$,

$$\frac{1}{2} v^2(y) = G(u(y); u_-, \lambda), \quad y \in \mathbb{R}. \tag{3.27}$$

Letting $y \to +\infty$ in (3.27) and using that $v(y) \to 0$ we obtain

$$G(u_+; u_-, \lambda) = 0$$

which, by (3.23), is equivalent to

$$E(u_+, u_-) = 0.$$

Using Lemma 3.2 we conclude that the right-hand state u_+ is uniquely determined, by the zero-entropy dissipation function:

$$u_+ = \varphi_0^\flat(u_-), \quad \lambda = \lambda_0(u_-). \tag{3.28}$$

Then, by assuming (3.28) and $u_- > 0$, Theorem 3.7 implies that the function $u \mapsto G(u; u_-, \lambda)$ remains strictly positive for all u (strictly) between u_+ and u_-. Since $v < 0$ we get from (3.27)

$$v(y) = -\sqrt{2\, G(u(y); u_-, \lambda)}. \tag{3.29}$$

In other words, we obtain the trajectory in the (u, v) plane:

$$v = \bar{v}(u) = -\sqrt{2\, G(u; u_-, \lambda)}, \quad u \in [u_+, u_-],$$

supplemented with the boundary conditions

$$\bar{v}(u_-) = \bar{v}(u_+) = 0.$$

Clearly, the function $\bar{v}$ is well-defined and satisfies $\bar{v}(u) < 0$ for all $u \in (u_+, u_-)$. Finally, based on the change of variable $y \in [-\infty, +\infty] \mapsto u = u(y) \in [u_+, u_-]$ given by

$$dy = \frac{c_2(u)}{\bar{v}(u)}\, du,$$

we immediately recover from the curve $v = \bar{v}(u)$ the (unique) trajectory

$$y \mapsto \left(u(y), v(y)\right).$$

This completes the proof of Theorem 3.8. $\qquad\qquad\qquad\qquad\qquad\square$

4. Traveling waves for a given speed

We prove in this section that, given u_0, u_2, and $\lambda = \bar{a}(u_0, u_2)$ in the range (see (3.21))

$$u_2 \in \left[\varphi_0^\flat(u_0), \varphi^\natural(u_0)\right), \quad \lambda \in \left(\lambda^\natural(u_0), \lambda_0(u_0)\right], \tag{4.1}$$

a nonclassical connection always exists if the ratio α *is chosen appropriately.* As we will show in the next section this result is the key step in the proof of Theorem 3.3. The main existence result proven in the present section is stated as follows.

THEOREM 4.1. (Nonclassical trajectories for a fixed speed.) *Consider two states $u_0 > 0$ and $u_2 < 0$ associated with a speed*

$$\lambda = \bar{a}(u_0, u_2) \in \left(\lambda^\natural(u_0), \lambda_0(u_0)\right].$$

Then, there exists a unique value $\alpha \geq 0$ such that u_0 is connected to u_2 by a diffusive-dispersive traveling wave solution.

By Lemma 3.4, u_0 is a saddle point and we have $\bar{\mu}(u_0) > 0$ and from Theorem 2.2 it follows that there are two trajectories leaving from u_0 at $y = -\infty$, both of them satisfying

$$\lim_{y \to -\infty} \frac{v(y)}{u(y) - u_0} = \bar{\mu}(u_0; \lambda, \alpha) \, c_2(u_0). \tag{4.2}$$

One trajectory approaches $(u_0, 0)$ in the quadrant $Q_1 = \{u > u_0, \, v > 0\}$, the other in the quadrant $Q_2 = \{u < u_0, \, v < 0\}$. On the other hand, u_2 is also a saddle point and there exist two trajectories reaching u_2 at $y = +\infty$, both of them satisfying

$$\lim_{y \to +\infty} \frac{v(y)}{u(y) - u_2} = \underline{\mu}(u_2; \lambda, \alpha) \, c_2(u_2). \tag{4.3}$$

One trajectory approaches $(u_2, 0)$ in the quadrant $Q_3 = \{u > u_2, \, v < 0\}$, the other in the quadrant $Q_4 = \{u < u_2, \, v > 0\}$.

LEMMA 4.2. *A traveling wave solution connecting u_0 to u_2 must leave the equilibrium $(u_0, 0)$ at $y = -\infty$ in the quadrant Q_2, and reach $(u_2, 0)$ in the quadrant Q_3 at $y = +\infty$.*

PROOF. Consider the trajectory leaving from the quadrant Q_1, that is, satisfying $u > u_0$ and $v > 0$ in a neighborhood of the point $(u_0, 0)$. By contradiction, suppose it would reach the state u_2 at $+\infty$. Since $u_2 < u_0$ by continuity there would exist y_0 such that

$$u(y_0) = u_0.$$

Multiplying (3.24b) by $u_y = v/c_2(u)$ we find

$$\left(v^2/2\right)_y + \alpha \, \frac{b(u)}{c_1(u) \, c_2(u)} \, v^2 = G'_u(u; u_0, \lambda) \, u_y.$$

Integrating over $(-\infty, y_0]$ we arrive at

$$\frac{v^2(y_0)}{2} + \alpha \int_{-\infty}^{y_0} v^2 \, \frac{b(u)}{c_1(u) \, c_2(u)} \, dy = G(u(y_0); u_0, \lambda) = 0. \tag{4.4}$$

Therefore $v(y_0) = 0$ and, since $u(y_0) = u_0$, a standard uniqueness theorem for the Cauchy problem associated with (3.24) implies that $u \equiv u_0$ and $v \equiv 0$ on $I\!\!R$. This contradicts the assumption that the trajectory would connect to u_2 at $+\infty$.

The argument around the equilibrium $(u_2, 0)$ is somewhat different. Suppose that the trajectory satisfies $u < u_2$ and $v > 0$ in a neighborhood of the point $(u_2, 0)$. There would exist some value y_1 achieving a *local minimum*, that is, such that

$$u(y_1) < u_2, \quad u_y(y_1) = 0, \quad u_{yy}(y_1) \geq 0.$$

From (3.24a) we would obtain $v(y_1) = 0$ and, by differentiation of (3.24a),

$$v_y(y_1) = u_{yy}(y_1)\, c_2(u(y_1)) \geq 0.$$

Combining the last two relations with (3.24b) we would obtain

$$G'_u(u(y_1); u_0, \lambda) \geq 0$$

which is in contradiction with Theorem 3.7 since $u(y_1) < u_2$ and $G'_u(u(y_1); u_0, \lambda) < 0$. $\square$

Next, we determine some intervals in which the traveling waves are always monotone.

LEMMA 4.3. *Consider a trajectory $u = u(y)$ leaving from u_0 at $-\infty$ and denote by $\underline{\xi}$ the largest value such that $u_1 < u(y) \leq u_0$ for all $y \in (-\infty, \underline{\xi})$ and $u(\underline{\xi}) = u_1$. Then, we have*

$$u_y < 0 \quad \text{on the interval } (-\infty, \underline{\xi}).$$

Similarly, if $u = u(y)$ is a trajectory connecting to u_2 at $+\infty$, denote by $\overline{\xi}$ the smallest value such that $u_2 \leq u(y) < u_1$ for all $y \in (\overline{\xi}, +\infty)$ and $u(\overline{\xi}) = u_1$. Then, we have

$$u_y < 0 \quad \text{on the interval } (\overline{\xi}, +\infty).$$

In other words, a trajectory cannot change its monotonicity before reaching the value u_1.

PROOF. We only check the first statement, the proof of the second one being similar. By contradiction, there would exist $y_1 \in (-\infty, \underline{\xi})$ such that

$$u_y(y_1) = 0, \quad u_{yy}(y_1) \geq 0, \quad u_1 < u(y_1) \leq u_0.$$

Then, using the equation (3.24b) would yield $G'_u(u(y_1); u_0, \lambda) \geq 0$, which is in contradiction with the monotonicity properties in Theorem 3.7. $\square$

PROOF OF THEOREM 4.1. For each $\alpha \geq 0$ we consider the orbit leaving from u_0 and satisfying $u < u_0$ and $v < 0$ in a neighborhood of $(u_0, 0)$. This trajectory reaches the line $\{u = u_1\}$ for the "first time" at some point denoted by $(u_1, V_-(\alpha))$. In view of Lemma 4.3 this part of trajectory is *the graph of a function*

$$[u_1, u_0] \ni u \mapsto v_-(u; \lambda, \alpha)$$

with of course $v_-(u_1; \lambda, \alpha) = V_-(\alpha)$. Moreover, by standard theorems on differential equations, v_- is a smooth function with respect to its argument $(u; \lambda, \alpha) \in [u_1, u_0] \times (\lambda^\natural(u_0), \lambda_0(u_0)] \times [0, +\infty)$.

Similarly, for each $\alpha \geq 0$ we consider the orbit arriving at u_2 and satisfying $u > u_2$ and $v < 0$ in a neighborhood of $(u_2, 0)$. This trajectory reaches the line $\{u = u_1\}$ for the "first time" *as y decreases from $+\infty$* at some point $(u_1, V_+(\alpha))$. By Lemma 4.3 this trajectory is the *graph of a function*

$$[u_2, u_1] \ni u \mapsto v_+(u; \lambda, \alpha).$$

The mapping v_+ depends smoothly upon $(u, \lambda, \alpha) \in [u_2, u_1] \times \left(\lambda^{\natural}(u_0), \lambda_0(u_0)\right] \times [0, +\infty)$.

For each of these curves $u \mapsto v_-(u)$ and $u \mapsto v_+(u)$ we derive easily from (3.24) a differential equation in the (u, v) plane:

$$v(u) \frac{dv}{du}(u) + \alpha \frac{b(u)}{c_1(u)} v(u) = G'_u(u, u_0, \lambda). \tag{4.5}$$

Clearly, the function

$$[0, +\infty) \ni \alpha \mapsto W(\alpha) := v_+(u_1; \lambda, \alpha) - v_-(u_1; \lambda, \alpha)$$
$$= V_+(\alpha) - V_-(\alpha)$$

measures the distance (in the phase plane) between the two trajectories at $u = u_1$. Therefore, the condition $W(\alpha) = 0$ characterizes the traveling wave solution of interest connecting u_0 to u_2. The existence of a root for the function W is obtained as follows.

Case 1: Take first $\alpha = 0$.

Integrating (4.5) with $v = v_-$ over the interval $[u_1, u_0]$ yields

$$\frac{1}{2}(V_-(0))^2 = G(u_1; u_0, \lambda) - G(u_0; u_0, \lambda) = G(u_1; u_0, \lambda),$$

while integrating (4.5) with $v = v_+$ over the interval $[u_2, u_1]$ gives

$$\frac{1}{2}(V_+(0))^2 = G(u_1; u_0, \lambda) - G(u_2; u_0, \lambda).$$

When $\lambda \neq \lambda_0(u_0)$, since $G(u_2; u_0, \lambda) > 0$ (Theorem 3.7) and $V_\pm(\alpha) < 0$ (Lemma 4.3) we conclude that $W(0) > 0$. When $\lambda = \lambda_0(u_0)$ we have $G(u_2; u_0, \lambda) = 0$ and $W(0) = 0$.

Case 2: Consider next the limit $\alpha \to +\infty$.

On one hand, since $v_- < 0$, for $\alpha > 0$ we get in the same way as in Case 1

$$\frac{1}{2}(V_-(\alpha))^2 < G(u_1; u_0, \lambda). \tag{4.6}$$

On the other hand, dividing (4.5) by $v = v_+$ and integrating over the interval $[u_2, u_1]$ we find

$$V_+(\alpha) = -\alpha \int_{u_2}^{u_1} \frac{b(u)}{c_1(u)} \, du + \int_{u_2}^{u_1} \frac{G'_u(u; u_0, \lambda)}{v_+(u)} \, du.$$

Since $v = c_2(u) u_y \leq 0$ and $G'_u(u) \geq 0$ in the interval $[u_2, u_1]$ we obtain

$$V_+(\alpha) \leq -\kappa \alpha (u_1 - u_2), \tag{4.7}$$

where $\kappa = \inf_{u \in [u_2, u_1]} b(u)/c_1(u) > 0$. Combining (4.6) and (4.7) and choosing α to be sufficiently large, we conclude that

$$W(\alpha) = V_+(\alpha) - V_-(\alpha) < 0.$$

Hence, by the intermediate value theorem there exists at least one value α such that

$$W(\alpha) = 0,$$

which establishes the existence of a trajectory connecting u_0 to u_2. Thanks to Lemma 4.3 it satisfies $u_y < 0$ *globally*.

The uniqueness of the solution is established as follows. Suppose that there would exist two orbits $v = v(u)$ and $v^* = v^*(u)$ associated with distinct values α and $\alpha^* > \alpha$, respectively. Then, Lemma 3.6 would imply that

$$\overline{\mu}(u_0; \lambda, \alpha^*) < \overline{\mu}(u_0; \lambda, \alpha), \quad \underline{\mu}(u_2; \lambda, \alpha^*) < \underline{\mu}(u_2; \lambda, \alpha).$$

So, there would exist $u_3 \in (u_2, u_0)$ satisfying

$$v(u_3) = v^*(u_3), \quad \frac{dv^*}{du}(u_3) \geq \frac{dv}{du}(u_3).$$

Comparing the equations (4.5) satisfied by both v and v^*, we get

$$v(u_3) \left(\frac{dv}{du}(u_3) - \frac{dv^*}{du}(u_3) \right) = (\alpha^* - \alpha) \frac{b(u_3)}{c_1(u_3)} v(u_3). \tag{4.8}$$

Now, since $v(u_3) \neq 0$ (the connection with the third critical point $(u_1, 0)$ is impossible) we obtain a contradiction, as the two sides of (4.8) have opposite signs. This completes the proof of Theorem 4.1.   □

REMARK 4.4. It is not difficult to see also that, in the proof of Theorem 4.1,

$$\alpha \mapsto V_-(\alpha) \text{ is non-decreasing} \tag{4.9i}$$

and

$$\alpha \mapsto V_+(\alpha) \text{ is decreasing.} \tag{4.9ii}$$

In particular, the function $W(\alpha) := V_+(\alpha) - V_-(\alpha)$ is decreasing.   □

THEOREM 4.5. (Threshold function associated with nonclassical shocks.) *Consider the function $A = A(u_0, u_2)$ which is the unique value α for which there is a nonclassical traveling wave connecting u_0 to u_2 (Theorem 4.1). It is defined for $u_0 > 0$ and $u_2 < 0$ with $u_2 \in [\varphi_0^\flat(u_0), \varphi^\natural(u_0))$ or, equivalently, $u_0 \in [\varphi_0^\flat(u_2), \varphi^{-\natural}(u_2))$. Then we have the following two properties:*

- *The function $A(u_0, u_2)$ is increasing in u_2 and maps $[\varphi_0^\flat(u_0), \varphi^\natural(u_0))$ onto some interval of the form $[0, A^\natural(u_0))$ where $A^\natural(u_0) \in (0, +\infty]$.*
- *The function A is also increasing in u_0 and maps the interval $[\varphi_0^\flat(u_2), \varphi^{-\natural}(u_2))$ onto the interval $[0, A^\natural(\varphi^{-\natural}(u_2)))$.*

Later (in Section 5) the function A will also determine the range in which *classical shocks* exist. From now on, we refer to the function A as the **critical diffusion-dispersion ratio.** On the other hand, the value $A^\natural(u_0)$ is called the **threshold diffusion-dispersion ratio** at u_0. Nonclassical trajectories leaving from u_0 exist if and only if $\alpha < A^\natural(u_0)$.

Observe that, in Theorem 4.5, we have $A(u_0, u_2) \to 0$ when $u_2 \to \varphi_0^\flat(u_0)$, which is exactly the desired property (3.18) in Theorem 3.3.

PROOF. We will only prove the first statement, the proof of the second one being completely similar. Fix $u_0 > 0$ and $u_2^* < u_2 < u_0$ so that

$$\lambda^\natural(u_0) < \lambda = \frac{f(u_2) - f(u_0)}{u_2 - u_0} < \lambda^* = \frac{f(u_2^*) - f(u_0)}{u_2^* - u_0} \leq \lambda_0(u_0).$$

Proceeding by contradiction we assume that

$$\alpha^* := A(u_0, u_2^*) \geq \alpha := A(u_0, u_2).$$

Then, Lemma 3.6 implies

$$\overline{\mu}(u_0; \lambda, \alpha) \geq \overline{\mu}(u_0; \lambda, \alpha^*) > \overline{\mu}(u_0; \lambda^*, \alpha^*).$$

Let $v = v(u)$ and $v^* = v^*(u)$ be the solutions of (4.5) associated with α and α^*, respectively, and connecting u_0 to u_2, and u_0 to u_2^*, respectively. Since $u_2^* < u_2$, by continuity there must exist some state $u_3 \in (u_2, u_0)$ such that

$$v(u_3) = v^*(u_3), \quad \frac{dv^*}{du}(u_3) \geq \frac{dv}{du}(u_3).$$

On the other hand, in view of (4.5) which is satisfied by both v and v^* we obtain

$$v(u_3) \left(\frac{dv^*}{du}(u_3) - \frac{dv}{du}(u_3) \right) + v(u_3)(\alpha^* - \alpha) \frac{b(u_3)}{c_1(u_3)} = (\lambda^* - \lambda)(u_0 - u_3) \frac{c_2(u_3)}{c_1(u_3)},$$

which leads to a contradiction since the left-hand side is non-positive and the right-hand side is positive. This completes the proof of Theorem 4.5. □

We complete this section with some important asymptotic properties (which will establish (3.16)-(3.17) in Theorem 3.3).

THEOREM 4.6. *The threshold diffusion-dispersion ratio satisfies the following two properties:*
- $A^\natural(u_0) < +\infty$ *for all* u_0.
- *There exists a traveling wave connecting* u_0 *to* $u_2 = \varphi^\natural(u_0)$ *for the value* $\alpha = A^\natural(u_0)$.

PROOF. Fix $u_0 > 0$. According to Theorem 4.1, given $\lambda \in (\lambda^\natural(u_0), \lambda_0(u_0)]$ there exists a nonclassical trajectory, denoted by $u \mapsto v(u)$, connecting u_0 to some u_2 with

$$\lambda = \frac{f(u_2) - f(u_0)}{u_2 - u_0}, \quad u_2 < \varphi^\natural(u_0), \quad \alpha = A(u_0, u_2). \tag{4.10}$$

On the other hand, choosing any state $u_0^* > u_0$ and setting

$$\lambda^* = \frac{f(u_0^*) - f(u_1^*)}{u_0^* - u_1^*}, \quad u_1^* = \varphi^\natural(u_0),$$

it is easy to check from (3.22) that, for all α^* *sufficiently large*, $\underline{\mu}(u_1^*; \lambda^*, \alpha^*)$ remains real with

$$\underline{\mu}(u_1^*; \lambda^*, \alpha^*) < 0.$$

Then, consider the trajectory $u \mapsto v^*(u)$ arriving at u_1^* and satisfying

$$\lim_{\substack{u \to u_1^* \\ u > u_1^*}} \frac{v^*(u)}{u - u_1^*} = \underline{\mu}(u_1^*; \lambda^*, \alpha^*) c_2(u_1^*) < 0.$$

Two different situations should be distinguished.

Case 1 : The curve $v^* = v^*(u)$ crosses the curve $v = v(u)$ at some point u_3 where

$$u_1^* < u_3 < u_0, \quad v(u_3) = v^*(u_3), \quad \frac{dv}{du}(u_3) \geq \frac{dv^*}{du}(u_3).$$

Using the equation (4.5) satisfied by both v and v^* we get

$$v(u_3) \left(\frac{dv^*}{du}(u_3) - \frac{dv}{du}(u_3) \right) + (\alpha^* - \alpha) \frac{b(u_3)}{c_1(u_3)} v(u_3) = G'_u(u_3; u_0^*, \lambda^*) - G'_u(u_3; u_0, \lambda)$$

$$< 0.$$

In view of our assumptions, since $v(u_3) < 0$ we conclude that $\alpha < \alpha^*$ in this first case.

Case 2 : $v^* = v^*(u)$ does not cross the curve $v = v(u)$ on the interval (u_1^*, u_0).

Then, the trajectory v^* crosses the u-axis at some point $u_4 \in (u_1^*, u_0]$. Integrating the equation (4.5) for the function v on the interval $[u_2, u_0]$ we obtain

$$\alpha \int_{u_0}^{u_2} \frac{b(u)}{c_1(u)} v(u) \, du = G(u_2; u_0, \lambda) - G(u_0; u_0, \lambda).$$

On the other hand, integrating (4.5) for the solution v^* over $[u_1^*, u_4]$ we get

$$\alpha^* \int_{u_4}^{u_1^*} \frac{b(u)}{c_1(u)} v^*(u) \, du = G(u_1^*; u_0^*, \lambda^*) - G(u_4; u_0^*, \lambda^*).$$

Since, by our assumption in this second case,

$$\int_{u_0}^{u_2} \frac{b(u)}{c_1(u)} v(u) du > \int_{u_4}^{u_1^*} \frac{b(u)}{c_1(u)} v^*(u) du,$$

we deduce from the former two equations that

$$\alpha \le \alpha^* \frac{G(u_2; u_0, \lambda) - G(u_0; u_0, \lambda)}{G(u_1^*; u_0^*, \lambda^*) - G(u_4; u_0^*, \lambda^*)} \le C\alpha^*,$$

where C is a constant *independent* of u_2. More precisely, u_2 describes a small neighborhood of $\varphi^\natural(u_0)$, while u_0^*, u_1^*, u_4, and λ^* remain fixed.

Finally, we conclude that in both cases

$$A(u_0, u_2) \le C' \alpha^*,$$

where α^* is sufficiently large (the condition depends on u_0 only) and C' is independent of the right-hand state u_2 under consideration. Hence, we have obtained an *upper bound* for the function $u_2 \mapsto A(u_0, u_2)$. This completes the proof of the first statement in the theorem.

The second statement is a consequence of the fact that $A(u_0, u_2)$ remains bounded as u_2 tends to $\varphi^\natural(u_0)$ and of the continuity of the traveling wave v with respect to the parameters λ and α, i.e., with obvious notation

$$v(.; \lambda^\natural(u_0), A^\natural(u_0)) = \lim_{u_2 \to \varphi^\natural(u_0)} v(.; \lambda(u_0, u_2), A(u_0, u_2)).$$

$\square$

The function $A^\natural = A^\natural(u_0)$ maps the interval $(0, +\infty)$ onto some interval $[\underline{A^\natural}, \overline{A^\natural}]$ where $0 \le \underline{A^\natural} \le \overline{A^\natural} \le +\infty$. The values $\underline{A^\natural}$ and $\overline{A^\natural}$ correspond to *lower* and *upper bounds* for the threshold ratio, respectively. The following theorem shows that the range of the function $A^\natural(u_0)$, in fact, has the form $[0, \overline{A^\natural}]$.

THEOREM 4.7. *With the notation in Theorem 4.5 the asymptotic behavior of $A^\natural(u_0)$ as $u_0 \to 0$ is given by*

$$A^\natural(u_0) \sim \kappa\, u_0, \quad \kappa := \frac{c_1(0)c_2(0)}{4\,b(0)}\, \sqrt{3 f'''(0)} > 0. \tag{4.11}$$

Note that of course (3.2) implies that $f'''(0) > 0$. In particular, Theorem 4.7 shows that $\underline{A}^\natural(0) = A^\natural(0) = 0$. Theorem 4.7 is the only instance where the assumption $f'''(0) \neq 0$ (see (3.2)) is needed. In fact, if this assumption is dropped one still have $A^\natural(u_0) \to 0$ as $u_0 \to 0$. (See the bibliographical notes.)

PROOF. To estimate $A^\natural$ near the origin we compare it with the corresponding critical function $A_\star^\natural$ determined explicitly from the third-order Taylor expansion $f^\star$ of $f = f(u)$ at $u = 0$. (See (4.16) below.) We rely on the results in Section 2, especially the formula (2.26) which provides the threshold ratio explicitly for the cubic flux.

Fix some value $u_0 > 0$ and the speed $\lambda = \lambda^\natural(u_0)$ so that, with the notation introduced earlier, $u_2 = u_1 = \varphi^\natural(u_0)$. Since $f'''(0) \neq 0$ it is not difficult to see that

$$u_2 = \varphi^\natural(u_0) = -(1 + O(u_0))\,\frac{u_0}{2}$$

(as is the case for the cubic flux $f(u) = u^3$). A straightforward Taylor expansion for the function

$$G(u) := G(u; u_0, \lambda^\natural(u_0))$$

yields

$$
\begin{aligned}
G(u) - G(u_2) &= G(u) - G(\varphi^\natural(u_0)) \\
&= \frac{(u - u_2)^3}{24}\left(f'''(0)\,\frac{c_2(0)}{c_1(0)}\,(3\,u_2 + u) + O(|u_2|^2 + |u|^2) \right).
\end{aligned}
$$

Since, for all $u \in [u_2, u_0]$

$$4\,u_2 < u + 3\,u_2 < u_0 + 3\,u_2 = u_2\,(1 + O(u_0)),$$

we arrive at

$$\left| G(u) - G(u_2) - f'''(0)\,\frac{c_2(0)}{c_1(0)}\,(u + 3\,u_2)\,\frac{(u - u_2)^3}{24} \right| \leq C\,u_0\,|u + 3\,u_2|\,(u - u_2)^3. \tag{4.12}$$

Now, given $\varepsilon > 0$, we can assume that u_0 is sufficiently small so that

$$(i) \quad -\frac{u_0}{2}\,(1 + \varepsilon) \leq u_2 \leq -\frac{u_0}{2}\,(1 - \varepsilon),$$

$$(ii) \quad (1 - \varepsilon)\,\frac{b(0)}{c_1(0)} \leq \frac{b(u)}{c_1(u)} \leq (1 + \varepsilon)\,\frac{b(0)}{c_1(0)}, \quad u \in [u_2, u_0], \tag{4.13}$$

$$(iii) \quad c_j(0)\,(1 - \varepsilon) \leq c_j(u) \leq c_j(0)\,(1 + \varepsilon), \quad u \in [u_2, u_0], \quad j = 1, 2.$$

Introduce next the flux-function

$$f_\star(u) = k\,\frac{u^3}{6}, \quad k = (1 + \varepsilon)\,f'''(0), \quad u \in \mathbb{R}. \tag{4.14}$$

Define the following (constant) functions

$$b^\star(u) = b(0), \quad c_1^\star(u) = c_1(0), \quad c_2^\star(u) = c_2(0).$$

To these functions we can associate a function G_* by the general definition in Section 3. We are interested in traveling waves associated with the functions f_*, b^*, c_1^*, and c_2^*, and connecting the left-hand state u_0^* given by

$$u_0^* = -2\,u_2$$

to the right-hand state u_2 (which will also correspond to the traveling wave associated with f).

The corresponding function

$$G_*(u) := G_*(u; u_0^*, \lambda^\natural(u_0^*))$$

satisfies

$$G_*(u) - G_*(u_2) = f'''(0)\,\frac{c_2(0)}{c_1(0)}\,\frac{(1+\epsilon)}{24}\,(u+3\,u_2)\,(u-u_2)^3. \qquad (4.15)$$

In view of Remark 2.6 the threshold function $A_*^\natural$ associated with f_*, b^*, c_1^*, and c_2^* is

$$A_*^\natural(u_0^*) = \frac{\sqrt{3k}\,c_1(0)c_2(0)}{4\,b(0)}\,u_0^*. \qquad (4.16)$$

By Theorem 4.6, for the value $\alpha^* := A_*^\natural(u_0^*)$ there exists also a traveling wave trajectory connecting u_0^* to $u_2^* := u_2$, which we denote by $v^* = v^*(u)$. By definition, in the phase plane it satisfies

$$v^*\,\frac{dv^*}{du}(u) + \alpha^*\,\frac{b^*(u)}{c_1^*(u)}\,v^*(u) = G_*'(u), \qquad (4.17)$$

with

$$G_*'(u) = \big(f_*(u) - f_*(u_0^*) - f_*'(u_2)\,(u-u_0^*)\big)\,\frac{c_2^*(u)}{c_1^*(u)}.$$

We consider also the traveling wave trajectory $u \mapsto v = v(u)$ connecting u_0 to u_2 which is associated with the data f, b, c_1, and c_2 and the threshold value $\alpha := A^\natural(u_0)$. We will now establish lower and upper bounds on $A^\natural(u_0)$; see (4.23) and (4.24) below.

Case 1 : First of all, in the easy case that $A^\natural(u_0)\,(1-\varepsilon) \le A_*^\natural(u_0^*)$, we immediately obtain by (4.16) and then (4.13)

$$A^\natural(u_0) \le (1+2\,\varepsilon)\,A_*^\natural(u_0^*) = (1+2\,\varepsilon)\,\sqrt{3k}\,\frac{c_1(0)c_2(0)}{4\,b(0)}\,u_0^*$$

$$\le (1+2\,\varepsilon)\,\sqrt{3k}\,\frac{c_1(0)c_2(0)}{4\,b(0)}\,u_0\,(1+\varepsilon)$$

$$\le (1+C\,\varepsilon)\,\sqrt{3f'''(0)}\,\frac{c_1(0)c_2(0)}{4\,b(0)}\,u_0,$$

which is the desired upper bound for the threshold function.

Case 2 : Now, assume that $A^\natural(u_0)\,(1-\varepsilon) > A_*^\natural(u_0^*)$ and let us derive a similar inequality on $A^\natural(u_0)$. Since $G'(u_2) = G_*'(u_2) = 0$, $G''(u_2) = G_*''(u_2) = 0$, and

$$v(u_2) = v(u_2^*) = 0, \quad \frac{dv}{du}(u_2) < 0, \quad \frac{dv^*}{du}(u_2) < 0,$$

it follows from the equation

$$\frac{dv}{du}(u) + \alpha \frac{b(u)}{c_1(u)} = \frac{G'_u(u; u_0, \lambda)}{v(u)}$$

by letting $u \to u_2$ that

$$\frac{dv}{du}(u_2) = -A^{\natural}_*(u_0) \frac{b(u_2)}{c_1(u_2)} < \frac{-1}{1-\varepsilon} A^{\natural}_*(u_0^*) \frac{b(0)}{c_1(0)} (1-\varepsilon) = \frac{dv^*}{du}(u_2).$$

This tells us that in a neighborhood of the point u_2 the curve v is locally below the curve v^*.

Suppose that the two trajectories meet for the "first time" at some point $u_3 \in (u_2, u_0]$, so

$$v(u_3) = v^*(u_3) \quad \text{with} \quad \frac{dv}{du}(u_3) \ge \frac{dv^*}{du}(u_3).$$

From the equations (4.5) satisfied by $v = v(u)$ and $v^* = v^*(u)$, we deduce

$$\frac{1}{2} v(u_3)^2 + \alpha \int_{u_2}^{u_3} v(u) \frac{b(u)}{c_1(u)} \, du = G(u_3) - G(u_2),$$

and

$$\frac{1}{2} v^*(u_3)^2 + \alpha^* \int_{u_2}^{u_3} v^*(u) \frac{b(u)}{c_1(u)} \, du = G_*(u_3) - G_*(u_2),$$

respectively. Subtracting these two equations and using (4.12) and (4.15), we obtain

$$\alpha \int_{u_2}^{u_3} v(u) \frac{b(u)}{c_1(u)} \, du - \alpha^* \int_{u_2}^{u_3} v^*(u) \frac{b^*(u)}{c_1^*(u)} \, du$$
$$= G(u_3) - G(u_2) - \big(G_*(u_3) - G_*(u_2)\big) \qquad (4.18)$$
$$\ge (O(u_0) - C\varepsilon) (u_3 + 3 u_2) (u_3 - u_2)^3.$$

But, by assumption the curve v is locally below the curve v^* so that the left-hand side of (4.18) is negative, while its right-hand side of (4.18) is positive if one chooses u_0 sufficiently small. We conclude that the two trajectories intersect only at u_2, which implies that $u_0^* \le u_0$ and thus

$$\int_{u_2}^{u_0} |v(u)| \, du > \int_{u_2}^{u_0^*} |v^*(u)| \, du. \qquad (4.19)$$

On the other hand we have by (4.13)

$$A^{\natural}(u_0) \frac{b(0)}{c_1(0)} (1-\varepsilon) \int_{u_2}^{u_0} |v(u)| \, du \le A^{\natural}(u_0) \int_{u_2}^{u_0} \frac{b(u)}{c_1(u)} |v(u)| \, du \qquad (4.20)$$
$$= G(u_2) - G(u_0).$$

Now, in view of the property (i) in (4.13) we have

$$|3 u_2 + u_0| \le \frac{u_0}{2} (1 + 3\varepsilon) \le |u_2| \frac{1 + 3\varepsilon}{1 - \varepsilon}, \quad |u_2 - u_0| \le \frac{u_0}{2} (3 + \varepsilon) \le |u_2| \frac{3 + \varepsilon}{1 - \varepsilon}.$$

Based on these inequalities we deduce from (4.12) that

$$G(u_2) - G(u_0) \le f'''(0) \frac{c_2(0)}{c_1(0)} \frac{9 |u_2|^4}{8} (1 + C\varepsilon). \qquad (4.21)$$

Concerning the second curve, $v^* = v^*(u)$, we have

$$A_*^\natural(u_0^*) \frac{b(0)}{c_1(0)} \int_{u_2}^{u_0^*} |v^*(u)| \, du = G_*(u_2) - G_*(u_0^*)$$

$$= f'''(0) \frac{c_2(0)}{c_1(0)} \frac{9 |u_2|^4}{8} (1 + \varepsilon) \qquad (4.22)$$

by using (4.15).

Finally, combining (4.19)–(4.22) we conclude that for every ε and for all sufficiently small u_0:

$$A^\natural(u_0) \le (1 + C \varepsilon) A_*^\natural(u_0^*)$$

$$\le (1 + C \varepsilon) \sqrt{3 f'''(0)} \frac{c_1(0) c_2(0)}{4 \, b(0)} u_0, \qquad (4.23)$$

which is the desired upper bound. Exactly the same analysis as before but based on the cubic function $f_*(u) = k \, u^3$ with $k = (1 - \varepsilon) f'''(0)$ (exchanging the role played by f_* and f, however) we can also derive the following inequality

$$A^\natural(u_0) \ge \sqrt{3 f'''(0)} \frac{c_1(0) c_2(0)}{4 \, b(0)} u_0 (1 - C \varepsilon). \qquad (4.24)$$

The proof of Theorem 4.7 is thus completed since ε is arbitrary in (4.23) and (4.24).
□

5. Traveling waves for a given diffusion-dispersion ratio

Fixing the parameter α, we can now complete the proof of Theorem 3.3 by identifying the set of right-hand state attainable from u_0 by classical trajectories. We rely here mainly on Theorem 4.1 (existence of the nonclassical trajectories) and Theorem 4.5 (critical function).

Given $u_0 > 0$ and $\alpha > 0$, a classical traveling wave must connect $u_- = u_0$ to $u_+ = u_1$ for some shock speed $\lambda \in (\lambda^\natural(u_0), f'(u_0))$. According to Theorem 4.5, to each pair of states (u_0, u_2) we can associate the critical ratio $A(u_0, u_2)$. Equivalently, to each left-hand state u_0 and each speed λ, we can associate a critical value $B(\lambda, u_0) = A(u_0, u_2)$. The mapping

$$\lambda \mapsto B(\lambda, u_0)$$

is defined and decreasing from the interval $\left[\lambda^\natural(u_0), \lambda_0(u_0)\right]$ onto $\left[0, A^\natural(u_0)\right]$. It admits an inverse

$$\alpha \mapsto \Lambda_\alpha(u_0),$$

defined from the interval $\left[0, A^\natural(u_0)\right]$ onto $\left[\lambda^\natural(u_0), \lambda_0(u_0)\right]$. By construction, given any $\alpha \in \left(0, A^\natural(u_0)\right)$ there exists a nonclassical traveling trajectory (associated with the shock speed $\Lambda_\alpha(u_0)$) leaving from u_0 and solving the equation with the prescribed value α.

It is natural to *extend the definition of the function* $\Lambda_\alpha(u_0)$ to arbitrary values α by setting

$$\Lambda_\alpha(u_0) = \lambda^\natural(u_0), \quad \alpha \ge A^\natural(u_0).$$

The nonclassical traveling waves are considered here when α is a fixed parameter. So, we define the **kinetic function** for nonclassical shocks,

$$(u_0, \alpha) \mapsto \varphi_\alpha^\flat(u_0) = u_2,$$

where u_2 denotes the right-hand state of the nonclassical trajectory, so that

$$\frac{f(u_0) - f(u_2)}{u_0 - u_2} = \Lambda_\alpha(u_0). \qquad (5.1)$$

Note that $\varphi_\alpha^\flat(u_0)$ makes sense for all $u_0 > 0$ but $\alpha < A^\natural(u_0)$.

THEOREM 5.1. *For all $u_0 > 0$ and $\alpha > 0$ and for every speed satisfying*

$$\Lambda_\alpha(u_0) < \lambda \leq f'(u_0),$$

there exists a unique traveling wave connecting $u_- = u_0$ to $u_+ = u_1$. Moreover, for $\alpha \geq A^\natural(u_0)$ there exists a traveling wave connecting $u_- = u_0$ to $u_+ = u_1$ for all

$$\lambda \in \left[\lambda^\natural(u_0), f'(u_0)\right].$$

PROOF. We first treat the case $\alpha \leq A^\natural(u_0)$ and $\lambda \in \left(\Lambda_\alpha(u_0), f'(u_0)\right)$. Consider the curve $u \mapsto v_-(u; \lambda, \alpha)$ defined on $[u_1, u_0]$ that was introduced earlier in the proof of Theorem 4.1. We have either $v_-(u_1; \lambda, \alpha) = 0$ and the proof is completed, or else $v_-(u_1; \lambda, \alpha) < 0$. In the latter case, the function v_- is a solution of (4.5) that extends further on the left-hand side of u_- in the phase plane. On the other hand, this curve cannot cross the nonclassical trajectory $u \mapsto v(u)$ connecting $u_- = u_0$ to $u_+ = \varphi_\alpha^\flat(u_0)$. Indeed, by Lemma 3.6 we have

$$\overline{\mu}(u_0; \lambda, \alpha) < \overline{\mu}(u_0; \Lambda_\alpha(u_0), \alpha).$$

If the two curves would cross, there would exist $u^* \in (\varphi_\alpha^\flat(u_0), u_1)$ such that

$$v(u^*) = v_-(u^*) \quad \text{and} \quad \frac{dv}{du}(u^*) \leq \frac{dv_-}{du}(u^*).$$

By comparing the equations (4.5) satisfied by these two trajectories we get

$$v(u^*) \left(\frac{dv}{du}(u^*) - \frac{dv_-}{du}(u^*)\right) = \left(\lambda - \Lambda_\alpha(u_0)\right)(u^* - u_0) \frac{c_2(u^*)}{c_1(u^*)}. \qquad (5.2)$$

This leads to a contradiction since the right-hand side of (5.2) is positive while the left-hand side is negative. We conclude that the function v_- must cross the u-axis at some point u_3 with $u_2 < \varphi_\alpha^\flat(u_0) < u_3 < u_1$. The curve $u \mapsto v_-(u, \lambda, \alpha)$ on the interval $[u_3, u_0]$ corresponds to a solution $y \mapsto u(y)$ in some interval $(-\infty, y_3]$ with $u_y(y_3) = 0$ and

$$u_{yy}(y_3) = \frac{g(u(y_3), \lambda) - g(u_0, \lambda)}{c_1(u(y_3)) c_2(u(y_3))} = \frac{G_u'(u_3; u_0, \lambda)}{c_2(u_3)^2}, \qquad (5.3)$$

which is positive by Theorem 3.7. Thus $u_{yy}(y_3) > 0$ and necessarily $u(y) > u_3$ for $y > y_3$. Indeed, assume that there exists $y_4 > y_3$, such that $u(y_4) = u(y_3) = u_3$. Then, multiplying (3.24b) by v_-/c_2 and integrating over $[y_3, y_4]$, we obtain

$$\frac{1}{2} v_-^2(y_4) + \alpha \int_{y_3}^{y_4} \frac{b(u)}{c_1(u) c_2(u)} v_-^2 \, dy = G(u_3; u_0, \lambda) - G(u_3; u_0, \lambda) = 0.$$

This would means that $u(y) = u_3$ for all y, which is excluded since $u_- = u_1$.

Now, since $u \leq u_0$ we see that u is bounded. Finally, by integration over the interval $(-\infty, y]$ we obtain

$$\frac{1}{2} v_-^2(y) + \alpha \int_{-\infty}^y \frac{b(u)}{c_1(u)\, c_2(u)} \, v_-^2 \, dy = G(u(y)) - G(u_0),$$

which implies that v is bounded and that the function u is defined on the whole real line $\mathbb{R}$. When $y \to +\infty$ the trajectory (u, v) converges to a critical point which can only be $(u_1, 0)$.

Consider now the case $\alpha > A^\natural(u_0)$. The proof is essentially same as the one given above. However, we replace the nonclassical trajectory with the curve $u \mapsto v_+(u)$ defined on the interval $[u_2, u_1]$. For each λ fixed in $(\lambda^\natural(u_0), f'(u_0))$ (since $\alpha > A^\natural(u_0)$) and thanks to Remark 4.4, the function, $W = V_+ - V_-$ (defined in the proof of Theorem 4.1, with $v_-(u; \lambda, \alpha)$ and $v_+(u; \lambda, \alpha)$ and extended to $\lambda \in (f'(u_2), f'(u_0))$) satisfies $W(\alpha) < 0$. On the left-hand side of u_1, with the same argument as in the first part above, we can prove that the extension of v_- does not intersect v_+ and must converge to $(u_1, 0)$. Finally, the case $\lambda = \lambda^\natural(u_0)$ is reached by continuity. This completes the proof of Theorem 5.1.   □

THEOREM 5.2. *If* $\lambda^\natural(u_0) < \lambda < \Lambda_\alpha(u_0)$ *there is no traveling wave connecting* $u_- = u_0$ *to* $u_+ = u_1$.

PROOF. Assume that there exists a traveling wave connecting u_0 to u_1. As in Lemma 4.2, we prove easily that such a curve must approach $(u_0, 0)$ from the quadrant Q_1 and coincide with the function v_- on the interval $[u_1, u_0]$. On the other hand, as in the proof of Theorem 5.1, we see that this curve does not cross the nonclassical trajectories. On the other hand, Lemma 3.7 gives

$$\overline{\mu}(u_0; \lambda, \alpha) \geq \overline{\mu}\big(u_0; \Lambda_\alpha(u_0), \alpha\big),$$

thus, the classical curve remains "under" the nonclassical one. So we have

$$v_-\big(\varphi_\alpha^\flat(u_0)\big) < v\big(\varphi_\alpha^\flat(u_0)\big),$$

where $u \mapsto (u, v(u))$ denotes the nonclassical trajectory. Assume now that the curve $(u, v_-(u))$ meets the u-axis for the first time at some point $(u_3, 0)$ with $u_3 < \varphi_\alpha^\flat(u_0) < u_2$. The previous curve defined on $[u_3, u_0]$ corresponds to a solution $y \mapsto u(y)$ defined on some interval $(-\infty, y_3]$ with $u_y(y_3) = 0$ and $u_{yy}(y_3) \geq 0$. Thus $v_y(y_3)$ satisfies (5.3) and is negative (Lemma 4.3). This implies that $u_{yy}(y_3) < 0$ which is a contradiction. Finally, the trajectory remains under the u-axis for $u < u_2$, and cannot converge to any critical point.   □

According to Theorem 5.1 the kinetic function can now be extended to all values of α by setting

$$\varphi_\alpha^\flat(u_0) = \varphi^\natural(u_0), \quad \alpha \geq A^\natural(u_0). \tag{5.4}$$

Finally we have:

THEOREM 5.3. (Monotonicity of the kinetic function.) *For each* $\alpha > 0$ *the mapping* $u_0 \mapsto \varphi_\alpha^\flat(u_0)$ *is decreasing.*

PROOF. Fix $u_0 > 0$, $\alpha > 0$, $\lambda = \Lambda_\alpha(u_0)$ and $u_2 = \varphi_\alpha^\flat(u_0)$. First suppose that $\alpha \geq A^\natural(u_0)$. Then, for all $u_0^* > u_0$, since $\varphi^\natural$ is known to be strictly monotone, it is clear that

$$\varphi_\alpha^\flat(u_0^*) \leq \varphi^\natural(u_0^*) < \varphi^\natural(u_0) = \varphi_\alpha^\flat(u_0).$$

Suppose now that $\alpha < A^\natural(u_0)$. Then, for $u_0^* > u_0$ in a neighborhood of u_0, the speed $\lambda^* = \frac{f(u_0^*) - f(u_2)}{u_0^* - u_2}$ satisfies $\lambda^* \in \left(\lambda^\natural(u_0^*), \lambda_0(u_0^*)\right)$. Then, there exists a nonclassical traveling wave connecting u_0^* to u_2 for some $\alpha^* = A(u_0^*, u_2)$. The second statement in Theorem 4.5 gives $\alpha^* > \alpha$. Since the function Λ_α is decreasing (by the first statement in Theorem 4.5) we have $\Lambda_{\alpha_*}(u_0^*) < \Lambda_\alpha(u_0^*)$ and thus $\varphi_\alpha^\flat(u_0^*) < u_2 = \varphi_\alpha^\flat(u_0)$ and the proof of Theorem 5.3 is completed.   □

PROOF OF THEOREM 3.3. Section 4 provides us with the existence of nonclassical trajectories, while Theorems 5.1 and 5.2 are concerned with classical trajectories. These results prove that the shock set is given by (3.14). By standard theorems on solutions of ordinary differential equations the kinetic function is smooth in the region $\left\{\alpha \leq A^\natural(u_0)\right\}$ while it coincides with the (smooth) function $\varphi^\natural$ in the region $\left\{\alpha \geq A^\natural(u_0)\right\}$. Additionally, by construction the kinetic function is continuous along $\alpha = A^\natural(u_0)$. This proves that $\varphi^\flat$ is Lipschitz continuous on each compact interval. On the other hand, the monotonicity of the kinetic function is provided by Theorem 5.3. The asymptotic behavior was the subject of Theorem 4.7.   □

REMARK 5.4. To a large extend the techniques developed in this chapter extend to *systems of equations*, in particular to the model of elastodynamics and phase transitions introduced in Examples I-4.7. With the notation of Examples I-4.7, the corresponding traveling wave solutions $(v, w) = (v(y), w(y))$ must solve

$$-s\,v_y - \Sigma\big(w, w_y, w_{yy}\big)_y = \big(\mu(w)\,v_y\big)_y,$$
$$s\,w_y + v_y = 0,$$

where s denotes the speed of the traveling wave, Σ is the total stress function, and $\mu(w)$ is the viscosity coefficient. When Σ is given by the law (I-4.20) and after some integration with respect to y we arrive at

$$-s\,(v - v_-) - \sigma(w) + \sigma(w_-) - \mu(w)\,v_y = \frac{\lambda'(w)}{2}\,w_y^2 - \big(\lambda(w)\,w_y\big)_y,$$
$$s\,(w - w_-) + v - v_- = 0,$$

where (v_-, w_-) denotes the upper left-hand limit and $\lambda(w)$ the capillarity coefficient. Using the second equation above we can eliminate the unknown $v(y)$, namely

$$\lambda(w)^{1/2}\left(\lambda(w)^{1/2}\,w_y\right)_y + \mu(w)\,v_y = s^2\,(w - w_-) - \sigma(w) + \sigma(w_-), \qquad (5.5)$$

which has precisely the structure of the equation (3.4) studied in the present chapter ! Additionally, the hypothesis (3.2) in this chapter is very similar to the hypotheses (I-4.8) and (I-4.12) in Examples I-4.4 and I-4.5, respectively. All the results in the present chapter extend to the equation (5.5) under the hypothesis (I-4.8) (monotonicity of the kinetic function, threshold diffusion-dispersion ratio, asymptotic properties) and most of them extend to (5.5) under the hypothesis (I-4.12). See the bibliographical notes.   □

CHAPTER IV

EXISTENCE THEORY
FOR THE CAUCHY PROBLEM

This chapter is devoted to the general *existence theory* for scalar conservation laws in the setting of functions with bounded variation. We begin, in Section 1, with an existence result for the Cauchy problem when the flux-function is *convex*. We exhibit a solution given by an explicit formula (Theorem 1.1) and prove the uniqueness of this solution (Theorem 1.3). The approach developed in Section 1 is of particular interest as it reveals important features of classical entropy solutions. However, it does not extend to non-convex fluxes or nonclassical solutions, and an entirely different strategy based on Riemann solvers and *wave front tracking* is developed in the following sections. In Sections 2 and 3, we discuss the existence of *classical* and of *nonclassical* entropy solutions to the Cauchy problem, respectively; see Theorems 2.1 and 3.2 respectively. Finally in Section 4, we derive *refined estimates* for the total variation of solutions (Theorems 4.1 to 4.3) which represent a preliminary step toward the forthcoming discussion of the Cauchy problem for systems (in Chapters VII and VIII).

1. Classical entropy solutions for convex flux

The main existence result in this section is:

THEOREM 1.1. (An explicit formula.) *Let $f : I\!R \to I\!R$ be a convex function satisfying*

$$f'' > 0 \text{ and } \lim_{\pm\infty} f' = \pm\infty,$$

and let u_0 be some initial data in $L^\infty(I\!R)$. Then, the Cauchy problem

$$\partial_t u + \partial_x f(u) = 0, \quad u = u(x,t) \in I\!R, \ x \in I\!R, \ t > 0,$$

$$u(x,0) = u_0(x), \quad x \in I\!R,$$

(1.1)

admits a weak solution $u \in L^\infty(I\!R \times I\!R_+)$ satisfying **Oleinik's one-sided inequality** *($x_1 \leq x_2$, $t > 0$):*

$$f'(u(x_2,t)) - f'(u(x_1,t)) \leq \frac{x_2 - x_1}{t}.$$

(1.2)

In particular, for almost all $t > 0$ the function $x \mapsto u(x,t)$ has locally bounded total variation.

Denote by $\tilde{f}$ *the Legendre transform of f and by g the inverse function of f'. Then, the solution of* (1.1) *is given by the* **explicit formula**

$$u(x,t) = g\left(\frac{x - y(x,t)}{t}\right),$$

(1.3a)

where $y(x,t)$ is a point that achieves the minimum value of the function

$$y \mapsto G(x,t;y) = \int_0^y u_0 \, dx + t \, \tilde{f}\left(\frac{x-y}{t}\right). \tag{1.3b}$$

A discussion of the initial condition at $t = 0$ is postponed to Theorem 1.2 below. Some important remarks are in order:

1. Recall that the **Legendre transform** of f is

$$\tilde{f}(m) = \sup_v \left(m \, v - f(v)\right), \quad m \in I\!R.$$

Since f is strictly convex and grows faster than any linear function, the supremum above is achieved at the (unique) point v such that

$$f'(v) = m \quad \text{or, equivalently,} \quad v = g(m).$$

Thus for all reals m

$$\tilde{f}(m) = m \, g(m) - f(g(m)), \quad \text{and so } \tilde{f}' = g, \tag{1.4}$$

which implies $\tilde{f}''(m) = 1/f''(g(m)) > 0$. Hence, the function $\tilde{f}$ is strictly convex and we also have $\lim_{\pm\infty} \tilde{f}' = \lim_{\pm\infty} g = \pm\infty$. For instance, if $f(u) = u^2/2$, then $g(u) = u$, $\tilde{f}(u) = u^2/2$, and

$$G(x,t;y) = \int_0^y u_0(z) \, dz + \frac{(x-y)^2}{2t}.$$

2. The formula in Theorem 1.1 should be regarded as a generalization to *discontinuous solutions* of the implicit formula (I-1.7) of Chapter I. The value $y(x,t)$ can be interpreted as the foot of the characteristic line passing through the point x at the time t. It is not difficult to deduce from the property that $y(x,t)$ minimizes G that, when the function u_0 has bounded variation,

$$u_{0-}(y(x,t)) \le u(x,t) \le u_{0+}(y(x,t)). \tag{1.5}$$

3. Setting $E := 1/\min f''$ where the minimum is taken over the range of the solution under consideration, (1.2) implies that ($x_1 \ne x_2$)

$$\frac{u(x_2,t) - u(x_1,t)}{x_2 - x_1} \le \frac{E}{t} \tag{1.2'}$$

and, by taking the limit $x_1 - x_2 \to 0$,

$$\partial_x u \le \frac{E}{t}.$$

In particular, the solution u has bounded variation in x (since the function $x \mapsto u(x,t) - E\,x/t$ is non-increasing) and u satisfies Lax shock inequality (see (II-1.9)):

$$u_-(x,t) \ge u_+(x,t),$$

where we use the notation $u_\pm(x,t) := u(x\pm,t)$.

PROOF. We consider the Cauchy problem (1.1) and, to begin with, we show that the formula (1.3) is valid for any *piecewise smooth solution with compact support* satisfying Lax shock inequality.

Define

$$w(x,t) = \int_{-\infty}^{x} u(y,t)\, dy, \quad w_0(x) = \int_{-\infty}^{x} u_0(y)\, dy,$$

and normalize the flux so that $f(0) = 0$. By integration of the equation in (1.1) we get

$$\partial_t w + f(\partial_x w) = 0.$$

Since f is convex we have for all $v \in \mathbb{R}$

$$f(v) + f'(v)\,(\partial_x w - v) \leq f(\partial_x w) = -\partial_t w,$$

thus by re-ordering the terms

$$\partial_t w + f'(v)\,\partial_x w \leq f'(v)\,v - f(v). \qquad (1.6)$$

Fix a point (x,t) and some real v. The line passing through (x,t) and with slope $f'(v)$ intersects the initial axis at some point y with, clearly,

$$y = x - t\,f'(v).$$

Integrating (1.6) along this straight line yields

$$w(x,t) - w_0(y) \leq t\left(f'(v)\,v - f(v)\right) = t\,\tilde{f} \circ f'(v),$$

since the right-hand side is a constant. When v describes the whole of $\mathbb{R}$ the parameter y also describes $\mathbb{R}$. We thus arrive at the fundamental inequality $(y \in \mathbb{R})$

$$w(x,t) \leq w_0(y) + t\,\tilde{f}\left(\frac{x-y}{t}\right) = \int_{-\infty}^{0} u_0\, dx + G(x,t;y). \qquad (1.7)$$

Since the left-hand side does not depend on y it is natural to minimize the right-hand side over all y.

Since u is piecewise smooth, from any point (x,t) (not on a shock curve) we can trace backward the (characteristic) line with slope

$$v = u(x,t).$$

Since u satisfies Lax shock inequality this line cannot meet a shock curve of u and, therefore, must eventually intersects the initial line at some point $y = y(x,t)$. Plugging this specific value in (1.6), we see that (1.6) and (1.7) become *equalities*. As a consequence, the minimum value of the right-hand side of (1.7) is achieved, precisely for the choice $v = u(x,t)$:

$$w(x,t) - \int_{-\infty}^{0} u_0\, dx = G(x,t;y(x,t)) = \min_{y \in \mathbb{R}} G(x,t;y).$$

Finally, in view of the relation

$$y(x,t) = x - t\,f'(u(x,t)),$$

the value $u(x,t)$ is recovered from $y(x,t)$ and (1.3a) holds. This establishes the explicit formula (1.3), at least for piecewise smooth solutions.

Conversely, consider now the function u given by the explicit formula (1.3). Note that the minimizer $y(x,t)$ always exists since $\tilde{f}$ grows faster than any linear function (indeed, $\lim_{\pm\infty} \tilde{f}' = \pm\infty$) while the term $\int_0^y u_0 \, dx$ grows at most linearly ($u_0 \in L^\infty(\mathbb{R})$). It need not be unique and, to begin with, we consider any (measurable) selection. However, we claim that the function $x \mapsto y(x,t)$ is increasing (but not necessarily strictly increasing). Indeed fix $x_1 < x_2$ and $y < y_1 := y(x_1, t)$ and let us check that

$$G(x_2, t; y_1) < G(x_2, t; y). \tag{1.8}$$

This will prove that $y(x_2, t) \geq y_1$. Indeed, we have

$$G(x_1, t; y_1) \leq G(x_1, t; y)$$

for all y, and especially for $y < y_1$. On the other hand, the function $\tilde{f}$ being strictly convex we have

$$\tilde{f}\left(\frac{x_2 - y_1}{t}\right) + \tilde{f}\left(\frac{x_1 - y}{t}\right) < \tilde{f}\left(\frac{x_1 - y_1}{t}\right) + \tilde{f}\left(\frac{x_2 - y}{t}\right).$$

Multiplying the latter inequality by t and adding to the former, we arrive at (1.8).

Therefore, for each t, the function $x \mapsto y(x,t)$ has locally bounded total variation and so is continuous at all but (at most) countably many points. Hence, for each t, excluding (at most) countably many x at most the minimizer $y(x,t)$ is uniquely defined.

Consider next the following approximation of u,

$$u_\varepsilon(x,t) = \frac{\displaystyle\int_{\mathbb{R}} g\left(\frac{x-y}{t}\right) e^{-\frac{1}{\varepsilon} G(x,t;y)} \, dy}{\displaystyle\int_{\mathbb{R}} e^{-\frac{1}{\varepsilon} G(x,t;y)} \, dy}$$

and similarly for $f(u)$

$$f_\varepsilon(x,t) = \frac{\displaystyle\int_{\mathbb{R}} (f \circ g)\left(\frac{x-y}{t}\right) e^{-\frac{1}{\varepsilon} G(x,t;y)} \, dy}{\displaystyle\int_{\mathbb{R}} e^{-\frac{1}{\varepsilon} G(x,t;y)} \, dy}.$$

Set also

$$v_\varepsilon(x,t) := \log \int_{\mathbb{R}} e^{-\frac{1}{\varepsilon} G(x,t;y)} \, dy.$$

(The functions under consideration are integrable, as can be checked easily from the estimates to be derived below.) A simple calculation using (1.3b) yields

$$\partial_t G = -(f \circ g)\left(\frac{x-y}{t}\right), \quad \partial_x G = g\left(\frac{x-y}{t}\right),$$

and thus $u_\varepsilon = -\varepsilon \, \partial_x v_\varepsilon$ and $f_\varepsilon = \varepsilon \, \partial_t v_\varepsilon$, from which it follows that

$$\partial_t u_\varepsilon + \partial_x f_\varepsilon = 0. \tag{1.9}$$

Consider a point (x,t) at which the function $y(.,t)$ is continuous in space (again, only countably many points are excluded). The minimizer is unique and the function $y \mapsto G(x,t;y)$ achieves its minimum solely at $y = y(x,t)$. We claim that, as $\varepsilon \to 0$, $u_\varepsilon(x,t)$ converges to the value of the integrand computed at a point that achieves the minimum of $G(x,t;.)$. We now provide a proof of this fact.

Normalize G by $G(x,t; y(x,t)) = 0$. (Adding a constant to G does not modify $u_\varepsilon(x,t)$.) Fix $\delta > 0$. Since G is Lipschitz continuous in y there exists $C_1 > 0$ (depending on the point (x,t)) such that

$$G(x,t;y) \leq C_1 \, |y - y(x,t)|, \quad y \in [y(x,t) - \delta, y(x,t) + \delta],$$

which implies

$$\int_{I\!R} e^{-G(x,t;y)/\varepsilon} \, dy \geq \int_{y(x,t)-\delta}^{y(x,t)+\delta} e^{-C_1 |y-y(x,t)|/\varepsilon} \, dy$$

$$= 2 \int_0^\delta e^{-C_1 y/\varepsilon} \, dy$$

$$= 2\varepsilon \int_0^{\delta/\varepsilon} e^{-C_1 y} \, dy \geq C_2 \, \varepsilon$$

for all $\varepsilon < \delta$ and for some C_2 independent of ε.

On the other hand, in the region $|y - y(x,t)| \geq \delta$ the function G is bounded away from zero (since $y(x,t)$ is the unique minimum of G). Since it tends to infinity at infinity there exists a constant $C_3 = C_3(\delta)$ (depending also on the point (x,t)) such that

$$e^{-G(x,t;y)/\varepsilon} \leq e^{-C_3 |y-y(x,t)|/\varepsilon}, \quad |y - y(x,t)| \geq \delta.$$

Collecting the above inequalities and denoting by C_4 the Lipschitz constant of $g(\cdot)/t$, we arrive at

$$|u_\varepsilon(x,t) - u(x,t)| \leq C_4 \frac{\displaystyle\int_{I\!R} |y - y(x,t)| \, e^{-G(x,t;y)/\varepsilon} \, dy}{\displaystyle\int_{I\!R} e^{-G(x,t;y)/\varepsilon} \, dy}$$

$$\leq C_4 \, \delta + \frac{C_4}{C_2 \, \varepsilon} \int_{\{|y-y(x,t)|\geq\delta\}} |y - y(x,t)| \, e^{-C_3 |y-y(x,t)|/\varepsilon} \, dy$$

$$= C_4 \, \delta + \frac{2 C_4}{C_2 \, \varepsilon} \int_0^{+\infty} y \, e^{-C_3 y/\varepsilon} \, dy$$

$$\leq C_4 \, \delta + C_5 \, \varepsilon,$$

for some constant $C_5 > 0$.

As $\varepsilon \to 0$ we find

$$\limsup_{\varepsilon \to 0} |u_\varepsilon(x,t) - u(x,t)| \leq C_4 \, \delta.$$

Since δ was arbitrary and the arguments for $f_\varepsilon(x,t)$ are completely similar we conclude that at each point (x,t) where the function $y(.,t)$ is continuous

$$\lim_{\varepsilon \to 0} u_\varepsilon(x,t) = u(x,t), \quad \lim_{\varepsilon \to 0} f_\varepsilon(x,t) = f(u(x,t)).$$

Passing to the limit in (1.9), we deduce that the function u is a weak solution of the conservation law in (1.1).

To show that u is an entropy solution, we derive the stronger statement (1.2). By using the explicit formula (1.3) and the monotonicity of the function $x \mapsto y(x,t)$ established earlier, we see that

$$f'(u(x_2,t)) - f'(u(x_1,t)) = \frac{x_2 - y(x_2,t)}{t} - \frac{x_1 - y(x_1,t)}{t} \leq \frac{x_2 - x_1}{t}.$$

This completes the proof of Theorem 1.1. $\qquad\qquad\qquad\qquad\qquad\qquad\square$

Concerning the initial condition we can now prove the following result.

THEOREM 1.2. (Initial condition.) *When $u_0 \in L^1(\mathbb{R}) \cap BV(\mathbb{R})$ (with bounded variation denoted by $TV(u_0)$), the solution $u = u(x,t)$ obtained in Theorem 1.1 assumes its initial data u_0 in the L^1 sense:*

$$\|u(t) - u_0\|_{L^1(\mathbb{R})} \to 0 \quad as \ t \to 0. \tag{1.10}$$

If the Legendre transform of f satisfies the lower bound

$$\tilde{f}(v) \geq C_0 \left(|v|^{1+\alpha} - 1\right), \quad v \in \mathbb{R} \tag{1.11}$$

for some constants $C_0, \alpha > 0$, then for each $T > 0$ there exists a constant $C = C(T)$ such that

$$\|u(t) - u_0\|_{L^1(\mathbb{R})} \leq C \, TV(u_0) \, t^{\alpha/(1+\alpha)}, \quad t \in [0,T]. \tag{1.12}$$

PROOF. Since y minimizes the function G we have

$$\begin{aligned}
G(x,t;y(x,t)) &= \int_0^{y(x,t)} u_0 \, dx + t \, \tilde{f}\left(\frac{x - y(x,t)}{t}\right) \\
&\leq \int_0^x u_0 \, dx + t \, \tilde{f}(0) \leq C_1.
\end{aligned} \tag{1.13}$$

Therefore, we have the following upper bound for $|y(x,t) - x|$:

$$t \, \tilde{f}\left(\frac{x - y(x,t)}{t}\right) \leq C_1 + \|u_0\|_{L^1(\mathbb{R})} =: C_2. \tag{1.14}$$

Under the assumption of Theorem 1.1 ($f'' > 0$ and $\lim_{\pm\infty} f' = \pm\infty$) it follows that for all x

$$y(x,t) \to x \quad as \ t \to 0.$$

In view of (1.5) we deduce that $u(x,t) \to u_0(x)$ at all x but the points of jump of u_0 and, by Lebesgue theorem, we conclude that (1.10) holds.

Now, with the stronger condition (1.11) we deduce from (1.14) that

$$\left|\frac{x - y(x,t)}{t}\right|^{1+\alpha} \leq \frac{C_2}{C_0 t} + 1$$

thus, for $C = C(T)$ and $t \in [0,T]$,

$$|y(x,t) - x| \leq C \, t^{\alpha/(1+\alpha)} =: C \, t^\beta.$$

Thus, from (1.5) it follows that

$$|u(x,t) - u_0(x)| \leq TV\left(u_0; (x - Ct^\beta, x + Ct^\beta)\right),$$

hence

$$\|u(t) - u_0\|_{L^1(\mathbb{R})} \leq \int_{\mathbb{R}} TV\left(u_0; (x - Ct^\beta, x + Ct^\beta)\right) dx.$$

When $u_0' \in L^1(\mathbb{R})$, by commuting the orders of integration we find

$$\int_{\mathbb{R}} \int_{x-Ct^\beta}^{x+Ct^\beta} |u_0'| \, dy dx = 2C \, t^\beta \int_{\mathbb{R}} |u_0'| \, dy.$$

Clearly, u_0 can be realized as the limit of functions whose derivatives are in L^1, and therefore $\|u(t) - u_0\|_{L^1(\mathbb{R})} \leq 2C \, TV(u_0) \, t^\beta$, which completes the proof of Theorem 1.2.
$\square$

We now turn to the uniqueness of solutions for conservation laws with convex flux. We provide a proof based on the inequality (1.2) discovered in Theorem 1.1, rather than on the (weaker) entropy conditions stated in Section II-1. A more general uniqueness result will be established in Chapter V using a different approach.

THEOREM 1.3. (Uniqueness of entropy solutions.) *Let* $f : \mathbb{R} \to \mathbb{R}$ *be a strictly convex flux-function. Let* u_1 *and* u_2 *be weak solutions of the problem (1.1) with*

$$u_1, u_2 \in L^\infty(\mathbb{R} \times \mathbb{R}_+) \cap L^\infty(\mathbb{R}_+, L^1(\mathbb{R})),$$

satisfying Oleinik one-sided entropy inequality $(x \neq y)$

$$\frac{u_1(x) - u_1(y)}{x - y} \le \frac{E}{t}, \quad \frac{u_2(x) - u_2(y)}{x - y} \le \frac{E}{t}, \tag{1.15}$$

where E *is a positive constant. Suppose that* u_1 *and* u_2 *share the same initial data* $u_0 \in L^\infty(\mathbb{R}) \cap L^1(\mathbb{R})$ *with*

$$\|u_1(t) - u_0\|_{L^1(\mathbb{R})} + \|u_2(t) - u_0\|_{L^1(\mathbb{R})} \le C t^\beta, \quad t > 0 \tag{1.16}$$

for some $C, \beta > 0$. *Then, we have*

$$u_2(t) = u_1(t) \quad \text{for all } t \ge 0.$$

In fact, in Chapter V we will derive the L^1 **contraction property** (Theorem V-5.2)

$$\|u_2(t) - u_1(t)\|_{L^1(\mathbb{R})} \le \|u_2(s) - u_1(s)\|_{L^1(\mathbb{R})}, \quad s \le t, \tag{1.17}$$

and see that the solution is actually Lipschitz continuous in time with values in L^1.

PROOF. Since u_1 and u_2 are weak solutions, the function $\varphi := u_2 - u_1$ satisfies the linear equation

$$\partial_t \varphi + \partial_x (a\,\varphi) = 0, \tag{1.18}$$

where $a = a(x, t)$ is defined by

$$f(u_2) - f(u_1) = \int_0^1 f'(\theta\,u_1 + (1 - \theta)\,u_2)\,d\theta\,(u_2 - u_1) =: a\,\varphi.$$

Setting $M := \max f''$ we deduce from (1.15) that

$$\frac{a(x) - a(y)}{x - y} = \int_0^1 \frac{f'(\theta\,u_1(x) + (1 - \theta)\,u_2(x)) - f'(\theta\,u_1(y) + (1 - \theta)\,u_2(y))}{x - y}\,d\theta$$

$$= \int_0^1 f''(v(x, y, \theta)) \left(\theta\,\frac{u_1(x) - u_1(y)}{x - y} + (1 - \theta)\,\frac{u_2(x) - u_2(y)}{x - y} \right) d\theta$$

for some point $v(x, y, \theta)$, thus $\dfrac{a(x) - a(y)}{x - y} \le \dfrac{ME}{t}$. So, letting $y \to x$ in the above inequality we arrive at

$$\partial_x a \le \frac{ME}{t}. \tag{1.19}$$

In view of (1.18) the function $\psi(x, t) := \displaystyle\int_{-\infty}^x \varphi(y, t)\,dy$ satisfies

$$\partial_t \psi + a\,\partial_x \psi = 0. \tag{1.20}$$

By definition, $\partial_x \psi$ is bounded and so is $\partial_t \psi$ in view of (1.20). Therefore, the function ψ is Lipschitz continuous and the calculations below make sense.

Next, given any even integer p we multiply (1.20) by ψ^{p-1} and obtain

$$\partial_t \psi^p + a\,\partial_x \psi^p = 0$$

or equivalently, using the weight t^{-m} ($m > 0$ being a real number),

$$\partial_t \left(t^{-m}\,\psi^p\right) + \partial_x \left(a\,t^{-m}\,\psi^p\right) = \left(-m\,t^{-m-1} + t^{-m}\,\partial_x a\right)\psi^p \leq 0. \tag{1.21}$$

The inequality above follows from (1.19) when m is chosen so that $m \geq M\,E$.

Fix now two constants $L, A > 0$ and integrate the equation (1.21) on the trapezoidal domain $-L + A\,t < x < L - A\,t$. Noticing that the boundary terms have a favorable sign when A is larger than the sup-norm of the coefficient a we arrive at

$$\frac{d}{dt}\left(t^{-m}\,\|\psi(t)\|^p_{L^p(-L+At,\,L-At)}\right) \leq 0,$$

that is,

$$t^{-m}\,\|\psi(t)\|^p_{L^p(-L+At,\,L-At)} \leq s^{-m}\,\|\psi(s)\|^p_{L^p(-L+As,\,L-As)}, \quad s < t. \tag{1.22}$$

Since, by (1.16), the two solutions assume the same initial data we find

$$
\begin{aligned}
\|\psi(s)\|_{L^p(-L+As,\,L-As)} &\leq (2L)^{1/p}\,\|\psi(s)\|_{L^\infty(\mathbb{R})} \\
&\leq (2L)^{1/p}\,\|u_2(s) - u_1(s)\|_{L^1(\mathbb{R})} \\
&\leq (2L)^{1/p}\left(\|u_1(s) - u_0\|_{L^1(\mathbb{R})} + \|u_2(s) - u_0\|_{L^1(\mathbb{R})}\right) \\
&\leq (2L)^{1/p}\,C\,s^\beta.
\end{aligned}
\tag{1.23}
$$

Finally, combining (1.22) with (1.23) we deduce that ($s \leq t$)

$$
\begin{aligned}
t^{-m}\|\psi(t)\|^p_{L^p(-L+At,\,L-At)} &\leq s^{-m}\|\psi(s)\|^p_{L^p(-L+As,\,L-As)} \\
&\leq C'\,s^{-m+p\beta} \to 0 \quad \text{when } s \to 0,
\end{aligned}
$$

provided p is chosen so large that $p\beta > m$. Therefore, we have proven that $\psi(x,t) = 0$ and thus $u_1(x,t) = u_2(x,t)$, for all $x \in (-L + At, L - At)$. Since L is arbitrary this establishes the desired uniqueness property. □

2. Classical entropy solutions for general flux

We now turn to the existence of classical entropy solutions for the Cauchy problem

$$\partial_t u + \partial_x f(u) = 0, \tag{2.1}$$

$$u(x,0) = u_0(x), \quad x \in \mathbb{R}, \tag{2.2}$$

where the flux $f : \mathbb{R} \to \mathbb{R}$ is a smooth function which need not be convex but, for simplicity, has only finitely many inflection points. The initial data $u_0 : \mathbb{R} \to \mathbb{R}$ are supposed to be integrable and with bounded total variation (denoted by $TV(u_0)$), that is, $u_0 \in L^1(\mathbb{R}) \cap BV(\mathbb{R})$. We restrict attention to weak solutions satisfying all of the entropy inequalities and we establish the existence of a classical entropy solution to the Cauchy problem (2.1) and (2.2). Later, in the following two sections, we shall extend the analysis to nonclassical entropy solutions selected by a kinetic relation. It will be convenient to use the following notation $(u_-, u_+ \in I\!R)$

$$\bar{a}(u_-, u_+) := \int_0^1 f'((1-\theta)\, u_- + \theta\, u_+)\, d\theta$$

$$= \begin{cases} \frac{f(u_+)-f(u_-)}{u_+ - u_-}, & u_- \neq u_+, \\ f'(u_-), & u_- = u_+. \end{cases}$$

The *classical Riemann solver* was defined explicitly in Theorem II-2.4. It corresponds to the solution of the Cauchy problem associated with (2.1) and

$$u(x,0) = \begin{cases} u_l, & x < 0, \\ u_r, & x > 0, \end{cases} \tag{2.3}$$

for constant data u_l and u_r. Recall that the classical entropy solution of the Riemann problem is easily determined by using Oleinik entropy inequalities (II-1.6)), and is defined from the convex (respectively, concave) hull of the flux f on the interval limited by u_l and u_r when $u_l < u_r$ (resp., $u_l > u_r$). Based on the Riemann solver, the **wave front tracking method** allows us to construct a sequence of *piecewise constant* approximate solutions of the Cauchy problem (2.1) and (2.2), as explained now.

Fixing a sequence $h \to 0+$ and the initial data u_0, we consider piecewise constant approximations $u_0^h : I\!R \to I\!R$ that have compact support and at most $1/h$ jump discontinuities and satisfy

$$\inf u_0 \leq u_0^h \leq \sup u_0,$$
$$TV(u_0^h) \leq TV(u_0), \tag{2.4}$$
$$u_0^h \to u_0 \text{ in the } L^1 \text{ norm, as } h \to 0.$$

For instance, one can choose finitely many points $x_1 < x_2 < \ldots < x_K$ and set

$$u_0^h(x) := \begin{cases} 0, & x < x_1, \\ \frac{1}{x_{k+1}-x_k} \int_{x_k}^{x_{k+1}} u_0 \, dx, & x \in (x_k, x_{k+1}) \quad (1 \leq k \leq K-1), \\ 0, & x > x_K. \end{cases} \tag{2.5}$$

Then, at each jump point x of u_0^h we can solve (at least locally in time) the Riemann problem associated with the initial data $u_0^h(x\pm)$. A Riemann solution is not truly piecewise constant and may contain both shock waves and rarefaction fans (Theorem II-2.4). Therefore, any rarefaction fan centered at some point $(x,t) = (x_0, 0)$ and connecting two states u_1 and u_2, say, will be replaced with a *single* **rarefaction front**, i.e.,

$$\begin{cases} u_1, & x - x_0 < t\,\bar{a}(u_1, u_2), \\ u_2, & x - x_0 > t\,\bar{a}(u_1, u_2), \end{cases}$$

if its strength $|u_2 - u_1|$ is less than or equal to h, while if $|u_2 - u_1| > h$ the rarefaction fan will be replaced with *several* rarefaction fronts with small strength, i.e.,

$$\begin{cases} u_1, & x - x_0 < t\,\bar{a}(u_1, w_1), \\ w_j, & t\,\bar{a}(w_{j-1}, w_j) < x - x_0 < t\,\bar{a}(w_j, w_{j+1}) \quad (1 \leq j \leq N-1), \\ u_2, & x - x_0 > t\,\bar{a}(w_{N-1}, u_2), \end{cases}$$

where $|u_2 - u_1|/N < h$ and

$$w_j := u_1 + \frac{j}{N}\,(u_2 - u_1) \quad \text{for } j = 0, \dots, N.$$

Each small jump travels with the speed determined by the Rankine-Hugoniot relation. We can patch together these local solutions and we obtain an approximate solution $u^h = u^h(x, t)$ defined up to the first interaction time t_1 when two waves from different Riemann solutions meet.

At the first interaction point we face again a Riemann problem which is solved by several shock waves and rarefaction fans. Again, rarefaction fans are replaced with small rarefaction fronts with strength less than or equal to h and traveling with the Rankine-Hugoniot speed. At the second interaction time t_2 we proceed similarly and continue the construction inductively.

We point out that the number of outgoing waves in each Riemann solution is finite, since f has finitely many inflection points so there are always finitely many rarefaction fans. However, it is not clear, at this stage, that our construction can be continued for all times since the number of waves may well increase at interactions and, in principle, could become infinite in finite time. The number of interaction points as well could be unbounded. In fact, we will show below that this is not the case and that the construction can be continued for all times.

By modifying slightly the initial approximation u_0^h if necessary, we can always assume that, at any given time, there are *at most one interaction point* and *only two waves interacting*. The condition is not essential but simplify our presentation.

Finally, since every front propagates at the Rankine-Hugoniot speed it is obvious that the functions u^h are *exact solutions* of (2.1) with the initial condition (2.2) replaced with

$$u^h(x, 0) = u_0^h(x), \quad x \in \mathbb{R}.$$

However, u^h does not quite satisfy the entropy inequalities since our construction introduces rarefaction fronts violating the entropy requirement (but having small strength). In fact, the correct solution is recovered in the limit $h \to 0$, as stated now.

We refer to u^h as the **sequence of wave front tracking approximations** generated by the sequence of initial data u_0^h and based on the classical Riemann solver.

THEOREM 2.1. (Existence of classical entropy solutions.) *Consider the Cauchy problem for a scalar conservation law, (2.1) and (2.2), associated with a flux-function f having finitely many inflection points and some initial data $u_0 \in L^1(\mathbb{R}) \cap BV(\mathbb{R})$.*

(i) *Then, the wave front tracking approximations $u^h = u^h(x, t)$, based on the classical Riemann solver determined by Oleinik entropy inequalities, are well defined globally in time. In particular, the total number of waves in $u^h(t)$ is uniformly bounded in t (but tends to infinity when h tends to 0).*

(ii) *The approximate solutions satisfy the uniform estimates*

$$
\begin{aligned}
&(a) \quad \inf u_0 \le u^h(x, t) \le \sup u_0, && x \in \mathbb{R},\, t > 0, \\
&(b) \quad TV(u^h(t)) \le TV(u_0), && t \ge 0, \\
&(c) \quad \|u^h(t) - u^h(s)\|_{L^1(\mathbb{R})} \le TV(u_0)\,\sup|f'|\,|t - s|, && s,\, t \ge 0,
\end{aligned}
\tag{2.6}
$$

where the sup-norm of f' is taken over the range determined by (2.6a).

(iii) *The sequence u^h (or a subsequence of it, at least) converges strongly to a classical entropy solution $u = u(x,t)$ of the Cauchy problem (2.1) and (2.2), i.e.,*

$$u^h(t) \to u(t) \quad \text{in } L^1_{loc} \text{ for all } t, \tag{2.7}$$

and for every convex entropy pair (U, F)

$$\partial_t U(u) + \partial_x F(u) \leq 0 \tag{2.8}$$

with, moreover,

(a) $\inf u_0 \leq u(x,t) \leq \sup u_0,$ $x \in \mathbb{R}, t > 0,$
(b) $TV(u(t)) \leq TV(u_0),$ $t \geq 0,$ (2.9)
(c) $\|u(t) - u(s)\|_{L^1(\mathbb{R})} \leq TV(u_0) \sup |f'| \, |t - s|,$ $s, t \geq 0,$

and, concerning the initial condition,

$$\|u(t) - u_0\|_{L^1(\mathbb{R})} \leq t \, TV(u_0) \sup |f'|, \quad t \geq 0.$$

PROOF. First of all, we check that the total number of waves in u^h remains finite (h being kept fixed). Consider an arbitrary interaction, involving a left-hand wave connecting u_l to u_m and a right-hand wave connecting u_m to u_r. We distinguish between two cases:

- monotone incoming patterns when $(u_m - u_l)(u_r - u_m) \geq 0,$
- non-monotone incoming patterns when $(u_m - u_l)(u_r - u_m) < 0.$

For each time t excluding interaction times we denote by $N_1(t)$ the total number of changes of monotonicity in $u^h(t)$. Observe that the function $N_1(t)$ diminishes at all interactions associated with a non-monotone pattern, precisely:

$$
\begin{aligned}
[N_1(t)] := N_1(t+) - N_1(t-) \\
= \begin{cases} 0 & \text{if there is a monotone incoming pattern at time } t, \\ -1 & \text{if there is a non-monotone incoming pattern at time } t. \end{cases}
\end{aligned}
$$

Since $N_1(0+)$ is obviously finite, this implies that the number of "non-monotone interactions" is finite. On the other hand, we observe that at each "monotone interaction" we have only the following three possibilities:

- Both incoming waves are shocks and the outgoing pattern is a single shock.
- The incoming pattern contains a shock and a rarefaction and the outgoing pattern contains a single shock.
- The incoming and outgoing patterns both contain exactly one shock and one rarefaction.

Therefore, "monotone interactions" cannot increase the number of waves. In turn, we deduce that the *total number of waves* is finite.

Call $N_2(t)$ the total number of waves in $u^h(t)$. Suppose that there exists a point (x_0, t_0) at which infinitely many interactions take place. As noted above, the total number of "non-monotone interactions" is also finite, so only "monotone interactions" take place in a backward neighborhood of the point (x_0, t_0). Since a Riemann solution contains at most two waves, it is clear geometrically that one wave must be cancelled and the number of waves must decrease strictly at that point, that is,

$$N_2(t_0) < \lim_{\substack{t \to t_0 \\ t < t_0}} N_2(t).$$

Therefore, there can be *at most finitely many points* at which infinitely many interactions take place. Finally, we can "pass through" any of these interactions by observing that since the singularity is localized at isolated points, on any given line $t = t_0$ we have

$$\lim_{\substack{t \to t_0 \\ t < t_0}} u^h(x, t) = u(x, t_0)$$

for all $x \neq x_0$. This completes the proof that the approximations $u^{\varepsilon,h}$ are well-defined globally in time.

We now derive uniform bounds on u^h. The properties (2.6a) and (2.6b) are clearly satisfied by the classical Riemann solver as well as by the approximate one. This is due to the facts that the classical Riemann solution is a monotone function of the space variable and that replacing the rarefaction fans by propagating fronts does not change the L^∞ nor the total variation norms.

The Lipschitz estimate (2.6c) is a consequence of the total variation estimate (2.6b) and the property of propagation at finite speed. Indeed, in any interval $[t_1, t_2]$ containing no interaction time let us denote by $y_k = y_k(t)$ $(k = 1, 2, \dots)$ the propagating fronts in u^h, which are in finite number. The speed $y'_k(t)$ is a constant in the time interval $[t_1, t_2]$ and it is not difficult to check, by decomposing the interval $[t_1, t_2]$ in smaller intervals if necessary,

$$\|u^h(t_2) - u^h(t_1)\|_{L^1(\mathbb{R})} \leq \sum_k |u_+^h(y_k(t_1), t_1) - u_-^h(y_k(t_1), t_1)| \, |y_k(t_2) - y_k(t_1)|$$

where $u_\pm^h(y_k(t_1), t_1)$ are the left- and right-traces, with

$$|y_k(t_2) - y_k(t_1)| = |y'_k| \, |t_2 - t_1| \leq \sup |f'| \, |t_2 - t_1|,$$

which yields (2.6c).

By Theorem A.3 in the appendix (Helly's compactness theorem) the conditions (2.6) imply the existence of a limit u and the convergence (2.7), as well as the properties (2.9). We rely here on the lower semi-continuity properties of the L^1 norm and total variation, for instance:

$$TV(u(t)) \leq \liminf_{h \to 0} TV(u^h(t)).$$

Since by construction

$$\partial_t u^h + \partial_x f(u^h) = 0,$$

it is obvious that the limit u satisfies the conservation law (2.1). The initial condition (2.2) follows from (2.3) and (2.6c), namely

$$\|u(t) - u_0\|_{L^1(\mathbb{R})} \leq \liminf_{h \to 0} \|u^h(t) - u_0^h\|_{L^1(\mathbb{R})}$$
$$\leq TV(u_0) \sup |f'| \, t \longrightarrow 0 \tag{2.10}$$

as $t \to 0$.

To check, finally, that u is the classical entropy solution of the Cauchy problem we rely on the fact that a propagating front either is a classical shock satisfying Oleinik entropy inequalities or else is a rarefaction front with small strength, that is, setting $u_k^\pm := u_\pm^h(y_k(t), t)$,

$$-y'_k \left(U(u_k^+) - U(u_k^-) \right) + F(u_k^+) - F(u_k^-) \leq 0 \quad \text{for shock fronts} \tag{2.11}$$

and
$$|u_k^+ - u_k^-| \le h \quad \text{for rarefaction fronts.} \tag{2.12}$$

Therefore, for every convex entropy pair (U, F) and every smooth function with compact support $\theta = \theta(x, t) \ge 0$ we have by (2.11)

$$\Omega(\theta) := -\int_{I\!R \times I\!R_+} \left(U(u^h)\, \partial_t \theta + F(u^h)\, \partial_x \theta \right) dx dt$$

$$\le \sum_{\text{rarefactions}} \int_{I\!R_+} \theta(y_k, t) \left(-y_k' \left(U(u_k^+) - U(u_k^-) \right) + F(u_k^+) - F(u_k^-) \right) dt.$$

Since
$$-y_k' \left(u_k^+ - u_k^- \right) + f(u_k^+) - f(u_k^-) = 0$$
and $F' := U' f'$, in view of (2.12) we obtain

$$\left| -y_k' \left(U(u_k^+) - U(u_k^-) \right) + F(u_k^+) - F(u_k^-) \right| = \left| \int_{u_k^-}^{u_k^+} U'(v) \left(-y_k' + f'(v) \right) dv \right|$$

$$\le \sup |U'| \, \sup f'' \, \left| u_k^+ - u_k^- \right|^2$$

$$\le C\, h \, \left| u_k^+ - u_k^- \right|,$$

so that
$$\Omega(\theta) \le C\, h \sum_k \int_{I\!R_+} \theta(y_k(t), t) \left| u_k^+(t) - u_k^-(t) \right| dt$$

$$\le C\, h \, \sup_t TV \left(u^h(t) \right) \int_{I\!R_+} \sup_{x \in I\!R} \theta(x, t)\, dt \longrightarrow 0.$$

This completes the proof of Theorem 2.1. $\square$

3. Nonclassical entropy solutions

The strategy described in Section 2 can be applied to the same Cauchy problem (2.1) and (2.2) but by replacing the classical Riemann solver with the nonclassical one discovered in Section II-4. As we will see, the corresponding approximate solutions are expected to converge toward weak solutions, which we will refer to as the *nonclassical entropy solutions* of the Cauchy problem. For a rigorous definition of nonclassical solutions we refer to Remark 3.3 and to Chapter X.

To be able to implement this approach we must overcome some new difficulties: As was pointed out in Chapter II, nonclassical entropy solutions of the Riemann problem do not satisfy the maximum principle ((2.6a) above) nor the total variation diminishing property ((2.6b) above). Namely, the total variation of a nonclassical solution may increase in time, especially at times when a nonclassical shock arises from the interaction between classical waves. To control the total variation on the approximate solutions uniformly, it will be necessary here to investigate carefully the geometric structure of the approximate solutions. In particular, we will have to keep track of certain wave fronts, that is, the "crossing shocks" defined below.

Let $f : I\!R \to I\!R$ be a concave-convex flux function satisfying

$$u\, f''(u) > 0 \quad (u \ne 0), \quad f'''(0) \ne 0,$$
$$\lim_{|u| \to \infty} f'(u) = +\infty. \tag{3.1}$$

As before, we associate with f the functions $\varphi^\natural, \varphi^{-\natural} : I\!R \to I\!R$; see (II-2.6). We want to consider the Cauchy problem (2.1)-(2.2) in a class of "nonclassical entropy solutions". In view of the results in Chapters II and III, a Lipschitz continuous *kinetic function* $\varphi^\flat : I\!R \to I\!R$ is now prescribed such that

$$\varphi^{-\natural}(u) < \varphi^\flat(u) \le \varphi^\natural(u), \quad u > 0,$$
$$\varphi^\natural(u) \le \varphi^\flat(u) < \varphi^{-\natural}(u), \quad u < 0, \tag{3.2a}$$

$$\varphi^\flat \text{ is monotone decreasing}, \tag{3.2b}$$

$\varphi^\flat$ satisfies the strict contraction property:

$$0 < \frac{\varphi^\flat \circ \varphi^\flat(u)}{u} < 1, \quad u \ne 0, \tag{3.2c}$$

the Lipschitz constant of $\varphi^\flat \circ \varphi^\flat$ near $u = 0$ is strictly less than 1:

$$\limsup_{\substack{u,v \to 0, \\ u \ne v}} \left| \frac{\varphi^\flat \circ \varphi^\flat(v) - \varphi^\flat \circ \varphi^\flat(u)}{v - u} \right| < 1. \tag{3.2d}$$

and the companion function $\varphi^\sharp : I\!R \to I\!R$ associated with $\varphi^\flat$ (see (II-4.3)) satisfies:

$$u\,\varphi^\sharp(u) \le 0, \quad u \in I\!R. \tag{3.2e}$$

REMARK 3.1.
- If a strictly convex entropy pair (U, F) is prescribed and $\varphi_0^\flat$ denotes the zero entropy dissipation function associated with f and U (Theorem II-3.1) and if a kinetic function satisfying the condition (II-4.1),

$$\varphi_0^\flat(u) < \varphi^\flat(u) \le \varphi^\natural(u), \quad u > 0,$$
$$\varphi^\natural(u) \le \varphi^\flat(u) < \varphi_0^\flat(u), \quad u < 0,$$

  is prescribed, then obviously (3.2a) holds true and since $\varphi_0^\flat \circ \varphi_0^\flat = id$, (3.2c) also holds. The setting proposed in the present section is more general than the one investigated in Section II-4 and in Chapter III. However, it is a simpler matter to observe that the Riemann solver is still well-defined under the conditions (3.2a)–(3.2c) and that the Riemann solution depends L^1-continuously upon its initial states. (See also Remark II-5.5.)
- Assumptions (3.2a) to (3.2d) are always satisfied by kinetic functions generated by nonlinear diffusive-dispersive limits in Chapter III. These kinetic functions are monotone decreasing and coincide with the classical value $\varphi^\natural$ on a neighborhood of 0. Since $\varphi^\natural(u) \sim -u/2$ at $u = 0$, the Lipschitz constant of $\varphi^\flat \circ \varphi^\flat$ near 0 is about 1/4.
- The assumption (3.2e), in fact, is the only genuine restriction made on the kinetic function in the present section. It implies that the Riemann solution is always classical when the Riemann data are in the same region of convexity or concavity of f. Of course, this assumption is fulfilled in most situations of interest. It is satisfied when the flux is $f(u) = u^3$ and the diffusion and dispersion functions are constant (Section III-2) or, more generally, when the regularization terms are consistent with the entropy $U(u) = u^2/2$. It is also

satisfied for a general concave-convex flux, provided the diffusion α is sufficiently large or provided u remains in a sufficiently small interval near 0. (See (III-3.17).)

$\square$

Now, to the kinetic function we can associate the nonclassical Riemann solver described in Theorem II-4.1. Recall that the Riemann solution contains *two waves, at most:* either a single rarefaction wave, or a single classical shock wave, or a nonclassical shock plus a classical one, or else a nonclassical shock plus a rarefaction wave. We can construct a sequence of piecewise constant, approximate solutions $u^h = u^h(x,t)$ of the Cauchy problem (2.1)-(2.2), as was done in the previous section but now relying on the nonclassical Riemann solver. Precisely, let us start with a sequence $h \to 0$ and some initial data u_0^h satisfying the usual convergence conditions (2.4). At the initial time $t = 0$, we decompose every rarefaction fan into small propagating jumps with strength less than h. At each interaction, we always replace a rarefaction fan with a *single rarefaction front* traveling with the Rankine-Hugoniot speed. For simplicity in the discussion we can assume that there is at most one interaction taking place at any given time and there are exactly two waves meeting at any interaction. Minor modifications are needed to cover the more general situation.

THEOREM 3.2. (Existence of nonclassical entropy solutions.) *Consider the Cauchy problem for the scalar conservation law (2.1) and (2.2) associated with a concave-convex flux-function f satisfying (3.1). Consider also a kinetic function $\varphi^\flat$ satisfying the assumptions (3.2).*

(i) *Then, for arbitrary initial data $u_0 \in L^\infty(I\!R) \cap BV(I\!R)$ the wave front tracking approximations determined from the nonclassical Riemann solver satisfy, for some constants $C_1, C_2 > 0$ depending only on $\|u_0\|_{L^\infty(I\!R)}$ and on the data f and $\varphi^\flat$,*

(a) $\|u^h(t)\|_{L^\infty(I\!R)} \leq C_1$, $t \geq 0$,

(b) $TV(u^h(t)) \leq C_2\, TV(u_0)$, $t \geq 0$,   (3.3)

(c) $\|u^h(t_2) - u^h(t_1)\|_{L^1(I\!R)} \leq C_2\, TV(u_0)\, \sup|f'|\, |t_2 - t_1|$, $t_1, t_2 \geq 0$.

(ii) *The sequence u^h (or a subsequence of it, at least) converges strongly to a weak solution $u = u(x,t)$ of (2.1) and (2.2), specifically*

$$u^h(t) \to u(t) \text{ in } L^1_{\text{loc}} \text{ for all times } t \qquad (3.4)$$

and $u^h(x,t) \to u(x,t)$ for almost every (x,t), with

(a) $\|u(t)\|_{L^\infty(I\!R)} \leq C_1$, $t \geq 0$,

(b) $TV(u(t)) \leq C_2\, TV(u_0)$, $t \geq 0$,   (3.5)

(c) $\|u(t_2) - u(t_1)\|_{L^1(I\!R)} \leq C_2\, TV(u_0)\, \sup|f'|\, |t_2 - t_1|$, $t_1, t_2 \geq 0$.

(iii) *If the kinetic function $\varphi^\flat$ satisfies the inequalities*

$$\varphi_0^\flat(u) < \varphi^\flat(u) \leq \varphi^\natural(u), \quad u > 0,$$

$$\varphi^\natural(u) \leq \varphi^\flat(u) < \varphi_0^\flat(u), \quad u < 0,$$

for the zero-entropy dissipation function $\varphi_0^\flat$ associated with some strictly convex entropy pair (U, F), then the solution u satisfies also the (single) entropy inequality

$$\partial_t U(u) + \partial_x F(u) \leq 0. \qquad (3.6)$$

Additionally, one can also establish that the solution satisfies the prescribed kinetic relation; see the forthcoming Section VIII-4. The solutions generated in Theorem 3.2 will be called **nonclassical entropy solutions** of the Cauchy problem (2.1)-(2.2).

REMARK 3.3. The solutions, in principle, could depend upon the approximation scheme under consideration or upon the discretization parameters. In Chapter X we will see that this is not the case. A general definition of nonclassical entropy solutions can be stated, independently of any approximation scheme, in the framework to be developed later in Chapter X: by definition, a **nonclassical entropy solution** of the Cauchy problem (2.1)-(2.2) is a (Φ, ψ)-admissible solution of (2.1)-(2.2) for the following families of admissible discontinuities and speeds:

$$\Phi := \left\{ (u_-, u_+) \ \middle/ \ \begin{array}{ll} u_+ \in [u_-, \varphi^\sharp(u_-)) \cup \{\varphi^\flat(u_-)\}, & u_- < 0 \\ u_+ \in \{\varphi^\flat(u_-)\} \cup (\varphi^\sharp(u_-), u_-], & u_- > 0 \end{array} \right\},$$

$$\psi(u_-, u_+) := \bar{a}(u_-, u_+) = \int_0^1 f'((1-\theta)u_- + \theta u_+)\, d\theta.$$

$\square$

The rest of this section is devoted to a proof of Theorem 3.2. We will now give a complete classification of all possible wave interaction patterns when a left-hand wave connecting two states u_l and u_m interacts with a right-hand wave connecting two states u_m and u_r. For definiteness, we restrict attention to positive left-hand states u_l, the other case being entirely similar. We use, for instance, the notation (RC)–(R') when the left-hand incoming wave is a rarefaction front, the right-hand incoming wave is a classical shock, and the outgoing wave pattern contains a single rarefaction front. The notation (RC)–$(N'C')$ is used when the outgoing pattern contains a nonclassical shock followed by a classical shock, etc. In each case, we indicate whether the incoming solution is locally monotone or not and we specify the relevant ranges for u_l, u_m, and u_r. We indicate whether the wave is increasing or decreasing by adding an up ($\uparrow$) or down ($\downarrow$) arrow. It is important to note that the notation N may also represent the *limiting case* when a left-hand state u_l is connected to $\varphi^\natural(u_l)$. So, our classification also covers the classical case. On the other hand, we exclude from our construction the classical shocks connecting u_l to $\varphi^\sharp(u_l)$ (when the latter is distinct from $\varphi^\flat(u_l)$) since such waves cannot be generated by interactions.

Our analysis below will keep track of the **crossing fronts,** that is, fronts connecting two states u_- and u_+ such that $u_- u_+ \leq 0$, $u_+ \neq 0$. So, in each case we will specify whether the incoming or outgoing pattern contains such fronts. Clearly, such waves are classical or nonclassical shock waves, but not rarefaction fronts. All nonclassical shocks are crossing shocks (by our assumption (3.2a)). So, we only have to distinguish between *crossing* classical shocks and *non-crossing* ones, the latter being referred to as classical shocks for short. We will use the notation C_+ for a shock connecting two non-negative states with $u_- \neq 0$, $C_\pm$ for a classical shock connecting a non-negative state to a negative one (so, here, $u_+ < 0$), R_+ for a rarefaction connecting two non-negative states, and so on. For the sake of completion we provide the classification when the assumptions (3.2a) to (3.2d) hold but without imposing (3.2e).

(i) *Interactions involving a rarefaction on the left-hand side:*

Case RC-1 : $(R_+^\uparrow C^\downarrow)$–$(C^{\downarrow\prime})$ (non-monotone and entirely classical) when

$$\max\big(\varphi^\sharp(u_l), \varphi^\sharp(u_m)\big) < u_r < u_l, \quad 0 < u_l < u_m.$$

The incoming classical shock survives the interaction with its strength decreased by an amount equal to the strength of the incoming rarefaction. The latter is completely cancelled. There are two-subcases: $(R_+^\uparrow C_+^\downarrow)$–$(C_+^{\downarrow\prime})$ if $u_r \geq 0$ and $(R_+^\uparrow C_\pm^\downarrow)$–$(C_\pm^{\downarrow\prime})$ otherwise.

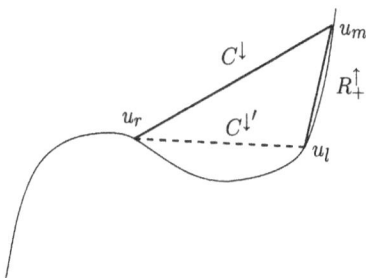

Case RC-2 : $(R_+^\uparrow C_\pm^\downarrow)$–$(N_\pm^{\downarrow\prime} R_-^{\downarrow\prime})$ (non-monotone and possibly entirely classical) when

$$\varphi^\sharp(u_m) < u_r \leq \varphi^b(u_l) < 0 < u_l < u_m.$$

The right-hand incoming crossing classical shock transforms into a nonclassical shock, while the left-hand incoming rarefaction passes through the crossing shock. Note that the wave $R_-^{\downarrow\prime}$ may be trivial, that is, the limiting case $u_r \leq \varphi^b(u_l)$ is possible.

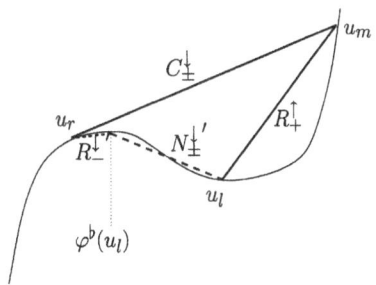

Case RC-3 : $(R_+^\uparrow C^\downarrow)$–$(N_\pm^{\downarrow\,'} C^{\uparrow\,'})$ (non-monotone and exclusively nonclassical) when

$$\max(\varphi^\flat(u_l), \varphi^\sharp(u_m)) < u_r < \varphi^\sharp(u_l), 0 < u_l < u_m.$$

The right-hand incoming crossing classical shock is transformed into a nonclassical shock, while the left-hand incoming rarefaction passes through the crossing shock and is transformed into a right-hand classical shock. There are two-subcases: $(R_+^\uparrow C_\pm^\downarrow)$–$(N_\pm^\downarrow C_-^{\uparrow\,'})$ if $u_r < 0$ and $(R_+^\uparrow C_+^\downarrow)$–$(N_\pm^{\downarrow\,'} C_\mp^{\uparrow\,'})$ otherwise. In the latter, the incoming pattern is entirely positive while a nonclassical shock and a crossing shock are generated; hence, the number of crossing shock increases.

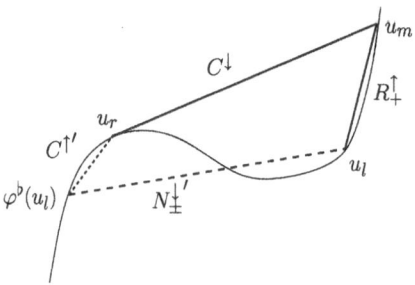

Case RN : $(R_+^\uparrow N_\pm^\downarrow)$–$(N_\pm^{\downarrow\,'} R_-^{\downarrow\,'})$ (non-monotone and exclusively nonclassical) when

$$0 < u_l < u_m \text{ and } u_r = \varphi^\flat(u_m).$$

The nonclassical shock survives the interaction, while the left-hand rarefaction passes through it and exits on its right-hand side.

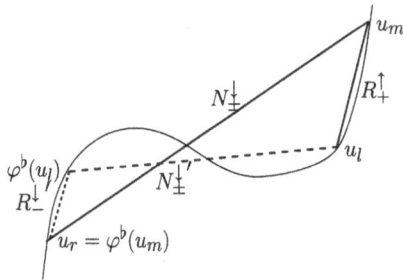

(ii) *Interactions involving a classical shock on the left-hand side:*

Case CR-1 : $(C_{\pm}^{\downarrow} R_{-}^{\downarrow})$–$(C_{\pm}^{\downarrow}{}')$ (monotone and entirely classical) when

$$\varphi^{\sharp}(u_l) < u_r < u_m \leq 0 < u_l.$$

The left-hand incoming classical crossing shock survives the interaction, its strength being increased by an amount equal to the strength of the right-hand incoming rarefaction. The latter is completely cancelled. In the special case $u_m = 0$ we have actually $(C_{+}^{\downarrow} R_{-}^{\downarrow})$–$(C_{\pm}^{\downarrow}{}')$.

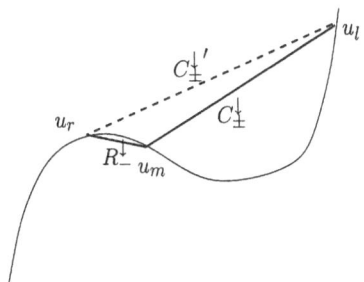

Case CR-2 : $(C_{+}^{\downarrow} R_{+}^{\uparrow})$–$(C_{+}^{\downarrow}{}')$ (non-monotone and entirely classical) when

$$\max(\varphi^{\sharp}(u_l), 0) \leq u_m < u_r < u_l.$$

The left-hand incoming classical shock survives the interaction, its strength being decreased by an amount equal to the strength of the right-hand incoming rarefaction. The latter is completely cancelled.

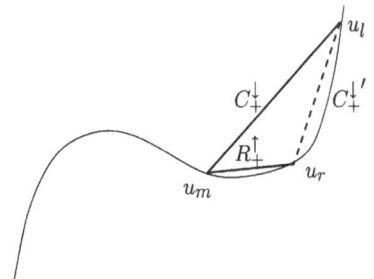

Case CR-3 : $(C_{\pm}^{\downarrow}R_{-}^{\downarrow})-(N_{\pm}^{\downarrow'}R_{-}^{\downarrow'})$ (monotone and possibly entirely classical) when

$$u_r \leq \varphi^b(u_l) < \varphi^\sharp(u_l) < u_m \leq 0 < u_l.$$

The incoming crossing classical shock is transformed into a nonclassical shock, while the right-hand incoming rarefaction bounces back on the right-side of the crossing shock. The wave $R_{-}^{\downarrow'}$ is trivial when $u_r = \varphi^b(u_l)$. In the special case $u_m = 0$ we have $(C_{+}^{\downarrow}R_{-}^{\downarrow})-(N_{\pm}^{\downarrow'}R_{-}^{\downarrow'})$.

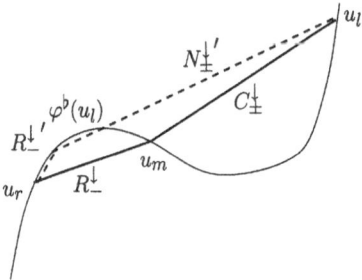

Case CR-4 : $(C_{\pm}^{\downarrow}R_{-}^{\downarrow})-(N_{\pm}^{\downarrow'}C_{-}^{\uparrow'})$ (monotone and exclusively nonclassical) when

$$\varphi^b(u_l) < u_r < \varphi^\sharp(u_l) < u_m \leq 0 < u_l.$$

The incoming crossing classical shock is transformed into a nonclassical shock, while the right-hand incoming rarefaction is transformed into a classical shock exiting on the same side. In the special case $u_m = 0$ we have $(C_{+}^{\downarrow}R_{-}^{\downarrow})-(N_{\pm}^{\downarrow'}C_{-}^{\uparrow'})$.

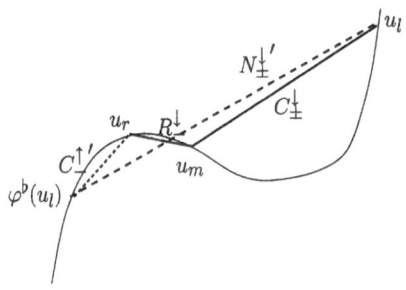

Case CC-1 : $(C_+^\downarrow C^\downarrow)$–$(C^{\downarrow\prime})$ (monotone and entirely classical) when

$$\max(\varphi^\sharp(u_l), \varphi^\sharp(u_m)) < u_r < u_m < u_l \quad \text{and } u_m \quad \geq 0.$$

Two incoming shocks join together to form a single classical shock. Either all waves are non-crossing or else there are exactly one incoming crossing shock and one outgoing crossing shock. There are two-subcases: $(C_+^\downarrow C_+^\downarrow)$–$(C_+^{\downarrow\prime})$ if $u_r \geq 0$ and $(C_+^\downarrow C_\pm^\downarrow)$–$(C_\pm^{\downarrow\prime})$ otherwise.

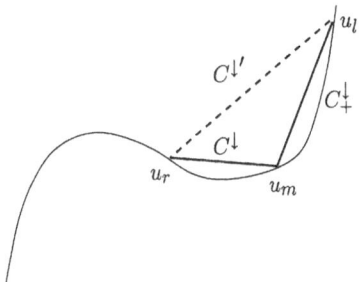

Case CC-2 : $(C_\pm^\downarrow C^\uparrow)$–$(C^{\downarrow\prime})$ (non-monotone and entirely classical) when

$$\varphi^\sharp(u_l) < u_m < u_r < \varphi^\sharp(u_m) < u_l \quad \text{and } u_m < 0.$$

Two incoming classical shocks cancel each other, and a single classical shock survives the interaction. There are two-subcases: $(C_\pm^\downarrow C_\mp^\uparrow)$–$(C_+^{\downarrow\prime})$ if $u_r \geq 0$ and $(C_\pm^\downarrow C_-^\uparrow)$–$(C_\pm^{\downarrow\prime})$ otherwise.

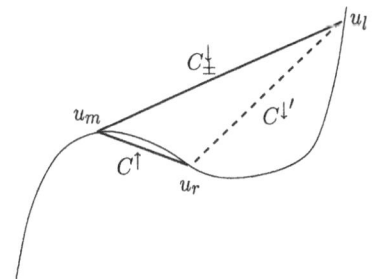

Case CC-3 : $(C_+^\downarrow C^\downarrow)$–$(N_\pm^{\downarrow'} C^{\uparrow'})$ (monotone and exclusively nonclassical) when

$$\varphi^\flat(u_l) < \varphi^\sharp(u_m) < u_r < \varphi^\sharp(u_l) < u_m < u_l, \quad u_m \geq 0.$$

The classical crossing shock is transformed into a nonclassical shock, while the classical shock passes through them from left to right. There are two subcases: $(C_+^\downarrow C_\pm^\downarrow)$–$(N_\pm^{\downarrow'} C_-^{\uparrow'})$ when $u_r < 0$ and $(C_+^\downarrow C_+^\downarrow)$–$(N_\pm^{\downarrow'} C_\mp^{\uparrow'})$ otherwise.

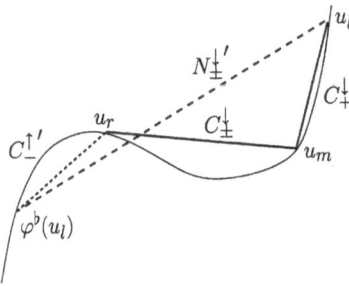

Case CN-1 : $(C_+^\downarrow N_\pm^\downarrow)$–$(C_\pm^{\downarrow'})$ (monotone and exclusively nonclassical) when

$$0 < u_m < u_l \text{ and } \varphi^\sharp(u_l) \leq u_r = \varphi^\flat(u_m).$$

The nonclassical shock is transformed into a crossing classical shock by combining its strength with the one of the incoming classical shock. The latter is completely cancelled.

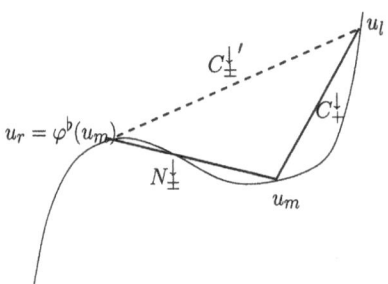

Case CN-2 : $(C^{\downarrow}_{\pm} N^{\uparrow}_{\mp})$-$(C^{\downarrow\,'}_{+})$ (non-monotone and exclusively nonclassical) when

$$\varphi^{\sharp}(u_l) < u_m < 0 \text{ and } u_r = \varphi^{\flat}(u_m).$$

The incoming classical and the nonclassical crossing shocks cancel each other, while only a classical shock survives the interaction.

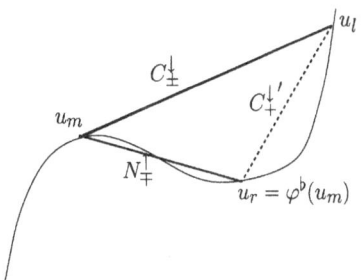

Case CN-3 : $(C^{\downarrow}_{+} N^{\downarrow}_{\pm})$-$(N^{\downarrow\,'}_{\pm} C^{\uparrow\,'}_{-})$ (monotone and exclusively nonclassical) when

$$0 < u_m < u_l \text{ and } u_r = \varphi^{\flat}(u_m) < \varphi^{\sharp}(u_l).$$

The classical shock passes from the left side to the right side of a nonclassical shock.

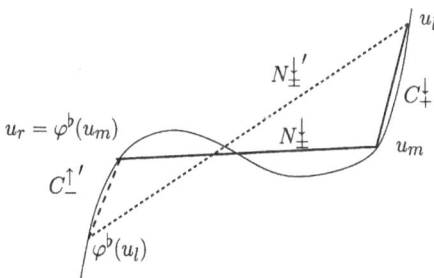

(iii) *Interactions involving a nonclassical shock on the left-hand side:*

Case NC : $(N_\pm^\downarrow C^\uparrow)$–$(C^{\downarrow'})$ (non-monotone and exclusively nonclassical) when

$$u_m = \varphi^\flat(u_l) \text{ and } \varphi^\sharp(u_l) < u_r < \varphi^\sharp(u_m) < u_l.$$

There are two possibilities. Either the incoming waves are nonclassical and classical crossing shocks, respectively, and cancel each other and a single classical shock leaves out. Or else, the incoming nonclassical shock is transformed into a crossing classical shock, while the incoming classical shock is completely cancelled. There are two subcases: $(N_\pm^\downarrow C_\mp^\uparrow)$–$(C_+^{\downarrow'})$ if $u_r \geq 0$ and $(N_\pm^\downarrow C_-^\uparrow)$–$(C_\pm^{\downarrow'})$ otherwise.

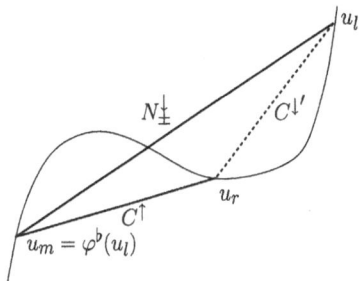

Case NN : $(N_\pm^\downarrow N_\mp^\uparrow)$–$(C_+^{\downarrow'})$ (non-monotone and exclusively nonclassical) when

$$u_m = \varphi^\flat(u_l) \text{ and } u_r = \varphi^\flat(u_m).$$

The two incoming nonclassical crossing shocks cancel each other, and generate a single classical shock.

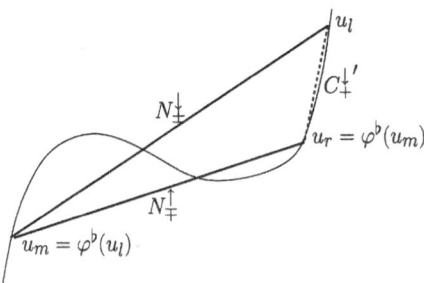

Some general comments concerning the interaction patterns are in order. Observe that Cases RC-1, CR-1, CR-2, CC-1, and CC-2 below involve classical waves only, while Cases RC-2 and CR-3 *may* involve classical waves only. The other cases require that there is one nonclassical shock, at least, in the incoming or outgoing wave pattern. The uniform norm and the total variation are non-increasing in all of the classical or monotone cases, at least. Actually, the uniform norm, as well as the total variation, increase in Cases RC-3, CR-4, CC-3, and CN-3. Note that in Case RC-3 there are two effects in competition: the decrease in total variation due to the cancellation of the incoming rarefaction R_+ and the increase due to the outgoing nonclassical shock $N_\pm$. The total variation may decrease if the former effect is stronger than the latter.

Next, we introduce the notation

$$I_m := [-m, m], \qquad J_m := I_m \cup \varphi^\flat(I_m). \tag{3.7}$$

We denote by $g^{[k]}$ the k-th iterate of a function g and by $\mathrm{Lip}_I(g)$ its Lipschitz constant on some interval I. Clearly, from the property (3.2c), for every interval I_m

$$\varphi^{\flat[2]}(I_m) \subset I_m, \quad \varphi^{\flat[3]}(I_m) \subset \varphi^\flat(I_m).$$

By (3.2d) the Lipschitz norm of $\varphi^\flat$ on the interval I_ε is less than 1. We fix $M > 0$, later on taken to be

$$M := \max\big(\|u_0\|_{L^\infty(\mathbb{R})}, \|\varphi^\flat(u_0)\|_{L^\infty(\mathbb{R})}\big),$$

and we estimate the Lipschitz norms on the interval I_M for iterates of $\varphi^\flat$ of arbitrary order. Note that, thanks to (3.2d), we can choose $\varepsilon > 0$ so small that

$$\eta := \mathrm{Lip}_{[-\varepsilon, \varepsilon]}\big(\varphi^\flat \circ \varphi^\flat\big) < 1. \tag{3.2d'}$$

LEMMA 3.4. (Estimates on the kinetic function.)
- *There exists an integer p such that*

$$\begin{aligned}
\varphi^{\flat[q]}(I_M) &\subset I_\varepsilon, & \text{even } q \geq 2p, \\
\varphi^{\flat[q]}(I_M) &\subset \varphi^\flat(I_\varepsilon), & \text{odd } q > 2p, \\
\varphi^{\flat[k]}(J_M) &\subset I_\varepsilon, & k \geq 2p.
\end{aligned} \tag{3.8}$$

- *There exists a constant $C_M > 0$ such that*

$$\mathrm{Lip}_{J_M}\big(\varphi^{\flat[q]}\big) \leq \begin{cases}
C_M, & \text{for all } q, \\
\eta^{(q-2p)/2}\, \mathrm{Lip}_{J_M}\big(\varphi^{\flat[2p]}\big), & \text{for all even } q \geq 2p, \\
\eta^{(q-1-2p)/2}\, \mathrm{Lip}_{J_M}\big(\varphi^{\flat[2p+1]}\big), & \text{for all odd } q > 2p.
\end{cases} \tag{3.9}$$

- *There exists a constant $C_M' > 0$ such that*

$$\sum_{k=0}^{\infty} \mathrm{Lip}_{J_M}\big(\varphi^{\flat[k]}\big) \leq C_M'. \tag{3.10}$$

PROOF. Combining the condition (3.2d') available near the origin with the global assumption (3.2c), we see that there exists a constant $\eta_0 \in (0, 1)$ (depending upon M) such that

$$\varphi^\flat \circ \varphi^\flat(u) \leq \eta_0 |u|, \quad |u| \leq M. \tag{3.11}$$

Then, it follows that for even exponents

$$\varphi^{\flat [2p]}(u) \leq \eta_0^p M, \quad |u| \leq M$$

and for odd exponents

$$\varphi^{\flat [2p+1]}(u) \leq \eta_0^p |\varphi^\flat(M)|, \quad |u| \leq M,$$

which establishes (3.8).

In view of (3.2d') and using the first property in (3.8) inductively with $\varphi^{\flat [q]} = \varphi^{\flat [2]} \circ \varphi^{\flat [q-2]}$, we obtain for any even integer $q \geq 2p$

$$\mathrm{Lip}_{J_M}\left(\varphi^{\flat [q]}\right) \leq \mathrm{Lip}_{I_\varepsilon}\left(\varphi^{\flat [2]}\right) \mathrm{Lip}_{J_M}\left(\varphi^{\flat [q-2]}\right)$$
$$\leq \eta^{(q-2p)/2} \mathrm{Lip}_{J_M}\left(\varphi^{\flat [2p]}\right)$$

and similarly for any odd integer $q \geq 2p + 1$

$$\mathrm{Lip}_{J_M}\left(\varphi^{\flat [q]}\right) \leq \eta^{(q-1-2p)/2} \mathrm{Lip}_{J_M}\left(\varphi^{\flat [2p+1]}\right).$$

We conclude that (3.9) holds with

$$C_M := \sup_{k=1,\dots,2p} \mathrm{Lip}_{J_M}\left(\varphi^{\flat [k]}\right).$$

The statement (3.10) is obvious from (3.9) since $\eta < 1$. This completes the proof of Lemma 3.4.   □

Relying on the technical estimates in Lemma 3.4, we arrive at the following:

LEMMA 3.5. (Basic properties of the wave front approximations.)
  (i) *The total number of fronts in $u^h(t)$ is less than or equal to the number of fronts in u_0^h. The total number of interaction points is finite.*
  (ii) *The range of the functions u^h is uniformly bounded: There exists some constant $M > 0$ depending only on $\|u_0\|_{L^\infty(\mathbb{R})}$ and $\varphi^\flat$ such that*

$$|u^h(x, t)| \leq M, \quad x \in \mathbb{R}, \, t \geq 0. \tag{3.12}$$

  (iii) *The strength of any rarefaction front is less than or equal to $C_M\, h$ where C_M was introduced in (3.10).*

The statement (ii) establishes (3.3a) in Theorem 3.2.

PROOF. The first property in (i) is obvious by construction since the interaction of two waves generates two outgoing waves at most. To estimate the number of interactions let us consider

$$B^h(t) := (A^h(0) + 1) A^h(t) + \sum_{C \cup R} A^h(t) + \sum_N A^h_{N,\mathrm{left}}(t), \tag{3.13}$$

where the sums are over all classical shocks and rarefaction fronts and over all non-classical shocks respectively (a wave connecting a state u with $\varphi^\natural(u)$ being counted as nonclassical.) Here, $A^h(t)$ is the total number of waves at the time t, and $A^h_{N,\text{left}}(t)$ is the total number of waves located on the right hand-side of the wave N. Clearly, since $A^h(t) \le A^h(0)$, we see that $B^h(t)$ is uniformly bounded by $(2\,A^h(0)+1)\,A^h(0)$. On the other hand, the function $B^h(t)$ decreases by at least 1 across each interaction having a single outgoing wave: the first term decreases by $A^h(t)+1$ while the sum of the last two terms increases by at most $A^h(t)$. In all of the other cases a left-hand outgoing nonclassical wave is generated: the first term remains constant since the number of waves does not change, while the sum of the last two terms decreases by at least 1. Hence, at each interaction time t, we have $B^h(t+) - B^h(t-) \le -1$, which implies that the number of interactions is finite.

The range of $u^h(t)$, denoted below by $\text{Range}\big(u^h(t)\big)$, may change only at interaction times. As is clear from the expression of the nonclassical Riemann solver, all of the new states created after an interaction are of the type $\varphi^\flat(v)$ where v is one of the left- or right-hand values at the interaction point, which are also values assumed before the interaction. Hence, a state v belongs to the range of the solution at time t only if it is some iterate of a state belonging to the range at the time 0. Setting $M_0 := \|u_0\|_{L^\infty(\mathbb{R})}$ we can write

$$\text{Range}\big(u^h(t)\big) \subset \bigcup_{k=0}^{+\infty} \varphi^{\flat\,[k]}\big(\text{Range}(u_0^h)\big)$$

$$\subset \bigcup_{k=0}^{+\infty} \varphi^{\flat\,[k]}\big(I_{M_0}\big) \subset \bigcup_{k=0}^{2p-1} \varphi^{\flat\,[k]}\big(I_{M_0}\big) \subset [-M, M] = I_M$$

for some (sufficiently large) M, where we have used (3.8) for the latter. This establishes (3.12).

To derive (iii), observe that the only interactions in which an outgoing rarefaction is produced are Cases RC-2, RN, and CR-3. In all of these cases, a rarefaction was already present before the interaction. This property allows us to keep track of all rarefaction fronts by starting at time $t = 0$, and there is also no ambiguity at interactions. Denote by $(y(t), t)$ a rarefaction front, defined on a bounded or unbounded time interval $[0, T]$ (T depending on h). By considering each one of the possible cases RC-2, RN, and CR-3, we see that at each interaction either the rarefaction strength

$$R^h(t_0-) := \lim_{t \to t_0-} |u_+^h(y(t), t) - u_-^h(y(t), t)|$$

decreases or else we have

$$R^h(t_0+) \le |\varphi^\flat\big(u_+^h\big) - \varphi^\flat\big(u_-^h\big)|,$$

where u_-^h and u_+^h are the left- and right-hand limits at the *incoming* rarefaction front. Iterating this argument we find

$$R^h(t) \le \sup_k \text{Lip}_{I_M}\big(\varphi^{\flat\,[k]}\big)\, R^h(0)$$

$$\le C_M\, h,$$

where we have used (3.9) and the fact that the initial strength of rarefactions is at most h. This completes the proof of Lemma 3.5. $\qquad\qquad\qquad\qquad\Box$

Our goal is to control the total variation of the approximate solutions u^h. To this end, we decompose the (x, t) plane using **crossing discontinuities** in u^h and therefore obtain regions in which u *keeps a constant sign*. In view of the expression of the nonclassical Riemann solver and thanks to the assumption (3.2e), the following properties are obvious:

- If the incoming pattern contains no crossing shock, then the outgoing pattern contains no crossing shocks.
- If the incoming pattern contains exactly one crossing shock, then the outgoing pattern contains exactly one crossing shock.
- If the incoming pattern contains two crossing shocks, then the outgoing pattern contains two crossing shocks or none.

New crossing shocks generated in principle in Case RC-3 (precisely, (R_+C_+)–$(N'_\pm C'_\mp)$) and in Case CC-3 (precisely, (C_+C_+)–$(N'_\pm C'_\mp)$) do not arise when the condition (3.2e) is satisfied. We can then keep track of the crossing fronts ordering them from left to right,

$$t \mapsto z_j(t), \quad j = 1, \ldots, m, \quad t \in (0, T_j)$$

with

$$u^h_-\big(z_j(t), t\big)\, u^h_+\big(z_j(t), t\big) \leq 0, \quad u^h_+\big(z_j(t), t\big) \neq 0,$$

and

$$z_1(t) \leq z_2(t) \leq \ldots \leq z_m(t).$$

In particular, the initial line is decomposed into a finite family of intervals I_j with

$$I_j := \big[z_j(0), z_{j+1}(0)\big], \quad j = 1, \ldots, m,$$

and $I_0 := \big(-\infty, z_1(0)\big]$ and $I_{m+1} := \big[z_m(0), +\infty\big)$. The initial data u^h_0 keep a constant sign in each interval I_j. A crossing path z_j may be defined for all times (when $T_j = +\infty$) or only on a finite time interval (when $T_j < +\infty$). The latter case happens when two crossing fronts meet and cancel each other; see Cases CC-2, CN-2, NC, and NN listed above. Note that some segments of a crossing shock may be classical while others are nonclassical. In Cases RC-2, RC-3, CC-3, CR-3, and CR-4, a classical crossing shock is transformed into a nonclassical shock. The opposite happens in Case CN-1.

Our analysis is also based on (generalized) characteristics, defined as follows. Given a point (x, t) (which, for simplicity in the presentation, is not a point of jump or interaction for u^h), we consider the **minimal backward characteristic** issuing from (x, t), i.e., by definition, a piecewise linear and continuous curve

$$s \mapsto X(s) = X(s; x, t), \quad s \in [0, t] \tag{3.14}$$

constructed as follows. Locally near $s = t$ the function u^h is constant and X coincides with the standard characteristic line passing through (x, t) and with slope $f'\big(u^h(x, t)\big)$. Continuing *backward* the construction we observe that the path X can meet only:

- a *rarefaction front,* by reaching it on its left- or right-hand side (since a rarefaction front propagates with the Rankine-Hugoniot speed by construction),
- or a *nonclassical front,* by reaching it on its *right-hand side* (since nonclassical shocks are slow undercompressive).

By definition, when the path meets a rarefaction it then coincides with it, at least until an interaction point is met. On the other hand, when the path meets a nonclassical shock it passes through it and continues again as a standard characteristic line propagating now on the left-hand side of the nonclassical shock.

It may happen that a characteristic path coinciding locally with a rarefaction front encounters an interaction point (x_0, t_0). Then, again by definition, the path propagates from the point of interaction by using the upper-left characteristic line exiting in a small backward neighborhood of (x_0, t_0). In this way, it is possible to define the backward characteristic $X(s; x, t)$ from the time $s = t$ down to the time $s = 0$ and encompass the whole interval $[0, t]$. Finally, our definition extends naturally to the case that (x, t) is an a point of jump or interaction for u^h.

LEMMA 3.6. (Properties of minimal backward characteristics.)
   (i) *Minimal backward characteristics cannot cross each other. Namely, given* $x_1 < x_2$, *we have* $X(s; x_1, t) \leq X(s; x_2, t)$ *for all* $s \in [0, t]$. *If two paths meet at some time* τ, *then they coincide for all* $s \in [0, \tau]$.
   (ii) *Given* (x, t) *and the backward characteristic* $s \mapsto X(s)$ *issuing from* (x, t), *we have*

$$u^h_-(x, t) = \varphi^{\flat[k]}\big(u^h_{0-}(X(0))\big), \qquad (3.15)$$

   where k is the number of nonclassical shocks encountered by X in the time interval $(0, t)$.
   (iii) *Given two points* (x_1, t) *and* (x_2, t) *with* $x_1 < x_2$, *suppose that the backward characteristics* $X(.; x_1, t)$ *and* $X(.; x_2, t)$ *satisfy*

$$X(0; x_1, t),\ X(0; x_2, t) \in I_j \quad \text{for some } j,$$

   then $k_1 \leq k_2$, where the integer k_i (i = 1, 2) is such that (3.15) holds with $X(.) = X(.; x_i, t)$ *and* $k = k_i$.

PROOF. The property (i) is clear: the backward characteristic may be non-unique at interaction points only, but then we have selected the *minimal* characteristic, making the construction unique.

Along a backward characteristic, u^h is piecewise constant and jumps only when the characteristic passes a shock. But, by construction, a minimal backward characteristic may pass through crossing fronts, only. Indeed, it is obvious that backward characteristics cannot pass through classical shocks and does pass through nonclassical shocks from right to left. When the characteristic passes through a nonclassical shock, say at $(x(\tau), \tau)$, the kinetic function is "acting" and, for ε sufficiently small, we find

$$u^h_-(x(\tau + \varepsilon), \tau + \varepsilon) = \varphi^\flat\big(u^h_-(x(\tau - \varepsilon), \tau - \varepsilon)\big). \qquad (3.16)$$

Additionally, from the list of interaction patterns, we can also see that the path may coincide locally with a rarefaction front emanating from an interaction point (x_0, t_0). When the path is traced backward, it may (but not always) passes through the nonclassical shock (if any), as happens in Cases RC-2, RN, and CR-3. In each of these interactions it can be checked that again the property (3.16) holds. Finally, by iterating the formula (3.16) we arrive at the statement (ii) of the theorem.

Since the interaction between crossing shocks and characteristic lines is always transversal from right to left *only*, and since characteristics can never cross each other, we obtain (iii). This completes the proof of Lemma 3.6. □

We are now ready to estimate the total variation of $u^h(t)$ at any given time $t > 0$. Let

$$-\infty = x_0 < x_1 < x_2 < \ldots < x_{N+1} = +\infty$$

be the discontinuity points in $u^h(t)$ and set also

$$u^h(x, t) = u_i, \quad x \in (x_i, x_{i+1}), \quad i = 0, \dots, N.$$

By using backward characteristics from any fixed point in each interval (x_i, x_{i+1}) and relying on Lemma 3.6, we see that there exists an integer k_i and a value v_i assumed by the initial data u_0^h such that

$$u_i = \varphi^{b\,[k_i]}(v_i). \tag{3.17a}$$

Precisely, for some interval j_i we have

$$v_i := u_0^h(X_i), \quad X_i := X(0; y_i, t) \in I_{j_i} \tag{3.17b}$$

where y_i has been chosen arbitrarily in the interval (x_i, x_{i+1}).

The total variation can be computed as follows:

$$
\begin{aligned}
TV\big(u^h(t)\big) &= \sum_{i=0}^{N-1} |u_{i+1} - u_i| = \sum_{i=0}^{N-1} \big|\varphi^{b\,[k_{i+1}]}(v_{i+1}) - \varphi^{b\,[k_i]}(v_i)\big| \\
&\leq \sum_{i=0}^{N-1} \big|\varphi^{b\,[k_{i+1}]}(v_{i+1}) - \varphi^{b\,[k_{i+1}]}(v_i)\big| + \sum_{i=0}^{N-1} \big|\varphi^{b\,[k_i]}(v_i) - \varphi^{b\,[k_{i+1}]}(v_i)\big| \\
&=: A_1 + A_2.
\end{aligned}
\tag{3.18}
$$

Estimating A_1 is easy by (3.10) in Lemma 3.4, indeed

$$
\begin{aligned}
A_1 &\leq \sup_k \big(\mathrm{Lip}_{J_M}(\varphi^{b\,[k]})\big) \sum_{i=0}^{N-1} |v_{i+1} - v_i| \\
&\leq C_M\, TV(u_0^h),
\end{aligned}
\tag{3.19}
$$

since $X_0 \leq X_1 \leq \dots \leq X_N$.

To estimate the term A_2 we consider the sets $\mathcal{A}^-$ and $\mathcal{A}^+$ made of all indices i such that

$$l_i := k_{i+1} - k_i$$

is strictly negative or strictly positive, respectively. (Obviously, there is nothing to estimate when $l_i = 0$.) In the expression of A_2 we can separate the summation over $\mathcal{A}^+$ and over $\mathcal{A}^-$, calling them A_2^+ and A_2^-, respectively. For each initial interval I_j consider the values of $u^h(t)$ that can be traced back to an initial state in the interval I_j, that is,

$$\mathcal{B}_j := \Big\{ i \,/\, 0 \leq i \leq N \text{ and } X_i \in I_j \Big\}.$$

By (iii) in Lemma 3.6 the map $i \mapsto k_i$ is increasing on $\mathcal{B}_j$ for every j. So, it is strictly increasing on $\mathcal{B}_j \cap \mathcal{A}^+$.

Now, using that $\mathcal{A}^+ = \bigcup_j (\mathcal{B}_j \cap \mathcal{A}^+)$ is a partition we find

$$
\begin{aligned}
A_2^+ &= \sum_{j=0}^{m} \sum_{i \in \mathcal{B}_j \cap \mathcal{A}^+} \big|\varphi^{b\,[k_i]}(v_i) - \varphi^{b\,[l_i]}\big(\varphi^{b\,[k_i]}(v_i)\big)\big| \\
&\leq \sup_l \big(\mathrm{Lip}_{J_M}(id - \varphi^{b\,[l]})\big) \sum_{j=0}^{m} \sum_{i \in \mathcal{B}_j \cap \mathcal{A}^+} \big|\varphi^{b\,[k_i]}(v_i)\big|,
\end{aligned}
$$

where id is the identity mapping. So, since $i \mapsto k_i$ is increasing on each subset under consideration and using (3.10) we obtain

$$
A_2^+ \leq \sup_l \left(1 + \mathrm{Lip}_{J_M}\left(\varphi^{\flat[l]}\right)\right) \sum_{j=0}^{m} \sum_{k=0}^{+\infty} \sup_{x \in I_j} \left|\varphi^{\flat[k]}(u_0^h(x))\right|
$$
$$
\leq \left(1 + C_M\right) C_M' \sum_{j=0}^{m} \sup_{x \in I_j} |u_0^h(x)| \leq C\,TV(u_0^h). \tag{3.20}
$$

The latter inequality holds since $\sup_{x \in I_j} |u_0^h(x)|$ is achieved at some value w_j of the initial data u_0^h with $w_j\,w_{j+1} < 0$ since the intervals I_j determine a decomposition of the real line into disjoint intervals in which u_0^h is alternatively positive and negative. Hence, $\sum_j |w_j|$ is less than $TV(u_0^h)$.

Finally, we estimate A_2^- for which $l_i < 0$. In view of (iii) in Lemma 3.6 and since characteristics cannot cross each other, those indices i must be associated with *distinct* intervals I_i', and since the characteristics cannot cross each other, we must have $j_i < j_{i+1}$. Hence, for every interval I' there exists *at most one index* $i \in \mathcal{A}^-$ such that $X_i \in I'$. It follows that

$$
\sum_{i \in \mathcal{A}^-} |v_i| \leq TV(u_0^h),
$$

thus

$$
A_2^- = \sum_{i \in \mathcal{A}^-} \left|\varphi^{\flat[k_i]}(v_i) - \varphi^{\flat[k_{i+1}]}(v_i)\right|
$$
$$
\leq \sup_{k,l} \mathrm{Lip}\left(\varphi^{\flat[k]} - \varphi^{\flat[l]}\right) \sum_{i \in \mathcal{A}^-} |v_i| \leq 2\,C_M\,TV(u_0^h). \tag{3.21}
$$

Finally, in view of (3.18)–(3.21) we conclude that there exists a constant $C > 0$ depending only on M and $\varphi^\flat$ such that

$$
TV(u^h(t)) \leq C\,TV(u_0^h), \quad t \geq 0.
$$

The derivation of the uniform bound on the total variation, (3.3b), is completed. The estimate (3.3c) is a consequence of (3.3b), as was already checked in Section 2. By Theorem A.3 in the appendix we conclude that a subsequence converges to a limit u satisfying (3.4) and (3.5). Since the functions u^h are exact solutions, it follows that u is a weak solution of the Cauchy problem. Similarly, one can see that the solution u satisfies the entropy inequality (3.6). This completes the proof of the properties (i) and (ii) of Theorem 3.2.

REMARK 3.7. When attention is restricted to *classical* solutions, the calculation made above simplifies drastically. Indeed, for classical solutions we have $k_i = 0$ for all i and, therefore, $A_2 = 0$, whereas the estimate (3.19) for A_1 can be replaced with

$$
A_1 = \sum_{i=0}^{N-1} |v_{i+1} - v_i| \leq TV(u_0^h), \tag{3.19'}
$$

which allows us to recover the total variation diminishing property (2.6b) in Theorem 2.1.   □

4. Refined estimates

In this last section we derive some additional properties of classical entropy solutions. In particular, we introduce the notion of interaction potential which will play a central role later in Chapters VII and VIII in our study of systems. Define the **interaction potential** of a function with bounded variation $u = u(x)$ by

$$Q(u) := \sum_{x<y} |u_+(x) - u_-(x)| \, |u_+(y) - u_-(y)|, \tag{4.1}$$

where the sum is over the points of discontinuity of the function u and $u_\pm$ denotes its left- and right-hand traces. Observe that

$$Q(u) \leq TV(u)^2 < \infty.$$

Furthermore, restricting now attention to piecewise constant functions $u = u(x)$ made of classical shock and rarefaction fronts only, we denote by $R(u)$ the maximal strength of rarefaction fronts in u.

For convex flux-functions we obtain immediately the following result.

THEOREM 4.1. (Refined estimates I) *Consider the Cauchy problem* (2.1) *and* (2.2) *where the flux f is either convex or concave. Let $u^h = u^h(x,t)$ be a sequence of wave front tracking approximations associated with the classical Riemann solver. Consider an interaction taking place at some time t_0 and involving two incoming fronts connecting the states u_l, u_m, and u_r. Then, we have*

$$\begin{aligned}
&[TV(u^h(t_0))] = |u_l - u_r| - |u_l - u_m| - |u_m - u_r| \leq 0, \\
&[R(u^h(t_0))] \leq 0, \\
&[Q(u^h(t_0))] \leq -|u_l - u_m| \, |u_m - u_r| \leq 0,
\end{aligned} \tag{4.2}$$

where, for instance, we use the notation

$$[TV(u^h(t_0))] := \lim_{\substack{\varepsilon \to 0 \\ \varepsilon > 0}} TV(u^h(t_0 + \varepsilon)) - TV(u^h(t_0 - \varepsilon)). \tag{4.3}$$

We omit the proof which is easy. From the local estimates (4.2) we deduce the global estimates $(t > 0)$

$$\begin{aligned}
TV(u^h(t)) &\leq TV(u^h(0+)) = TV(u_0^h), \\
R(u^h(t)) &\leq R(u^h(0+)) \leq 1/h, \\
Q(u^h(t)) &\leq Q(u^h(0+)) \leq TV(u_0^h)^2.
\end{aligned}$$

Recall that $TV(u_0^h)$ is uniformly bounded in view of (2.4).

We now discuss the interaction potential for concave-convex flux-functions, using the standard notation $\varphi^\natural$, $\varphi^{-\natural}$, etc. A technical inequality will be needed, which we introduce first. Given two points u and v with $v \neq u, \varphi^\natural(u)$ we denote by $\rho(u,v)$ the solution of

$$\frac{f(\rho(u,v)) - f(u)}{\rho(u,v) - u} = \frac{f(v) - f(u)}{v - u}, \quad \rho(u,v) \neq u, v. \tag{4.4}$$

The function ρ is extended by continuity, so that, in particular, $\rho(u, \varphi^\natural(u)) = \varphi^\natural(u)$. Expanding the relation above, it is not difficult to see that, in the neighborhood of the origin,

$$\rho(u,v) \sim -u - v.$$

In turn, we see that for some $\theta \in (0,1)$ and in the neighborhood of the origin,

$$\mathrm{Lip}(\varphi^\natural) < 1,$$
$$|u - \varphi^\natural(u)| \, |v - \varphi^\natural(u)| \leq \theta \, \min\big(|u| \, |v|, |u - \rho(u,v)| \, |v - \rho(u,v)|\big) \tag{4.5}$$

for all $u > 0$ and $\rho(u,0) \leq v \leq \varphi^\natural(u)$, as well as an analogous property for $u < 0$. In fact, (4.5) holds globally for all u and v if $f(u) = u^3$, since then $\varphi^\natural(u) = -u/2$ and $\rho(u,v) = -u - v$:

$$\frac{3}{2} u \left| v + \frac{u}{2} \right| \leq \theta \, \min\big(|u| \, |v|, |2u + v| \, |2v + u|\big),$$

which holds in the range given above if, for instance, $\theta = 3/4$.

THEOREM 4.2. (Refined estimates II) *Consider the Cauchy problem* (2.1) *and* (2.2) *when the flux f is a concave-convex function and let us restrict attention to solutions whose range is included in a small neighborhood of the origin (or, more precisely, assume that the flux satisfies* (4.5)*). Let $u^h = u^h(x,t)$ be a sequence of classical wave front tracking approximations. Consider an interaction taking place at some time t_0 and involving two incoming waves connecting the states u_l, u_m, and u_r and (at most) two outgoing waves connecting u_l, u'_m, and u_r (with possibly $u'_m = u_r$). Then, we have*

$$\big[TV(u^h(t_0))\big] = |u_l - u'_m| + |u'_m - u_r| - |u_l - u_m| - |u_m - u_r| \leq 0,$$
$$\big[R(u^h(t_0))\big] \leq 0, \tag{4.6}$$
$$\big[Q(u^h(t_0))\big] \leq |u_l - u'_m| \, |u'_m - u_r| - |u_l - u_m| \, |u_m - u_r|$$
$$\leq -c \, |u_l - u_m| \, |u_m - u_r|$$

for some uniform constant $c > 0$.

PROOF. We distinguish between several interactions, following the general classification given in Section 3. Since only classical shocks are allowed here, only seven different cases may arise. Note that the decreasing property of the interaction potential is obvious when the outgoing pattern contains a "single wave" since then

$$\big[Q(u^h(t_0))\big] = -|u_l - u_m| \, |u_m - u_r| \leq 0.$$

So, we omit this calculation in the single wave cases below. Additionally, in our calculations of the maximal rarefaction strength and of the potential we focus on those waves involved in the interaction, or in other words we assume that the solution under consideration contains only two waves interacting and no other waves. It is obvious that the contribution to the potential due to "other" waves would diminish since the total variation diminishes at interactions.

Case RC-1 : That is, $(R_+ C)$–(C') when $0 < u_l < u_m$ and $\varphi^\natural(u_l) \leq u_r < u_l$. There is only one outgoing wave, and the incoming pattern is non-monotone so some waves strength is cancelled. We find

$$\big[TV(u^h(t_0))\big] = |u_r - u_l| - |u_m - u_l| - |u_r - u_m| = -2 \, |u_m - u_l| \leq 0$$

and

$$\big[R(u^h(t_0))\big] = -|u_l - u_m| \leq 0.$$

Case RC-2 : That is, $(R_+C_\pm)$–$(C'_\pm R'_-)$ when $0 < u_l < u_m$ and $\varphi^\natural(u_m) \leq u_r < \varphi^\natural(u_l)$. The outgoing pattern contains here two waves and some cancellation is taking place. We have

$$\left[TV(u^h(t_0))\right] = |\varphi^\natural(u_l) - u_l| + |u_r - \varphi^\natural(u_l)| - |u_m - u_l| - |u_r - u_m|$$
$$= -2\,|u_m - u_l| \leq 0,$$

and

$$\left[R(u^h(t_0))\right] = |\varphi^\natural(\varphi^{-\natural}(u_r)) - \varphi^\natural(u_l)| - |u_m - u_l|$$
$$\leq -\left(1 - \mathrm{Lip}(\varphi^\natural)\right)|u_m - u_l|.$$

Using $u_l < \varphi^{-\natural}(u_r) \leq u_m$ we find also that for every $\kappa \in (0,1)$

$$\left[Q(u^h(t_0))\right]$$
$$\leq |\varphi^\natural(u_l) - u_l|\,|u_r - \varphi^\natural(u_l)| - |u_m - u_l|\,|u_r - u_m|$$
$$\leq -(1 - \kappa)\,|u_m - u_l|\,|u_r - u_m| + \mathrm{Lip}(\varphi^\natural)\,|\varphi^\natural(u_l) - u_l|\,|\varphi^{-\natural}(u_r) - u_l|$$
$$\quad - \kappa\,|\varphi^{-\natural}(u_r) - u_l|\,|u_r - \varphi^{-\natural}(u_r)|$$
$$\leq -(1 - \kappa)\,|u_m - u_l|\,|u_r - u_m| - \left(\kappa - \mathrm{Lip}(\varphi^\natural)\right)|\varphi^\natural(u_l) - u_l|\,|\varphi^{-\natural}(u_r) - u_l|$$
$$\leq -(1 - \kappa)\,|u_m - u_l|\,|u_r - u_m| \leq 0,$$

since $u_r < \varphi^\natural(u_l) < u_l < \varphi^{-\natural}(u_r)$ and provided we choose κ such that

$$1 > \kappa > \mathrm{Lip}(\varphi^\natural),$$

which is possible by the first assumption in (4.5).

Case CR-1 : That is, $(C_\pm R_-)$–$(C'_\pm)$ when $\varphi^\natural(u_l) \leq u_r < u_m < 0$. There is only one outgoing wave and the incoming solution is monotone. We find here

$$\left[TV(u^h(t_0))\right] = 0, \quad \left[R(u^h(t_0))\right] = -|u_r - u_m| \leq 0.$$

Case CR-2 : That is, (C_+R_+)–(C'_+) when $0 < u_m < u_r < u_l$. There is only one outgoing wave and some cancellation is taking place. This case is analogous to (RC-1).

Case CR-3 : That is, $(C_\pm R_-)$–$(C'_\pm R'_-)$ when $\rho(u_l, u_m) < u_r < \varphi^\natural(u_l) \leq u_m < 0$. The outgoing pattern contains two waves and the incoming solution is monotone. We obtain

$$\left[TV(u^h(t_0))\right] = |\varphi^\natural(u_l) - u_l| + |u_r - \varphi^\natural(u_l)| - |u_m - u_l| - |u_r - u_m| = 0$$

and

$$\left[R(u^h(t_0))\right] = |u_r - \varphi^\natural(u_l)| - |u_r - u_m| = -|\varphi^\natural(u_l) - u_m| \leq 0.$$

For every $\kappa \in (0,1)$ we have

$$\left[Q(u^h(t_0))\right] \leq |\varphi^\natural(u_l) - u_l|\,|u_r - \varphi^\natural(u_l)| - |u_m - u_l|\,|u_r - u_m|$$
$$= |\varphi^\natural(u_l) - u_l|\,|u_r - \varphi^\natural(u_l)| - \kappa\,|u_m - u_l|\,|u_r - u_m| - (1 - \kappa)\,|u_m - u_l|\,|u_r - u_m|.$$

The polynomial function $u_m \mapsto |u_m - u_l|\,|u_r - u_m|$ over the interval determined by $\rho(u_l, u_r) \leq u_m \leq 0$ satisfies the inequality

$$|u_m - u_l|\,|u_r - u_m| \geq \min\left(|u_l|\,|u_r|, |u_l - \rho(u_l, u_r)|\,|u_r - \rho(u_l, u_r)|\right).$$

Therefore, we conclude that

$$
\begin{aligned}
&[Q(u^h(t_0))] \\
&\leq |\varphi^\natural(u_l) - u_l| \, |u_r - \varphi^\natural(u_l)| - \kappa \min\big(|u_l| \, |u_r|, |u_l - \rho(u_l, u_r)| \, |u_r - \rho(u_l, u_r)|\big) \\
&\quad - (1 - \kappa) \, |u_m - u_l| \, |u_r - u_m| \\
&\leq -(1 - \kappa) \, |u_m - u_l| \, |u_r - u_m|
\end{aligned}
$$

by the assumption (4.5), provided κ is chosen such that $\theta \leq \kappa < 1$.

Case CC-1 : That is, (C_+C)–(C') when $0 \leq u_m < u_l$ and $\varphi^\natural(u_m) \leq u_r < u_m$. This case is analogous to (CR-1).

Case CC-2 : That is, $(C_\pm C)$–(C') when $\varphi^\natural(u_l) \leq u_m \leq 0$ and $u_m < u_r \leq \varphi^\natural(u_m)$. This case is similar to (RC-1).

This completes the proof of Theorem 4.2. $\qquad\square$

Finally, we consider the **weighted interaction potential**

$$
\tilde{Q}(u) = \sum_{x<y} q(u_-(x), u_+(x)) \, |u_+(x) - u_-(x)| \, |u_+(y) - u_-(y)|, \qquad (4.7)
$$

where the weight q is determined so that *a right-contact located at x is regarded as non-interacting with all waves located at $y > x$.* Precisely, setting $u_\pm := u_\pm(x)$ we define the function $q(u_-, u_+)$ by

$$
q(u_-, u_+) := \begin{cases}
u_+ - \rho(u_-, u_+), & \varphi^\natural(u_-) \leq u_+ \leq u_- \text{ and } u_- > 0, \\
\rho(u_-, u_+) - u_+, & u_- \leq u_+ \leq \varphi^\natural(u_-) \text{ and } u_- < 0, \\
1, & \text{otherwise.}
\end{cases} \qquad (4.8)
$$

Recall that $\rho(u_-, u_+)$ was defined in (4.4).

THEOREM 4.3. (Refined estimates III) *With the same assumptions and notation as in Theorem 4.2, we have at each interaction time t_0*

$$
[\tilde{Q}(u^h(t_0))] \leq \begin{cases}
C \, TV(u^h(0)) \, |u_l - u_m| \, |u_m - u_r| & \text{in Case CC-2,} \\
-c \, |u_l - u_m| \, |u_m - u_r| & \text{in all other cases,}
\end{cases} \qquad (4.9)
$$

for some uniform constants $C, c > 0$.

We observe that the rate of decrease in (4.9) is weaker than the one obtained in Theorem 4.2 since the coefficient q may vanish. Furthermore, in one case (Case CC-2) the potential may increase whereas the total variation functional decreases:

$$
[TV(u^h(t_0))] = -2 \, |u_r - u_m| \leq 0.
$$

In turn, for in a range of constants $C_* > 0$ at least, the functional $TV(u^h(t_0)) + C_* \, \tilde{Q}(u^h(t_0))$ decreases strictly in all interaction cases. The weighted potential will be of particular interest in Chapter VIII to study nonclassical solutions.

PROOF. The jump of the interaction potential can be decomposed in three parts:

$$
[\tilde{Q}(u^h(t_0))] = P_1 + P_2 + P_3,
$$

where P_1 contains products between the waves involved in the interaction, P_2 between waves which are not involved in the interaction, and P_3 products between these two sets of waves.

On one hand, we have immediately

$$P_1 = -q(u_l, u_m) |u_l - u_m| |u_m - u_r|,$$

since there is only one outgoing wave or else the two outgoing waves are regarded as non-interacting in view of the definition (4.8). If u^h would contain only the two interacting waves and no other wave then (4.9) would follow by combining the above result with (4.6). On the other hand, clearly, $P_2 = 0$ since the waves which are not involved in the interaction are not modified.

We now concentrate on the contribution P_3 of "other waves". Denote by W_l and W_r the total strength of waves located on the left- and right-hand sides of the interaction point, respectively. Let us decompose P_3 accordingly, say $P_3 = P_{3l} + P_{3r}$. Waves located on the left-hand side are dealt with by relying on the decrease of the total variation:

$$P_{3l} = W_l \left(|u_r - u_l| - |u_m - u_l| - |u_m - u_r| \right) \le 0.$$

To deal with waves located on the right-hand side of the interaction point we set $P_{3r} = \Omega_r W_r$ with

$$\Omega_r := q(u_l, u'_m) |u_l - u'_m| + q(u_l, u'_m) |u_l - u'_m| - q(u_l, u_m) |u_l - u_m| - q(u_m, u_r) |u_r - u_m|,$$

and we now estimate Ω_r by distinguishing between several cases, as in the proof of Theorem 4.2.

Case RC-1 :

$$\Omega_r = \left(u_r - \rho(u_l, u_r) \right) (u_l - u_r) - (u_m - u_l) - \left(u_r - \rho(u_m, u_r) \right) (u_m - u_r)$$
$$= -\left| \rho(u_l, u_r) - \rho(u_m, u_r) \right| |u_l - u_r| - \left(1 + |u_r - \rho(u_m, u_r)| \right) |u_m - u_l| \le 0.$$

Case RC-2 :

$$\Omega_r = \left(\varphi^\natural(u_l) - u_r \right) - (u_m - u_l) - \left(u_r - \rho(u_m, u_r) \right) (u_m - u_r)$$
$$\le \left| \varphi^\natural(u_l) - \varphi^\natural(u_m) \right| - |u_m - u_l| \le 0,$$

since $\mathrm{Lip}(\varphi^\natural) < 1$ near the origin.

Case CR-1 :

$$\Omega_r = \left(u_r - \rho(u_l, u_r) \right) (u_l - u_r) - \left(u_m - \rho(u_l, u_m) \right) (u_l - u_m) - (u_m - u_r)$$
$$= -\left(|u_m - u_r| + |\rho(u_l, u_r) - \rho(u_l, u_m)| \right) |u_l - u_m|$$
$$\quad - \left(1 + |u_r - \rho(u_l, u_r)| \right) |u_m - u_r| \le 0.$$

Case CR-2 : We have

$$\Omega_r = \left(u_r - \rho(u_l, u_r) \right) (u_l - u_r) - \left(u_m - \rho(u_l, u_m) \right) (u_l - u_m) - (u_r - u_m)$$
$$= -\left(1 + |u_m - \rho(u_l, u_m)| - |u_l - u_r| \left(1 + \frac{|\rho(u_l, u_m) - \rho(u_l, u_r)|}{|u_r - u_m|} \right) \right) |u_r - u_m|$$
$$\le 0,$$

provided the term $|u_l - u_r|$ is sufficiently small.

Case CR-3 :

$$\begin{aligned}\Omega_r &= (\varphi^\natural(u_l) - u_r) - (u_m - \rho(u_l, u_m))\,(u_l - u_m) - (u_m - u_r)\\ &\leq - |u_m - \varphi^\natural(u_l)| \leq 0.\end{aligned}$$

Case CC-1 :

$$\begin{aligned}\Omega_r &= (u_r - \rho(u_l, u_r))\,(u_l - u_r)\\ &\quad - (u_m - \rho(u_l, u_m))\,(u_l - u_m) - (u_r - \rho(u_m, u_r))\,(u_m - u_r)\\ &= -|\rho(u_l, u_r) - \rho(u_l, u_m)|\,|u_l - u_m|\\ &\quad - |u_m - u_r|\,\big|u_l - u_m - \rho(u_m, u_r) + \rho(u_l, u_r)\big| \leq 0,\end{aligned}$$

since the function ρ satisfies when $u_r \leq u_m \leq u_l$

$$0 \leq \rho(u_l, u_r) - \rho(u_l, u_m) \leq \big(1 + O(\delta)\big)\,|u_m - u_r|,$$
$$\big|u_l - u_m - \rho(u_m, u_r) + \rho(u_l, u_r)\big| \leq C\,\delta\,|u_l - u_m|.$$

Indeed, when $f(u) = u^3$ these inequalities are obvious since $\rho(u, v) = -u - v$. For a general concave-convex flux-function, it can be checked that they hold in a sufficiently small neighborhood of 0.

Case CC-2 :

$$\begin{aligned}\Omega_r &= \big(u_r - \rho(u_l, u_r)\big)\,(u_l - u_r) - \big(u_m - \rho(u_l, u_m)\big)\,(u_l - u_m)\\ &\quad - \big(\rho(u_m, u_r) - u_r\big)\,(u_r - u_m)\\ &= O(1)\,|u_r - u_m|\,|u_m - u_l|.\end{aligned}$$

This completes the proof of Theorem 4.3. $\square$

CHAPTER V

CONTINUOUS DEPENDENCE OF SOLUTIONS

In the present chapter, we investigate the *continuous dependence* of solutions to scalar conservation laws. In Section 2, we study a class of hyperbolic equations with discontinuous coefficient, and we establish a *general stability result* in L^1 when the coefficient does not contain *rarefaction-shocks;* see Definition 1.1 and the main result in Theorem 1.7. Next, in Section 2 we apply this setting to conservation laws with convex flux; see Theorems 2.2 and 2.3. The proofs in Section 2 are based on the key observation that no rarefaction-shock (in the sense of Section 1) can arise from comparing two entropy solutions. In Section 3, we derive a *sharp estimate* in a *weighted L^1 norm,* which provides a quantitative bound on the decrease of the L^1 norm; see Theorem 3.1. Finally, in Section 4 we state the generalization to nonclassical solutions.

1. A class of linear hyperbolic equations

Consider the Cauchy problem associated with the scalar conservation law

$$\partial_t u + \partial_x f(u) = 0, \quad u = u(x,t) \in I\!R, \ x \in I\!R, \ t > 0, \qquad (1.1)$$

in which the flux $f : I\!R \to I\!R$ is a given smooth function. We want to establish the L^1 *continuous dependence property*

$$\|u(t) - v(t)\|_{L^1(I\!R)} \le \|u(0) - v(0)\|_{L^1(I\!R)}, \quad t \ge 0, \qquad (1.2)$$

for solutions u and v of (1.1). Our general strategy can be sketched as follows.
    After introducing the notation

$$\psi := v - u, \quad \overline{a}(u,v) = \int_0^1 f'\big(\theta\, u + (1-\theta)\, v\big)\, d\theta, \qquad (1.3)$$

where $\overline{a}(u,v)$ will be called the **averaging speed**, for any two solutions $u = u(x,t)$ and $v = v(x,t)$ of (1.1) we have

$$\partial_t \psi + \partial_x\big(\overline{a}(u,v)\,\psi\big) = 0. \qquad (1.4)$$

Therefore, to establish (1.2) it is sufficient to establish the L^1 *stability property*

$$\|\psi(t)\|_{L^1(I\!R)} \le \|\psi(0)\|_{L^1(I\!R)}, \quad t \ge 0, \qquad (1.5)$$

for a class of equations and solutions covering the situation (1.3) and (1.4). In the present section, we discuss precisely the derivation of (1.5) and leave for the following section the applications to the situation (1.3) and (1.4).
    Given a piecewise constant speed $a : I\!R \times I\!R_+ \to I\!R$ we consider piecewise constant solutions $\psi : I\!R \times I\!R_+ \to I\!R$ of the linear hyperbolic equation with *discontinuous coefficient*

$$\partial_t \psi + \partial_x\big(a\,\psi\big) = 0, \quad x \in I\!R, \ t > 0, \qquad (1.6)$$

and we aim at deriving the inequality (1.5). The set of *points of continuity,* where a is locally constant, is denoted by $\mathcal{C}(a)$. The (finitely many) polygonal lines of jump discontinuities in the function a determine a set of *jump points* $\mathcal{J}(a) \subset \mathbb{R} \times \mathbb{R}_+$. The finite set of interaction times when these lines intersect is denoted by $\mathcal{I}(a) \subset \mathbb{R}_+$. To each propagating discontinuity $(x,t) \in \mathcal{J}(a)$ we associate a *shock speed* $\lambda^a = \lambda^a(x,t)$ and *left- and right-hand traces* $a_\pm = a_\pm(x,t) = a(x\pm,t)$. As we will see later, it will be convenient to extend the definition of the shock speed by setting

$$\lambda^a(x,t) = a(x,t), \quad (x,t) \in \mathcal{C}(a).$$

Finally, observe that if ψ is a solution of (1.6) (in the weak sense of distributions) and admits a jump discontinuity at a point (x,t) propagating at the shock speed λ^a, then the traces $\psi_\pm = \psi_\pm(x,t)$ and $a_\pm = a_\pm(x,t)$ satisfy the *jump condition*

$$-\lambda^a \left(\psi_+ - \psi_- \right) + a_+ \, \psi_+ - a_- \, \psi_- = 0.$$

The geometric properties of the speed a will be critical in the forthcoming discussion, and it will be useful to adopt the following terminology (illustrated by Figure V-1).

DEFINITION 1.1. A point $(x,t) \in \mathcal{J}(a)$ is called:

• A **Lax** discontinuity if

$$a_-(x,t) \geq \lambda^a(x,t) \geq a_+(x,t).$$

• A **slow undercompressive** discontinuity if

$$\lambda^a(x,t) < \min\big(a_-(x,t), a_+(x,t)\big).$$

• A **fast undercompressive** discontinuity if

$$\lambda^a(x,t) > \max\big(a_-(x,t), a_+(x,t)\big).$$

• A **rarefaction-shock** discontinuity if

$$a_-(x,t) \leq \lambda^a(x,t) \leq a_+(x,t).$$

Observe that the cases listed in Definition 1.1 are disjoint (since $a_-(x,t) \neq a_+(x,t)$ when $(x,t) \in \mathcal{J}(a)$). Introduce the partition

$$\mathcal{J}(a) =: \mathcal{L}(a) \cup \mathcal{S}(a) \cup \mathcal{F}(a) \cup \mathcal{R}(a),$$

where $\mathcal{L}(a), \mathcal{S}(a), \mathcal{F}(a)$, and $\mathcal{R}(a)$ are the sets of Lax, slow undercompressive, fast undercompressive, and rarefaction-shock discontinuities, respectively. We first illustrate a fundamental feature of the equation (1.6) with an example.

EXAMPLE 1.2. When

$$a(x,t) = \begin{cases} -1, & x < 0, \\ 1, & x > 0, \end{cases}$$

the Cauchy problem associated with (1.6) with the initial condition

$$\psi(x,0) = 0, \quad x \in \mathbb{R},$$

admits a large class of solutions (including the trivial one $\psi \equiv 0$), namely

$$
\psi(x,t) = \begin{cases}
0, & x < -t, \\
\psi_0(t+x), & -t \leq x < 0, \\
-\psi_0(t-x), & 0 \leq x < t, \\
0, & x \geq t,
\end{cases}
$$

where the arbitrary function ψ_0 is Lipschitz continuous or, more generally, of bounded variation. (See Figure V-2.) $\qquad\qquad\qquad\qquad\qquad\qquad\qquad\qquad\quad$ $\square$

Example 1.2 shows that we cannot expect the uniqueness of weak solutions for the Cauchy problem associated with (1.6). In the above example, the characteristics happen to be moving away from the discontinuity at $x = 0$ from both sides, and thus the discontinuity in the coefficient a is a rarefaction-shock in the sense of Definition 1.1. The fact that rarefaction-shocks are the *only* source of non-uniqueness is a consequence of the following theorem. (Throughout this chapter, we always tacitly restrict attention to solutions ψ with compact support.)

THEOREM 1.3. (L^1 stability for linear hyperbolic equations.) *Consider a coefficient $a = a(x,t)$ and a solution $\psi = \psi(x,t)$ of (1.6), both being piecewise constant. Then, for all $t \geq 0$ we have the identity*

$$
\|\psi(t)\|_{L^1(\mathbb{R})} + \int_0^t \sum_{(x,\tau)\in\mathcal{L}(a)} 2\left|\lambda^a(x,\tau) - a_-(x,\tau)\right| |\psi_-(x,\tau)|\, d\tau
$$
$$
= \|\psi(0)\|_{L^1(\mathbb{R})} + \int_0^t \sum_{(x,\tau)\in\mathcal{R}(a)} 2\left|\lambda^a(x,\tau) - a_-(x,\tau)\right| |\psi_-(x,\tau)|\, d\tau. \tag{1.7}
$$

At this juncture, observe that:
- Lax discontinuities contribute to the *decrease* of the L^1 norm.
- Rarefaction-shocks *increase* the L^1 norm.
- Undercompressive discontinuities *keep constant* the L^1 norm.

Hence, in the special case that a has no rarefaction shocks, Theorem 1.3 implies

$$
\|\psi(t)\|_{L^1(\mathbb{R})} \leq \|\psi(0)\|_{L^1(\mathbb{R})}, \quad t \geq 0, \tag{1.8}
$$

which is the desired estimate (1.5). In particular, if the coefficient a has no rarefaction-shocks, then the Cauchy problem for (1.6) admits a *unique* solution (in the class of piecewise constant solutions under consideration here, at least).

PROOF. It is sufficient to check that for all $t \notin \mathcal{I}(a)$

$$
\frac{d}{dt}\int_{\mathbb{R}} |\psi(x,t)|\, dx = -\sum_{(x,t)\in\mathcal{L}(a)} 2\left|\lambda^a(x,t) - a_-(x,t)\right| |\psi_-(x,t)|
$$
$$
+ \sum_{(x,t)\in\mathcal{R}(a)} 2\left|\lambda^a(x,t) - a_-(x,t)\right| |\psi_-(x,t)|.
$$

The estimate (1.7) then follows by using that the mapping $t \mapsto \|\psi(t)\|_{L^1(\mathbb{R})}$ is Lipschitz continuous. So, in the rest of this proof, we restrict attention to any time interval in which a contains *no interaction point*.

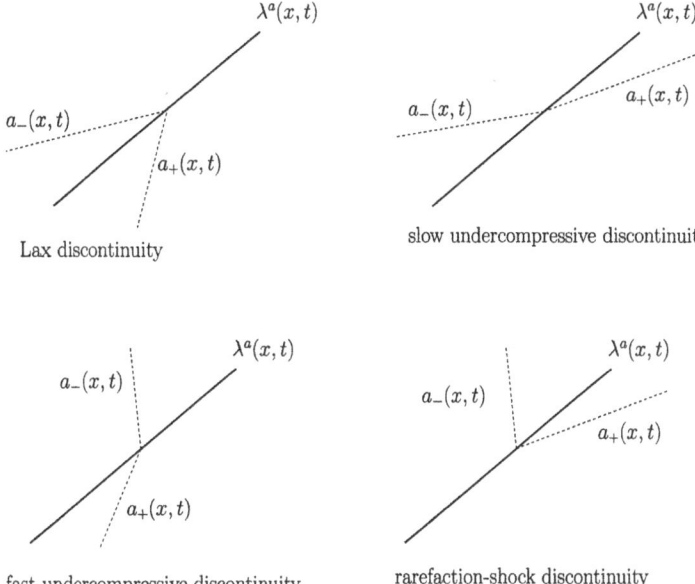

Figure V-1 : Classification : $\mathcal{L}(a), \mathcal{S}(a), \mathcal{F}(a)$, and $\mathcal{R}(a)$.

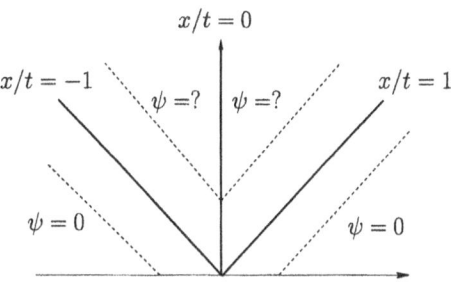

Figure V-2 : Non-uniqueness of solutions.

We denote by $t \mapsto x_j(t)$ $(0 \le j \le m+1)$ the straight lines (or polygonal lines if interaction times would be included) along which the function $x \mapsto \psi(x,t)$ changes sign with the convention that, for $0 \le j \le m$,

$$(-1)^j \, \psi(x,t) \ge 0, \quad x \in [x_j(t), x_{j+1}(t)], \tag{1.9}$$

$x_0(t) = -\infty$, and $x_{m+1}(t) = +\infty$. For all $1 \le j \le m$ we set

$$\psi_j^\pm(t) := \psi_\pm(x_j(t), t), \quad a_j^\pm(t) := a_\pm(x_j(t), t), \quad \lambda_j(t) := \dot{x}_j(t),$$

$\lambda_{m+1}(t) = \lambda_0(t) = 0$, and $\psi_{m+1}^-(t) = \psi_0^+(t) = 0$. Since ψ solves (1.6) we find (the terms $\partial_t \psi$ and $\partial_x(a\,\psi)$ being measures)

$$\frac{d}{dt} \int_{I\!\!R} |\psi(x,t)| \, dx = \frac{d}{dt} \sum_{j=0}^m (-1)^j \int_{x_j(t)}^{x_{j+1}(t)} \psi(x,t) \, dx$$

$$= \sum_{j=0}^m (-1)^j \left(x_{j+1}'(t) \, \psi_-(x_{j+1}(t),t) - x_j'(t) \, \psi_+(x_j(t),t) + \int_{x_j(t)}^{x_{j+1}(t)} \partial_t \psi(x,t) \right)$$

$$= \sum_{j=0}^m (-1)^j \left(\lambda_{j+1}(t) \, \psi_{j+1}^-(t) - \lambda_j(t) \, \psi_j^+(t) + \int_{x_j(t)}^{x_{j+1}(t)} -\partial_x(a(x,t) \, \psi(x,t)) \right)$$

$$= \sum_{j=1}^m (-1)^j \left((a_j^+(t) - \lambda_j(t)) \, \psi_j^+(t) + (a_j^-(t) - \lambda_j(t)) \, \psi_j^-(t) \right).$$

The Rankine-Hugoniot relation associated with (1.6) reads

$$(a_j^+(t) - \lambda_j(t)) \, \psi_j^+(t) = (a_j^-(t) - \lambda_j(t)) \, \psi_j^-(t), \tag{1.10}$$

therefore with (1.9)

$$\frac{d}{dt} \int_{I\!\!R} |\psi(x,t)| \, dx = \pm \sum_{j=1}^m 2 \left(a_j^\pm(t) - \lambda_j(t) \right) |\psi_j^\pm(t)|. \tag{1.11}$$

(In the right-hand side of (1.11), one can use either the $+$ or the $-$ sign.)

Consider any of the points $x_j(t)$:

- Suppose $x_j(t) \in \mathcal{C}(a)$, and thus $a_j(t) := a_j^-(t) = a_j^+(t)$. Since $\psi_j^-(t)$ and $\psi_j^+(t)$ have opposite signs, it follows from the Rankine-Hugoniot relation (1.10) that either $\lambda_j(t) - a_j(t) = 0$ or else $\psi_j^-(t) = \psi_j^+(t) = 0$. In both cases, the terms $\left(a_j^\pm(t) - \lambda_j(t) \right) |\psi_j^\pm(t)|$ vanish and so the points $x_j(t) \in \mathcal{C}(a)$ do not contribute to the right-hand side of (1.11).

- Suppose $x_j(t) \in \mathcal{L}(a)$, and thus $\lambda_j(t) = \lambda^a(x_j(t),t)$. Since, for a Lax discontinuity, $a_j^-(t) \ge \lambda_j(t) \ge a_j^+(t)$, we see that both coefficients $\pm(a_j^\pm(t) - \lambda_j(t))$ are negative. Therefore, the points $x_j(t) \in \mathcal{L}(a)$ contribute "favorably" to the right-hand side of (1.11).

- Suppose $x_j(t) \in \mathcal{R}(a)$, and thus $\lambda_j(t) = \lambda^a(x_j(t),t)$. Since, for a rarefaction-shock, $a_j^-(t) \le \lambda_j(t) \le a_j^+(t)$, we see that both coefficients $\pm(a_j^\pm(t) - \lambda_j(t))$ are positive. Therefore, the points $x_j(t) \in \mathcal{R}(a)$ contribute with a "wrong" sign to the right-hand side of (1.11).

- Suppose $x_j(t) \in \mathcal{S}(a) \cup \mathcal{F}(a)$, and thus $\lambda_j(t) = \lambda^a(x_j(t), t)$. By definition of an undercompressive discontinuity, the two sides of (1.10) have different signs and, therefore,

$$\left(a_j^+(t) - \lambda_j(t)\right)\psi_j^+(t) = \left(a_j^-(t) - \lambda_j(t)\right)\psi_j^-(t) = 0.$$

Hence, undercompressive discontinuities $x_j(t) \in \mathcal{S}(a) \cup \mathcal{F}(a)$ do not contribute to the right-hand side of (1.11).

This completes the derivation of (1.7).   □

REMARK 1.4. Observe that at rarefaction shocks we have

$$\left|\lambda^a(x,\tau) - a_-(x,\tau)\right| \leq \left|a_+(x,\tau) - a_-(x,\tau)\right|$$

and since ψ changes sign (see (1.10) above)

$$\left|\psi_-(x,\tau)\right| \leq \left|\psi_+(x,\tau) - \psi_-(x,\tau)\right|.$$

This implies that the last terms in the right-hand side of (1.7) is bounded by the total variation of ψ and the maximum strength of rarefaction-shocks, namely

$$2 \sup_{\substack{(x,\tau)\in\mathcal{R}(a)\\ \tau\in(0,t)}} \left|a_+(x,\tau) - a_-(x,\tau)\right| \int_0^t TV(\psi(\tau))\, d\tau.$$

This estimate will play an important role later, in Section 2.   □

REMARK 1.5. Of course, the left-hand traces appearing in (1.7) are chosen for the sake of definiteness, only. It is obvious from (1.10) that

$$\left|\lambda^a(x,\tau) - a_-(x,\tau)\right|\left|\psi_-(x,\tau)\right| = \left|\lambda^a(x,\tau) - a_+(x,\tau)\right|\left|\psi_+(x,\tau)\right|.$$

By (1.10), the function ψ changes sign at Lax and rarefaction shocks only, and keeps a constant sign at undercompressive discontinuities. Therefore, *if a solution ψ keeps a constant sign*, the two sums in (1.7) vanish and we find

$$\|\psi(t)\|_{L^1(\mathbb{R})} = \|\psi(0)\|_{L^1(\mathbb{R})}, \quad t \geq 0.$$

This implies that the Cauchy problem associated with (1.6) admits *at most one* solution ψ of a given sign.   □

We now discuss the properties of weighted norms of the form

$$\|\psi(t)\|_{w(t)} := \int_{\mathbb{R}} |\psi(x,t)|\, w(x,t)\, dx, \quad t \geq 0,$$

where $w = w(x,t) > 0$ is a piecewise constant function. We search for conditions on the weight w guaranteeing that $\|\psi(t)\|_{w(t)}$ is non-increasing in time when ψ is a solution of (1.6). Our goal is to quantify the rate of decrease of these weighted L^1 norms. Following the notation introduced for the speed a, we use the obvious notation λ^w, $\mathcal{C}(w)$, $\mathcal{J}(w)$, and $\mathcal{I}(w)$. It is convenient also to denote by $\mathcal{I}(a,w)$ the set consisting of all interaction times for a or w (that is, $\mathcal{I}(a) \cup \mathcal{I}(w)$) as well as all times when two polygonal lines of discontinuity for a and for w intersect in the (x,t)-plane.

We will impose that the discontinuity lines of w coincide with discontinuity lines of a or with characteristic lines associated with a, in other words, *with the exception of all times* $t \in \mathcal{I}(a, w)$,

$$\lambda^w(x, t) = \lambda^a(x, t), \quad (x, t) \in \mathcal{J}(w). \tag{1.12}$$

Recall that we extended the definition of the shock speed λ^a by setting $\lambda^a(x, t) := a(x, t)$ when $(x, t) \in \mathcal{C}(a)$. Attempting to generalize the calculation made in the proof of Theorem 1.3, we arrive at the following identity for the weighted norm.

LEMMA 1.6. *Consider a coefficient* $a = a(x, t)$, *a solution* $\psi = \psi(x, t)$ *of* (1.6), *and a weight* $w = w(x, t)$, *all of them being piecewise constant. If the weight satisfies the constraint* (1.12), *then for all* $t \geq 0$ *we have*

$$\|\psi(t)\|_{w(t)} = \|\psi(0)\|_{w(0)} + \int_0^t \sum_{(x,\tau) \in \mathcal{J}(a)} (\gamma_- w_- + \gamma_+ w_+) \left| \lambda^a - a_- \right| |\psi_-|(x, \tau) \, d\tau, \tag{1.13}$$

where

$$\begin{aligned} \gamma_-(x, \tau) &:= \operatorname{sgn}\big(\lambda^a(x, \tau) - a_-(x, \tau)\big), \\ \gamma_+(x, \tau) &:= \operatorname{sgn}\big(a_+(x, \tau) - \lambda^a(x, \tau)\big), \end{aligned} \tag{1.14}$$

and

$$\operatorname{sgn}(\xi) := \left\{ \begin{array}{ll} -1, & \xi < 0, \\ 0, & \xi = 0, \\ 1, & \xi > 0. \end{array} \right.$$

Observe that the contribution in the right-hand side of (1.13) vanishes when the argument of the sign function vanishes. So, in practice, the specific value of $\operatorname{sgn}(\xi)$ at $\xi = 0$ does not matter. In the forthcoming statements (for instance in (1.16), below) we will restrict attention to points (x, t) where $\lambda^a(x, t) - a_\pm(x, t) \neq 0$.

PROOF. Denote by $t \mapsto x_j(t)$ $(1 \leq j \leq m)$ the polygonal lines of discontinuity for any of the functions a, ψ, or w. That is, include all of the points in $\mathcal{J}(a) \cup \mathcal{J}(\psi) \cup \mathcal{J}(w)$. Let us exclude the set $\mathcal{I}(a, \psi, w)$ of all of the interaction times when two lines of discontinuity for either ones of the functions a, ψ, or w intersect in the (x, t)-plane. To derive (1.13) it is sufficient to consider any time interval disjoint from $\mathcal{I}(a, \psi, w)$. In such a time interval the discontinuity lines are straight lines and the following calculation makes sense.

In each interval $\big(x_j(t), x_{j+1}(t)\big)$ all of the functions are constant and we can write, with completely similar notations as in the proof of Theorem 1.3,

$$\begin{aligned} &\int_{x_j(t)}^{x_{j+1}(t)} |\psi(x, t)| \, w(x, t) \, dx \\ &= \big(x_{j+1}(t) - x_j(t)\big) |\psi| \, w \\ &= \big(x_{j+1}(t) - a_{j+1}^-(t) \, t\big) |\psi_{j+1}^-(t)| \, w_{j+1}^-(t) + \big(a_j^+(t) \, t - x_j(t)\big) |\psi_j^+(t)| \, w_j^+(t), \end{aligned}$$

by observing for instance that

$$a_j^+(t) := a_+(x_j(t), t) = a_-(x_{j+1}(t), t) =: a_{j+1}^-(t).$$

After a summation over the index j and a re-ordering of the terms obtained by collecting two contributions at each discontinuity, we arrive at

$$\|\psi(t)\|_{w(t)} = \sum_{j=1}^{m} \big(x_j(t) - a_j^-(t)\,t\big)\,|\psi_j^-(t)|\,w_j^-(t) + \big(a_j^+(t)\,t - x_j(t)\big)\,|\psi_j^+(t)|\,w_j^+(t).$$

Differentiating the above identity with respect to t and noting that all terms but $x_j(t)$ are constant, we find

$$\frac{d}{dt}\|\psi(t)\|_{w(t)}$$
$$= \sum_{j=1}^{m} \big(\dot{x}_j(t) - a_j^-(t)\big)\,|\psi_j^-(t)|\,w_j^-(t) + \big(a_j^+(t) - \dot{x}_j(t)\big)\,|\psi_j^+(t)|\,w_j^+(t). \tag{1.15}$$

We now rely on the jump relation (1.10) and distinguish between the following possibilities:

- Case $(x_j(t), t) \in \mathcal{J}(a)$: Here, $x_j(t)$ is a jump point for the function a we have $\dot{x}_j(t) = \lambda^a(x_j(t), t)$.
- Case $(x_j(t), t) \in \mathcal{C}(a) \cap \mathcal{J}(\psi)$: The function a is locally constant near $x_j(t)$ but ψ has a jump, and, therefore from (1.10), $\dot{x}_j(t) = a(x_j(t), t) = a_j^\pm(t)$.
- Case $(x_j(t), t) \in \mathcal{C}(a) \cap \mathcal{C}(\psi) \cap \mathcal{J}(w)$: Both a and ψ are continuous but w is discontinuous and, thanks to the constraint (1.12), $\dot{x}_j(t) = a(x_j(t), t) = a_j^\pm(t)$.

Clearly, in the last two cases above, the contribution in the right-hand side of (1.15) vanishes. Therefore, only the points $(x, t) \in \mathcal{J}(a)$ are relevant, and using once more (1.10) we obtain

$$\frac{d}{dt}\|\psi(t)\|_{w(t)} = \sum_{(x,t)\in\mathcal{J}(a)} \Big(\mathrm{sgn}(\lambda^a(x,t) - a_-(x,t))\,w_-(x,t)$$
$$+ \mathrm{sgn}(a_+(x,t) - \lambda^a(x,t))\,w_+(x,t)\Big)\,|\lambda^a(x,t) - a_-(x,t)|\,|\psi_-(x,t)|,$$

which establishes (1.13). □

In view of Definition 1.1, the coefficient introduced in (1.14) depend whether the discontinuity is of type Lax, undercompressive, or rarefaction:

$$\gamma_\pm(x,t) = \begin{cases} -1, & (x,t) \in \mathcal{L}(a), \\ \pm1, & (x,t) \in \mathcal{S}(a), \\ \mp1, & (x,t) \in \mathcal{F}(a), \\ 1, & (x,t) \in \mathcal{R}(a), \end{cases} \tag{1.16}$$

(provided $\lambda^a(x,t) - a_\pm(x,t) \neq 0$). Consider one pair of terms

$$\big(\gamma_- \, w_- + \gamma_+ \, w_+\big)\,|\lambda^a - a_-|\,|\psi_-|$$

in the right-hand side of (1.13). By (1.16), for each Lax or rarefaction-shock discontinuity the two terms are both non-positive or both non-negative, respectively, while for each undercompressive discontinuity there is exactly one non-positive and one non-negative terms.

To take advantage of the property (1.16) within the identity (1.13) it is natural to determine the weight w in order that

$$w_+(x,t) - w_-(x,t) \begin{cases} \leq 0, & (x,t) \in \mathcal{S}(a), \\ \geq 0, & (x,t) \in \mathcal{F}(a), \end{cases} \qquad (1.17)$$

while no condition need be imposed at Lax or rarefaction-shock discontinuities. In conclusion, the following "weighted" version of Theorem 1.3 follows immediately from Lemma 1.6.

THEOREM 1.7. (Weighted L^1 stability for linear hyperbolic equations.) *Consider a coefficient $a = a(x,t)$ and a solution $\psi = \psi(x,t)$ of (1.6), both being piecewise constant. Suppose that there exists a weight-function $w > 0$ satisfying the conditions (1.12) and (1.17). Then, for all $t \geq 0$ we have the identity*

$$\|\psi(t)\|_{w(t)} + \int_0^t \Big(\mathbf{D}_2(s) + \mathbf{D}_3(s) \Big) \, ds = \|\psi(0)\|_{w(0)} + \int_0^t \mathbf{R}(s) \, ds, \qquad (1.18)$$

where

$$\begin{aligned}
\mathbf{D}_2(s) &:= \sum_{(x,s) \in \mathcal{L}(a)} \big(w_-(x,s) + w_+(x,s) \big) \big| a_-(x,s) - \lambda^a(x,s) \big| \, |\psi_-(x,s)|, \\
\mathbf{D}_3(s) &:= \sum_{(x,s) \in \mathcal{S}(a) \cup \mathcal{F}(a)} \big| w_+(x,s) - w_-(x,s) \big| \big| a_-(x,s) - \lambda^a(x,s) \big| \, |\psi_-(x,s)|,
\end{aligned} \qquad (1.19)$$

and

$$\mathbf{R}(s) := \sum_{(x,s) \in \mathcal{R}(a)} \big(w_-(x,s) + w_+(x,s) \big) \big| a_-(x,s) - \lambda^a(x,s) \big| \, |\psi_-(x,s)|.$$

$\square$

The **dissipation terms** D_2 and D_3 are *quadratic* and *cubic* in nature, respectively. To apply Theorem 1.7, we need construct a weight w fulfilling the constraints (1.12) and (1.17). We will rely on certain geometrical conditions of the coefficient a related to the distribution of slow and fast undercompressive shocks. On the other hand, in the applications the **remainder R** either will vanish identically or will tend to zero together with some approximation parameter.

2. L^1 continuous dependence estimate

The framework in Section 1 is now applied to approximate classical entropy solutions of the conservation law (1.1), which, by construction, satisfy Oleinik entropy inequalities up to some approximation error. In the rest of this chapter, the coefficient a and the solution ψ have the form (1.3). We begin in Theorem 2.2 by deriving the L^1 contraction estimate for conservation laws with general flux, relying here on Theorem 1.3. Next in Theorem 2.3 we derive a weighted L^1 estimate when the flux-function is convex.

The theory in Section 1 unveiled the fact that rarefaction shocks are the source of non-uniqueness in linear hyperbolic equations. So, we start by pointing out the following key property.

THEOREM 2.1. (Fundamental property of the averaging coefficient.) *If u and v are two piecewise constant, classical entropy solutions of the conservation law (1.1) defined in some region of the (x,t) plane, then the averaging speed*

$$a = a(x,t) = \bar{a}(u(x,t), v(x,t))$$

contains no rarefaction shocks.

PROOF. The discontinuity lines in $\bar{a}(u,v)$ are determined by superimposing the discontinuity lines in u and v. At a jump point $(x,t) \in \mathcal{J}(\bar{a}(u,v))$ we have

(i) either $(x,t) \in \mathcal{J}(u)$ while v is locally constant,

(ii) either $(x,t) \in \mathcal{J}(v)$ while u is locally constant,

(iii) or else $(x,t) \in \mathcal{J}(u) \cap \mathcal{J}(v)$ and $\lambda^u(x,t) = \lambda^v(x,t)$. (If the latter would not hold, (x,t) would be an interaction point of $\bar{a}(u,v)$.)

Case (iii) above is not "generic" and can be avoided by an arbitrary small perturbation of one of the solution. Alternatively we can decompose the discontinuity in Case (iii) into two discontinuities, one in Case (i), and one in Case (ii). On the other hand, Case (ii) being completely similar to Case (i). So we only consider Case (i), when u has a discontinuity connecting some states u^- and u^+ satisfying Oleinik entropy inequalities. Recall that the graph of the flux f on the interval limited by u^- and u^+ remains below (above, respectively) the chord connecting the two points with coordinates u^- and u^+, when $u^+ < u^-$ ($u^- < u^+$, respectively).

Consider for definiteness the case $u^+ < u^-$. By extending to the real line the chord connecting u^- to u^+ on the graph of f, we determine finitely many points of intersection

$$u_p^+ < u_{p-1}^+ < \ldots < u_1^+ < u_0^+ = u^+ < u^- = u_0^- < u_1^- < \ldots < u_{q-1}^- < u_q^-,$$

such that f is above the line connecting u_{2j}^- to u_{2j+1}^- and the one connecting u_{2j+1}^+ to u_{2j}^+, while it below the line connecting u_{2j+1}^- to u_{2j+2}^- and the one connecting u_{2j+2}^+ to u_{2j+1}^+ for $j = 0, 1, \ldots$

Observe that the nature of a discontinuity in $\bar{a}$ depends on the location of the constant value (say v) taken by the function v. Precisely, in view of Definition 1.1 it is clear geometrically that the discontinuity is

- a Lax shock when $v \in (u^+, u^-)$,
- slow undercompressive when $v \in \left(u_{2j}^-, u_{2j+1}^-\right)$ and when $v \in \left(u_{2j+2}^|, u_{2j+1}^|\right)$,
- or fast undercompressive when $v \in \left(u_{2j+1}^-, u_{2j+2}^-\right)$ and when $v \in \left(u_{2j+1}^+, u_{2j}^+\right)$.

In particular, there are no rarefaction-shocks. This completes the proof of Theorem 2.1. $\square$

Combining Theorems 2.1 and 1.3 we arrive at the following important property for scalar conservation laws.

THEOREM 2.2. (L^1 *contraction property for general flux.*) *Let $f : \mathbb{R} \to \mathbb{R}$ be a smooth flux-function. Then, the classical entropy solutions of (1.1) satisfy the L^1* **contraction property**

$$\|u(t) - v(t)\|_{L^1(\mathbb{R})} \leq \|u(0) - v(0)\|_{L^1(\mathbb{R})}, \quad t \geq 0. \tag{2.1}$$

For definiteness, we restrict attention in the present chapter to solutions of (1.1) arising as limits of the (piecewise constant) approximate solutions described earlier (in Chapter IV). A general *uniqueness theory* will be developed later on (in Chapter X).

PROOF. Starting from some initial values $u(0)$ and $v(0)$, we construct piecewise approximate solutions $u^h = u^h(x,t)$ and $v^h = v^h(x,t)$ by wave front tracking, as described in Section IV-2. Recall that $u^h(0) \to u(0)$ and $u^h \to u$ almost everywhere with

$$\|u^h(t)\|_{L^\infty(\mathbb{R})} \leq \|u(0)\|_{L^\infty(\mathbb{R})}, \quad t \geq 0, \tag{2.2}$$

and

$$TV(u^h(t)) \leq TV(u(0)), \quad t \geq 0. \tag{2.3}$$

Similar statements hold for $v^h \to v$. The functions u^h and v^h are actually exact solutions of (1.1) but do not fulfill exactly Oleinik entropy inequalities. The rarefaction fronts in the solution u^h have small strength:

$$|u_+^h(x,t) - u_-^h(x,t)| \leq h. \tag{2.4}$$

Furthermore, by slightly modifying the initial data $u^h(0)$ or $v^h(0)$ if necessary, we can always guarantee that the discontinuity lines of u^h and v^h cross at finitely many points (x,t) only. Furthermore, we can assume that within a compact region in space we always have $u^h(x,t) \neq v^h(x,t)$ (so that ψ^h does not vanish) while $u^h(x,t) = v^h(x,t) = 0$ outside this region. As the analysis is completely trivial in the latter we only treat the former region. These conditions will be tacitly *assumed in the rest of this chapter*.

Following the general strategy sketched in the beginning of Section 1, we set $\psi^h := v^h - u^h$ and $\overline{a}^h := \overline{a}(u^h, v^h)$. Since both u^h and v^h are solutions of (1.1) we have

$$\partial_t \psi^h + \partial_x(\overline{a}^h \psi^h) = 0. \tag{2.5}$$

In view of Theorem 1.3 and Remark 1.4 and using the total variation bound (2.3) we get

$$\|\psi^h(t)\|_{L^1(\mathbb{R})} \leq \|\psi^h(0)\|_{L^1(\mathbb{R})}$$
$$+ 2t\left(TV(u(0)) + TV(v(0))\right) \sup_{\substack{(x,\tau)\in\mathcal{R}(\overline{a}^h)\\\tau\in(0,t)}} \left(a_+^h(x,\tau) - a_-^h(x,\tau)\right),$$

thus

$$\|v^h(t) - u^h(t)\|_{L^1(\mathbb{R})} \leq \|v^h(0) - u^h(0)\|_{L^1(\mathbb{R})} + C_1\, t \sup_{\substack{(x,\tau)\in\mathcal{R}(\overline{a}^h)\\\tau\in(0,t)}} \left(a_+^h(x,\tau) - a_-^h(x,\tau)\right).$$

Since $u^h(0) \to u(0)$ and $v^h(0) \to v(0)$, by lower semi-continuity of the L^1 norm it follows that

$$\|v(t) - u(t)\|_{L^1(\mathbb{R})} \leq \liminf_{h\to 0} \|v^h(t) - u^h(t)\|_{L^1(\mathbb{R})}$$
$$\leq \|v(0) - u(0)\|_{L^1(\mathbb{R})} + C_1\, t \liminf_{h\to 0} \sup_{\substack{(x,\tau)\in\mathcal{R}(\overline{a}^h)\\\tau\in(0,t)}} \left(a_+^h(x,\tau) - a_-^h(x,\tau)\right).$$

To conclude we show that

$$\sup_{\substack{(x,\tau)\in\mathcal{R}(\overline{a}^h)\\\tau\in(0,t)}} \left(a_+^h(x,\tau) - a_-^h(x,\tau)\right) \leq C_2\, h. \tag{2.6}$$

Suppose that $v^h(t)$ is continuous at some point x, i.e., is identically equal to some constant v, while u^h has a shock connecting u^- to u^+, say. If the front in u^h is an entropy-satisfying shock, then by Theorem 2.1 the discontinuity in $\overline{a}^h$ cannot be a rarefaction shock. If the front in u^h is a rarefaction-shock, then we estimate the jump of $\bar{a}^h$ as follows

$$\bar{a}_+^h(x,\tau) - \bar{a}_-^h(x,\tau) = \int_0^1 f'\big(u_+ + \theta\,(v - u_+)\big)\,d\theta - \int_0^1 f'\big(u_- + \theta\,(v - u_-)\big)\,d\theta$$
$$\leq \|f''\|_{L^\infty}\,|u_+ - u_-| \leq C_2\,h,$$

where we used the uniform bound (2.4). This completes the proof of Theorem 2.2.

$\square$

We now rely on Theorem 1.7 and derive a *weighted L^1 estimate* when the flux-function is convex. This new estimate provides us with additional information beyond the L^1 contraction property (2.1), specifically a lower bound on the decrease of the L^1 distance between two solutions of (1.1). Later on in Chapter IX in the context of systems, it will be essential to work with a weighted norm in order to cope with "wave interaction" error terms and derive a generalization of (2.1) for systems.

THEOREM 2.3. (Weighted L^1 estimates.) *Consider the scalar conservation law* (1.1) *where the flux f is a convex function. Given sequences of front tracking approximations u^h and v^h and a constant $K > 0$ satisfying*

$$2\,K\left(TV(u^h(0)) + TV(v^h(0))\right) < 1, \tag{2.7}$$

there exists a (piecewise constant) weight $w^h = w^h(x,t)$ satisfying for all (x,t) the uniform bounds $w_{\min}^h \leq w^h(x,t) \leq w_{\max}^h$ with

$$w_{\min}^h := 1 - 2K\,TV(u^h(0)) - 2K\,TV(v^h(0)),$$
$$w_{\max}^h := 1 + 2K\,TV(u^h(0)) + 2K\,TV(v^h(0)),$$

such that the **weighted L^1 continuous dependence estimate**

$$\int_{I\!R} |v^h(t) - u^h(t)|\,w^h(t)\,dx + \int_0^t \left(\mathbf{D}_2^h(s) + \mathbf{D}_3^h(s)\right) ds$$
$$= \int_{I\!R} |v^h(0) - u^h(0)|\,w^h(0)\,dx + \int_0^t \mathbf{R}^h(s)\,ds \tag{2.8}$$

holds for all $t \geq 0$, where $\bar{a}^h := \bar{a}(u^h, v^h)$ and

$$\mathbf{D}_2^h(s) := \sum_{(x,s)\in\mathcal{L}(\bar{a}^h)} \left(w_-^h + w_+^h\right)\big|a_-^h - \lambda^{\bar{a}^h}\big|\,|v_-^h - u_-^h|,$$
$$\mathbf{D}_3^h(s) := \sum_{(x,s)\in\mathcal{S}(\bar{a}^h)\cup\mathcal{F}(\bar{a}^h)} K\left(|u_+^h - u_-^h| + |v_+^h - v_-^h|\right)\big|a_-^h - \lambda^{\bar{a}^h}\big|\,|v_-^h - u_-^h|, \tag{2.9a}$$

and

$$\mathbf{R}^h(s) := \sum_{(x,s)\in\mathcal{R}(\bar{a}^h)} \left(w_-^h + w_+^h\right)\big|a_-^h - \lambda^{\bar{a}^h}\big|\,|v_-^h - u_-^h|. \tag{2.9b}$$

Interestingly, by letting $K \to 0$ in (2.8) and neglecting the (non-negative) term D_2^h we recover the L^1 *contraction estimate* (2.1) derived in Theorem 2.2. Of course, when $K > 0$ the statement (2.8) is *much stronger* than (2.1). The passage to the limit $h \to 0$ in (2.8) and (2.9) is discussed in Section 3 below. Of course, as was already observed (Remark 1.4 and the proof of Theorem 2.2), the last term in (2.8) vanishes when $h \to 0$.

It will be convenient to denote by $\mathcal{S}(\bar{a}^h; u^h)$ and $\mathcal{F}(\bar{a}^h; u^h)$ the slow and the fast undercompressive discontinuities in $\bar{a}^h$ associated with jumps of u^h. A similar notation is used for v^h. Recall that the difference between two solutions, $\psi^h := v^h - u^h$, keeps a constant sign at each undercompressive discontinuity. (See for instance Remark 1.5.)

Using that f is convex it is not difficult to check that:

LEMMA 2.4. *The nature of the fronts is determined in each region where $\psi^h = v^h - u^h$ keeps a constant sign, as follows:*

$$
\psi^h_{\pm}(x,t) \begin{cases} > 0 & \text{if and only if } (x,t) \in \mathcal{S}(\bar{a}^h; u^h) \cup \mathcal{F}(\bar{a}^h; v^h), \\[2mm] < 0, & \text{if and only if } (x,t) \in \mathcal{F}(\bar{a}^h; u^h) \cup \mathcal{S}(\bar{a}^h; v^h). \end{cases} \tag{2.10}
$$

PROOF OF THEOREM 2.3. **Step 1 : Preliminaries.** To simplify the notation we omit the upper index h. In view of (2.10) we can rewrite the required conditions (1.17) on the weight, in the following strengthened form:

$$
w_+(x,t) - w_-(x,t) = \begin{cases} -K\,|u_+(x,t) - u_-(x,t)|, & (x,t) \in \mathcal{J}(u), \quad \psi_{\pm}(x,t) > 0, \\[1mm] K\,|u_+(x,t) - u_-(x,t)|, & (x,t) \in \mathcal{J}(u), \quad \psi_{\pm}(x,t) < 0, \\[1mm] K\,|v_+(x,t) - v_-(x,t)|, & (x,t) \in \mathcal{J}(v), \quad \psi_{\pm}(x,t) > 0, \\[1mm] -K\,|v_+(x,t) - v_-(x,t)|, & (x,t) \in \mathcal{J}(v), \quad \psi_{\pm}(x,t) < 0. \end{cases} \tag{2.11}
$$

When such a weight exists, Theorem 1.7 provides us immediately with the desired identity (2.8).

Let us also point out the following property associated with convex-flux functions. Suppose that two fronts in the solution u interact at some point (x,t), and generate a (single) front. Call u_l, u_m, u_r the constant states achieved before the interaction, so that u_l and u_r are the left- and right-hand states after the interaction. The strength of the outgoing front in u is less than or equal to the total sum of the strengths before the interaction, and the corresponding decrease is measured by the **amount of cancellation at the point** (x,t):

$$
\theta^u(x,t) := |u_l - u_m| + |u_r - u_m| - |u_r - u_l| \geq 0. \tag{2.12}
$$

Suppose that $\theta^u(x,t) > 0$ and, for definiteness, that $u_m < u_r < u_l$. Then, we will say that the solution u contains, *from the interaction time on*, a **past (or cancelled) wave front** with strength $2\,|u_r - u_m|$. For clarity, we will sometime refer to the fronts in the solution u as **present wave fronts**. The same notation and terminology apply also to the solution v.

Step 2 : Main argument of proof. By setting (away from interaction times and discontinuities)

$$
w(x,t) = 1 + \sum_{y < x} [w](y,t),
$$

the (piecewise constant) weight w is uniquely determined if we prescribe its derivative w_x in the form of finitely many Dirac masses propagating along polygonal lines and referred to as **particles** or **anti-particles** (see below for the precise definition) which may propagate along characteristic lines or discontinuity lines of a in agreement with the earlier requirement (1.12). By convention the (signed) mass of a particle may be positive or negative. Particles and anti-particles will be generated from the same location and the sum of their masses be zero. The jump $[w](y,t)$ is the **total mass** of all of the particles and anti-particles located at (y,t).

Our construction will associate a *train of particles and anti-particles* with every front in the solutions u and v as well as with every cancelled front. To proceed we will decompose the (x,t)-plane in finitely many regions in which the function ψ keeps a constant sign. We will ensure that the following property holds:

(P): With every present wave front with strength ε and within every region Ω in which ψ keeps a constant sign, we can associate a set of particles and a set of anti-particles located within the region Ω such that the total mass of particles is equal to $\pm\varepsilon$ while the total mass of anti-particles is equal to $\mp\varepsilon$ (the sign depending on the region).

An analogous property will hold for past waves. Based on (P) we conclude the proof as follows.

Denoting by $E_1(t)$ the set of locations of particles and anti-particles associated with present fronts, we find $(x_1 < x_2)$

$$-K\,TV(u(t)) - K\,TV(v(t)) \le \sum_{\substack{x_1<y<x_2 \\ y\in E_1(t)}} [w](y,t) \le K\,TV(u(t)) + K\,TV(v(t)). \quad (2.13)$$

On the other hand, observe that the total amount of cancellation can be bounded by the initial total variation

$$\sum_{\text{interactions } (x,t)} \theta^u(x,t) + \theta^v(x,t)$$

$$= \sum_{\text{interactions } t} TV(u(t-)) - TV(u(t+) + \sum_{\text{interactions } t} TV(v(t-)) - TV(v(t+))$$

$$\le TV(u(0)) + TV(v(0)).$$

So, calling $E_2(t)$ the set of locations of particles and anti-particles associated with past fronts, we find now

$$-K\,TV(u(0)) - K\,TV(v(0)) \le \sum_{\substack{x_1<y<x_2 \\ y\in E_2(t)}} [w](y,t) \le K\,TV(u(0)) + K\,TV(v(0)). \quad (2.14)$$

Finally, collecting the terms in (2.13) and (2.14) we arrive at

$$-2K\,TV(u(0)) - 2K\,TV(v(0)) \le \sum_{x_1<y<x_2} [w](y,t) \le 2K\,TV(u(0)) + 2K\,TV(v(0)),$$

which leads us to the conclusion of the theorem.

Step 3 : Constructing the weight. It remains to explain how particles and anti-particles are generated and propagate in time. Recall first that the interaction of two incoming fronts in u (v, respectively) generates a single outgoing front in u (v, respectively), while the crossing of two fronts in u and v can be regarded as an interaction point with two outgoing fronts. The former are referred to as u/u **interactions** and v/v **interactions**, respectively. The latter are referred to as u/v **interactions** if the front in u travels faster than the front in v, and as v/u **interactions** otherwise. Let us decompose the (x,t)-plane in finitely many regions in which the function ψ

keeps a constant sign. According to (2.10), the waves within these regions can only be slow or fast undercompressive, while the boundaries are made of Lax shocks or rarefaction-shocks.

In each region where ψ keeps a constant sign, say a region denoted by Ω_+ where $\psi \geq 0$, the weight is defined locally near the initial line $t = 0$, as follows. With each front leaving from a point $(x, 0)$ we associate a *particle* with strength $\pm\varepsilon$ determined by (2.11) and propagating along with the wave front, that is with the speed λ^a. We also introduce a corresponding *anti-particle* with opposite strength $\mp\varepsilon$ propagating with the local characteristic speed associated with the speed a. Then, (2.11) is satisfied for small times.

Next, the dynamics of anti-particles within Ω_+ is straightforward: Any jump in the speed a is slow or fast undercompressive and, therefore, all characteristic lines cross these jumps *transversally*. A characteristic can never exit Ω_+ since its boundary is made of Lax and rarefaction-shock fronts, only. Hence, the anti-particles simply propagate within Ω_+, passing through fronts of a until they possibly reach the boundary Ω_+ and stick with it. At this juncture observe that, within Ω_+, all of the waves in the function a associated with jumps in u are slow undercompressive, while the waves associated with v are fast undercompressive.

For small times we have associated with each wave one particle and one anti-particle. More generally, for arbitrary times and to each wave front it is necessary to associate a train made of several pairs of particles and anti-particles, in agreement with the property (P) above. Furthermore, waves may be cancelled and to each cancelled front we also associate a train of particles and anti-particles satisfying (P). So, to complete the definition of the weight w we now distinguish between several interaction cases, depending whether the waves under consideration interact within, enter, or leave the region Ω_+:

- u/v or v/u *Interactions within* Ω_+ : Suppose, for instance, that a line of discontinuity in the solution u crosses a line of discontinuity in the solution v, at some point (x, t). Since the two fronts cross each other without changing their respective strengths we impose that the particles associated with the incoming fronts propagate forward without change.

- u/u or v/v *Interactions within* Ω_+ : Suppose, for instance, that two fronts in the solution u interact at some point (x, t) and generate a (single) front in the solution u. Call u_l, u_m, and u_r the constant states achieved before the interaction. Here, we replace the two incoming particles (propagating together with the incoming fronts) with masses $-K|u_m - u_l|$ and $-K|u_r - u_m|$ respectively with a single particle (propagating together with the outgoing front) with mass $-K|u_r - u_l|$. The discrepancy in taken care of by introducing a *new particle* associated with the *cancelled* front (propagating at the local characteristic speed and) carrying the mass $-K\theta^u(x, t)$. When $\theta^u(x, t) = 0$, all particles and anti-particles associated with the incoming fronts are naturally associated with the outgoing front. When $\theta^u(x, t) \neq 0$, the incoming front which is completely cancelled is refer to a *past front* and remains associated with its anti-particles, while the mass of the anti-particles associated with the surviving front is decomposed in such a way that the property (P) holds.

- *Leaving/entering fronts* : Consider first a front in the solution u leaving Ω_+ on the right-hand side. Calling u_l, u_r and v_l, v_r the corresponding left- and right-hand states, we have here $v_r < \min(u_l, u_r) < \max(u_l, u_r) < v_l$. It is an interaction of the type $(S_u L_v)$–$(L'_v F'_u)$, that is, a slow undercompressive front in u meeting with a Lax front in v and generating a Lax front in v and a fast compressive front in u. We impose here that the particle associated with the front S_u sticks with the boundary L'_v of Ω_+. (In fact, here it cancels together with its corresponding anti-particle.) On the other hand, with the outgoing wave F'_u (within a new region where now $\psi < 0$) we associate a new pair of particle and anti-particle, in agreement with the constraints (2.11): The particle with mass $K\,|u_r - u_l|$ propagates along with the front F'_u while the anti-particle sticks with the boundary L'_v.

Consider next a front in the solution u leaving Ω_+ on the left-hand side. We now have $v_l < \min(u_l, u_r) < \max(u_l, u_r) < v_r$ and an interaction of the type $(R_v S_u)$–$(F'_u R'_v)$. We impose that the particle associated with S_u sticks with the boundary R'_v of the region Ω_+. On the other hand, with the wave F'_u and in a region where now $\psi < 0$, we associate a particle along F'_u and an anti-particle along R'_v, in agreement with the constraints (2.11).

- u/v or v/u *Interactions closing the region* Ω_+ : At an interaction involving two boundaries of the region Ω_+ it turns out that all particles and anti-particles within that region cancel each other.

This completes the construction of the weight and, therefore, the proof of Theorem 2.3. □

REMARK 2.5. It is also possible to initialize the weight near $t = 0$ without introducing anti-particles. Then, all anti-particles (which, anyway, must be introduced when waves pass from a region $\psi > 0$ to a region $\psi < 0$ and vice-versa) remain "stuck" along the boundaries $\psi_- \, \psi_+ < 0$. Only particles generated by cancellation propagate along characteristics in the interior of the regions. □

3. Sharp version of the continuous dependence estimate

Our next purpose is to pass to the limit $(h \to 0)$ in the statement established in Theorem 2.3 for piecewise constant approximate solutions. The proofs are omitted here as they will be given in Section IX-3 for *systems* of equations.

For each function with bounded variation $u = u(x)$ we denote by $\mathcal{C}(u)$ and $\mathcal{J}(u)$ its points of continuity and jump discontinuity, respectively, and by $u_\pm(x)$ its left- and right-hand limits at $x \in \mathcal{J}(u)$. Let V^u denote the total variation function $x \mapsto TV_{-\infty}^x(u(t))$ associated with $u(t)$.

To any three functions of bounded variation u, v, w in x we associate the **non-conservative product**

$$\left(\overline{a}(u, v) - f'(u)\right)(v - u)\, dw := \mu$$

which is defined as the unique measure on $\mathbb{R}$ characterized by the following two conditions:

- If B is a Borel set included in the set $\mathcal{C}(w)$ of continuity points of w

$$\mu(B) := \int_B \left(\overline{a}(u, v) - f'(u)\right)(v - u)\, dw, \qquad (3.1a)$$

where the integral is defined in a classical sense;

- If $x \in \mathcal{J}(w)$ is a jump point of w, then

$$
\mu(\{x\}) := \frac{1}{2} \Big(\big(\overline{a}(u_+, v_+) - \overline{a}(u_-, u_+)\big) (v_+ - u_+) \\
+ \big(\overline{a}(u_-, v_-) - \overline{a}(u_-, u_+)\big) (v_- - u_-) \Big) |w_+ - w_-|
\tag{3.1b}
$$

with $u_\pm = u(x\pm)$, etc.

The following theorem provides a sharp estimate for the decrease of the weighted distance between two solutions at time t:

$$
\|v(t) - u(t)\|_{w(t)} := \int_{I\!R} |v(x,t) - u(x,t)| \, w(x,t) \, dx.
$$

THEOREM 3.1. *Let f be a strictly convex function and u and v be two entropy solutions of bounded variation of the conservation law (1.1). Then, for each K satisfying*

$$
3K \Big(TV(u) + TV(v) \Big) < 1,
$$

there exists a function $w = w(x,t)$ which is bounded and remains bounded away from zero such that the **sharp L^1 continuous dependence estimate**

$$
\|v(t) - u(t)\|_{w(t)} + \int_0^t \Big(\mathbf{D}_2(u(s), v(s)) + \mathbf{D}_2(v(s), u(s)) \\
+ \mathbf{D}_3(u(s), v(s)) + \mathbf{D}_3(v(s), u(s)) \Big) ds
\tag{3.2}
$$
$$
\leq \|v(0) - u(0)\|_{w(0)}
$$

holds for all $t \geq 0$, where the dissipation terms are defined for each given time by

$$
\mathbf{D}_2(u,v) := 2T(u,v) \sum_{x \in \mathcal{L}(u,v)} \big|\overline{a}\big(u_-(x), v_-(x)\big) - \overline{a}\big(u_-(x), u_+(x)\big)\big| \, |v_-(x) - u_-(x)|,
$$
$$
\mathbf{D}_3(u,v) := K \int_{I\!R} \big(\overline{a}(u,v) - f'(u)\big) (v - u) \, dV^u.
\tag{3.3}
$$

with

$$
T(u,v) := 1 - 2K \Big(TV(u) + TV(v) \Big),
\tag{3.4}
$$

and the set of Lax *discontinuities $\mathcal{L}(u,v) \subset I\!R$ in the function u is the set of points x satisfying $\big(v_-(x) - u_-(x)\big)\big(v_-(x) - u_+(x)\big) \leq 0$.* □

The terms $\mathbf{D}_2$ and $\mathbf{D}_3$ are quadratic and cubic in nature, respectively. Observe that the integral term in the left-hand side of (3.2) is positive and, therefore, do contribute to the decrease of weighted distance between the two solutions. To see this, let us decompose the cubic dissipation term in continuous and atomic parts:

$$
\mathbf{D}_3(u,v) = K \int_{I\!R} \big(\overline{a}(u,v) - f'(u)\big) (v - u) \, dV_c^u \\
+ K \sum_{x \in \mathcal{J}(u)} \big(\overline{a}(u_-, v_-) - \overline{a}(u_+, u_-)\big) (v_- - u_-) |u_+ - u_-|,
\tag{3.5}
$$

where V_c^u denotes the continuous part of the total variation measure V^u. The set of jump points in u can be decomposed as follows:

$$
\mathcal{J}(u) = \mathcal{L}(u,v) \cup \mathcal{S}(u,v) \cup \mathcal{F}(u,v),
\tag{3.6}
$$

where $\mathcal{L}(u,v)$, $\mathcal{S}(u,v)$, $\mathcal{F}(u,v)$ are the subsets of Lax, slow undercompressive, and fast undercompressive discontinuities in the solution u, respectively, with

$$
\begin{aligned}
\big(v_-(x) - u_-(x)\big)\big(v_-(x) - u_+(x)\big) &\leq 0 &&\text{when } x \in \mathcal{L}(u,v), \\
\big(v_-(x) - u_-(x)\big) \geq 0 \text{ and } \big(v_-(x) - u_+(x)\big) &\geq 0 &&\text{when } x \in \mathcal{S}(u,v), \\
\big(v_-(x) - u_-(x)\big) \leq 0 \text{ and } \big(v_-(x) - u_+(x)\big) &\leq 0 &&\text{when } x \in \mathcal{F}(u,v).
\end{aligned} \tag{3.7}
$$

(Note that exact entropy solutions cannot have rarefaction-shock discontinuities.) Then the sharp L^1 estimate (3.2) implies

$$
\begin{aligned}
\|v(t) - u(t)\|_{w(t)} + \int_0^t \Big(&\widehat{\mathbf{D}}_2(u(s),v(s)) + \widehat{\mathbf{D}}_2(v(s),u(s)) \\
&+ \widehat{\mathbf{D}}_3(u(s),v(s)) + \widehat{\mathbf{D}}_3(v(s),u(s)) \Big)\, ds \\
\leq \|v(0) - u(0)\|_{w(0)},&
\end{aligned} \tag{3.8}
$$

where

$$
\widehat{\mathbf{D}}_2(u,v) := \widehat{T}(u,v) \sum_{x \in \mathcal{L}(u,v)} \big| \bar{a}(u_-,v_-) - \bar{a}(u_-,u_+) \big| \, |v_- - u_-|,
$$

$$
\begin{aligned}
\widehat{\mathbf{D}}_3(u,v) := K \sum_{x \in \mathcal{S}(u,v) \cup \mathcal{F}(u,v)} &\big| \bar{a}(u_-,v_-) - \bar{a}(u_-,u_+) \big| \, |v_- - u_-| \, |u_+ - u_-| \\
&+ K \int_{I\!R} \big| \bar{a}(u,v) - f'(u) \big| \, |v - u| \, dV_c^u.
\end{aligned}
$$

with

$$
\widehat{T}(u,v) := 1 - 3K \Big(TV(u) + TV(v) \Big).
$$

The following is a simplified version of Theorem 3.1 in which the weight equals $w = 1$ with $K = 0$ and $\widehat{T}(u,v) = 1$.

THEOREM 3.2. *Under the assumptions in Theorem 3.1, for all $t \geq 0$ we have*

$$
\begin{aligned}
\|v(t) - u(t)\|_{L^1(I\!R)} + \int_0^t \Big(&\widetilde{\mathbf{D}}_2(u(s),v(s)) + \widetilde{\mathbf{D}}_2(v(s),u(s)) \Big)\, ds \\
&\leq \|u(0) - v(0)\|_{L^1(I\!R)},
\end{aligned} \tag{3.9}
$$

where

$$
\widetilde{\mathbf{D}}_2(u,v) := \sum_{x \in \mathcal{L}(u,v)} 2 \big(\bar{a}(u_-(x),v_-(x)) - \bar{a}(u_-(x),u_+(x)) \big) \, |v_-(x) - u_-(x)|.
$$

$\square$

4. Generalizations

By taking $v = 0$ in the continuous dependence results derived in Sections 2 and 3, for instance in Theorem 2.2, we find immediately that the norm $\|u(t)\|_{L^1(I\!R)}$ is non-increasing in time for every classical entropy solution u. In fact, this result holds also for *nonclassical* entropy solutions, as shown now.

THEOREM 4.1. (L^1 stability of nonclassical solutions.) *Let $f : \mathbb{R} \to \mathbb{R}$ be a concave-convex flux-function and $\varphi^\flat$ be a kinetic function satisfying the conditions (IV-3.2). Then, any nonclassical entropy solution $u = u(x,t)$ (generated by wave front tracking and) based on the corresponding kinetic relation satisfies the L^1* **stability property**

$$\|u(t)\|_{L^1(\mathbb{R})} \leq \|u(0)\|_{L^1(\mathbb{R})}, \quad t \geq 0. \tag{4.1}$$

Theorem 4.1 applies, for instance, to the cubic flux-function and kinetic functions determined by diffusive-dispersive limits (Section III-2).

PROOF. Given a sequence of piecewise constant approximations u^h converging to the nonclassical solution u (see Section IV-3), we can follow the approach developed in Section 1 (Theorem 1.3) with now the function

$$\psi^h := u^h. \tag{4.2}$$

Thanks to (IV-3.2e), all of the discontinuities of the averaging coefficient

$$\overline{a}^h := \overline{a}(u^h, 0) = \frac{f(u^h) - f(0)}{u^h}$$

are *Lax or undercompressive discontinuities* except for those associated with rarefaction fronts in u^h, that is, an analogue of Theorem 2.1 holds true ! Therefore, the last term in the right-hand side of (1.7) tends to zero with h. Neglecting the last term in the left-hand side we obtain

$$\|u^h(t)\|_{L^1(\mathbb{R})} \leq \|u^h(0)\|_{L^1(\mathbb{R})} + o(h), \tag{4.3}$$

which yields (4.1).   □

In fact, nonclassical entropy solutions should satisfy an analogue of the L^1 continuous dependence results obtained in Sections 2 and 3. We conjecture the following stability result, which supplements the existence result in Theorem IV-3.2.

THEOREM 4.2. (L^1 continuous dependence of nonclassical solutions.) *Under the notations and assumptions in Theorem IV-3.2, any two nonclassical entropy solutions $u = u(x,t)$ and $v = v(x,t)$ generated by wave front tracking and based on a given kinetic relation satisfy the L^1* **continuous dependence property**

$$\|u(t) - v(t)\|_{L^1(\mathbb{R})} \leq C_* \|u(0) - v(0)\|_{L^1(\mathbb{R})}, \quad t \geq 0. \tag{4.4}$$

where the constant $C_* > 0$ depends on the kinetic function and the L^∞ norm of the solutions under consideration.

PART 2

SYSTEMS OF CONSERVATION LAWS

CHAPTER VI

THE RIEMANN PROBLEM

In this first chapter on systems we explicitly construct the *classical* and the *nonclassical* entropy solutions to the Riemann problem associated with a strictly hyperbolic system of conservation laws. The initial data consist of single jump discontinuities of sufficiently small strength. As was already observed with scalar conservation laws (Chapter II), solutions can be obtained by combining shock waves and rarefaction waves together. Motivated by the applications (Sections I-3 and I-4) we are *primarily* interested in systems endowed with a strictly convex entropy pair and in solutions satisfying a single entropy inequality.

In Section 1 we discuss general properties of shock and rarefaction waves. In the (comparatively easier) case of systems whose all characteristic fields are *genuinely nonlinear* or *linearly degenerate,* a single entropy inequality is sufficient to select a unique solution to the Riemann problem; see Theorem 1.6. Then, in the following sections we focus on characteristic fields that are not globally genuinely nonlinear. In Section 2 we prove that Lax shock inequalities select a unique (classical) entropy solution when the characteristic fields of the system are *concave-convex* or *convex-concave* (or genuinely nonlinear or linearly degenerate); see (2.2) and Theorem 2.1. On the other hand, in Sections 3 and 4 we show that a single entropy inequality leads to *undercompressive* shock waves and to a *multi-parameter* family of solutions (one for each concave-convex or convex-concave characteristic field) and imposing a *kinetic relation* we arrive at a unique nonclassical entropy solution to the Riemann problem; see Theorems 3.4, 3.5, and 4.3.

1. Shock and rarefaction waves

We consider the *Riemann problem* for a system of conservation laws

$$\partial_t u + \partial_x f(u) = 0, \quad u = u(x,t) \in \mathcal{U}, \tag{1.1}$$

$$u(x,0) = \begin{cases} u_l, & x < 0, \\ u_r, & x > 0, \end{cases} \tag{1.2}$$

where $\mathcal{U} \subset I\!\!R^N$ is an open set, the function $f : \mathcal{U} \to I\!\!R^N$ is a smooth mapping, and u_l, u_r are constant states in $\mathcal{U}$. Since this problem is invariant under the transformation $(x,t) \mapsto (\theta x, \theta t)$ (for any $\theta > 0$), it is natural to search for a **self-similar solution** $u = u(\xi)$, with $\xi = x/t$. In view of (1.1) and (1.2) the solution must satisfy the ordinary differential equation

$$-\xi u' + f(u)' = 0, \quad u = u(\xi), \tag{1.3}$$

and the boundary conditions

$$u(-\infty) = u_l, \quad u(+\infty) = u_r. \tag{1.4}$$

Throughout, we restrict attention to values u in a neighborhood of a constant state normalized to be $0 \in I\!R^N$ and, so, we set $\mathcal{U} := \mathcal{B}(\delta_0)$, the ball with center 0 and radius $\delta_0 > 0$. Moreover, the flux-function f is assumed to be strictly hyperbolic, that is, the Jacobian matrix $A(u) := Df(u)$ has N real and distinct eigenvalues,

$$\lambda_1(u) < \ldots < \lambda_N(u),$$

and basis of left- and right-eigenvectors $l_i(u)$ and $r_i(u)$, $1 \leq i \leq N$, respectively. Recall that $l_i(u)\, r_j(u) = 0$ if $i \neq j$ and, after normalization, we can always assume that $l_i(u)\, r_i(u) \equiv 1$. Finally, we assume that δ_0 is sufficiently small so that the wave speeds are separated, in the sense that

$$\sup_{u \in \mathcal{U}} \lambda_{j-1}(u) < \inf_{u \in \mathcal{U}} \lambda_j(u) \quad (j = 2, \ldots, N).$$

Here, we are primarily interested in a system of conservation laws (1.1) endowed with a strictly convex, mathematical entropy pair $(U, F) : \mathcal{U} \to I\!R^2$. Following the discussion in Section I-3 we constrain the weak solutions to satisfy the *single entropy inequality*

$$\partial_t U(u) + \partial_x F(u) \leq 0, \tag{1.5}$$

which, for self-similar solutions, reads

$$-\xi\, U(u)' + F(u)' \leq 0. \tag{1.6}$$

Shock waves and Hugoniot curves.

The so-called *shock waves* provide an important family of elementary solutions of (1.1). Those solutions take only two constant values, u_- and $u_+ \in \mathcal{U}$, and have the form

$$u(x,t) = \begin{cases} u_-, & x < \lambda t, \\ u_+, & x > \lambda t, \end{cases} \tag{1.7}$$

where λ is called the *shock speed*. We know from Theorem I-2.3 that (1.7) is a weak solution of (1.1) if and only if the states u_-, u_+ and the speed λ satisfy the *Rankine-Hugoniot relation*

$$-\lambda\, (u_+ - u_-) + f(u_+) - f(u_-) = 0. \tag{1.8}$$

To study (1.8), we fix the left-hand state u_- and we study the local structure of the **Hugoniot set** consisting of all right-hand states u_+ satisfying (1.8) for some λ.

THEOREM 1.1. (Hugoniot curves.) *There exist $\delta_1 < \delta_0$ and $\varepsilon > 0$ such that for each $u_- \in \mathcal{B}(\delta_1)$ the following holds. The Hugoniot set can be decomposed into N curves $s \mapsto v_i(s; u_-)$ $(1 \leq i \leq N)$ defined for $s \in (-\varepsilon, \varepsilon)$ and depending smoothly on s and u_-. Moreover, we have*

$$v_i(s; u_-) = u_- + s\, r_i(u_-) + \frac{s^2}{2} \left(Dr_i\, r_i\right)(u_-) + O(s^3) \tag{1.9}$$

and the corresponding shock speed $\lambda = \overline{\lambda}_i(s; u_-)$ satisfies

$$\begin{aligned}
\overline{\lambda}_i(s; u_-) =&\, \lambda_i(u_-) + \frac{s}{2} \left(\nabla \lambda_i \cdot r_i\right)(u_-) \\
&+ \frac{s^2}{6} \left(\left(\nabla(\nabla \lambda_i \cdot r_i) \cdot r_i\right) + \frac{\nabla \lambda_i \cdot r_i}{2}\, l_i(Dr_i\, r_i) \right)(u_-) + O(s^3).
\end{aligned} \tag{1.10}$$

We will use the notation

$$\mathcal{H}_i(u_-) = \left\{ v_i(s; u_-) \,/\, s \in (-\varepsilon, \varepsilon) \right\}$$

and refer to $\mathcal{H}_i(u_-)$ as the i-**Hugoniot curve issuing from** u_-. Note in passing that, by (1.10), when the *genuine nonlinearity* condition (Definition I-1.4)

$$\nabla \lambda_i(u_-) \cdot r_i(u_-) > 0$$

holds, the shock speed $\overline{\lambda}_i(s; u_-)$ is an *increasing* function of s. (If necessary, replace here ε with a smaller value.)

Since the matrix $Df(u_-)$ is strictly hyperbolic for all $u_- \in \mathcal{B}(\delta_0)$, so is the **averaging matrix**

$$\overline{A}(u_-, u_+) := \int_0^1 Df\big(\theta \, u_- + (1 - \theta) \, u_+\big) \, d\theta \tag{1.11}$$

for all $u_-, u_+ \in \mathcal{B}(\delta_1)$, where $\delta_1 < \delta_0$ is sufficiently small. We denote by $\overline{\lambda}_i(u_-, u_+)$, $\overline{r}_i(u_-, u_+)$, $\overline{l}_i(u_-, u_+)$ its eigenvalues and eigenvectors which we normalize so that $\overline{l}_i(u_-, u_+) \overline{r}_i(u_-, u_+) = 1$. Observe that $\overline{\lambda}_i(u_-, u_-) = \lambda_i(u_-)$, etc.

PROOF. The Rankine-Hugoniot relation (1.8) is equivalent to

$$\big(\lambda - \overline{A}(u_-, u_+)\big) (u_+ - u_-) = 0, \tag{1.12}$$

which shows that, for *some* index $i = 1, \ldots, N$, the vector $(u_+ - u_-)$ is an i-eigenvector and $\lambda = \overline{\lambda}_i(u_-, u_+)$. In particular, we get

$$\overline{l}_j(u_-, u_+) (u_+ - u_-) = 0, \quad j \neq i, \quad j = 1, \ldots, N,$$

which is a nonlinear algebraic system of $N - 1$ equations for the unknown N-vector u_+. Let us apply the implicit function theorem to the mapping

$$u_+ \in \mathcal{U} \mapsto G(u_+) = \big(\overline{l}_j(u_-, u_+) (u_+ - u_-)\big)_{j \neq i} \in I\!R^{N-1}$$

in a neighborhood of $u_+ = u_-$, which is a trivial solution of

$$G(u_+) = 0.$$

Since the Jacobian matrix at $u_+ = u_-$,

$$DG(u_-) = \big(l_j(u_-)\big)_{j \neq i},$$

has maximal rank, $N - 1$, there exists a smooth curve of solutions u_+ of (1.12), depending smoothly also upon the base value u_-. By continuity, when $u_+ \to u_-$ we have $\lambda = \overline{\lambda}_i(u_-, u_+) \to \lambda_i(u_-)$.

Regarding u_+ and λ as functions of some parameter s, say

$$u_+ = v_i(s; u_-) = v(s), \quad \lambda = \overline{\lambda}_i(s; u_-) = \overline{\lambda}(s),$$

let us differentiate (1.8) with respect to s:

$$\overline{\lambda}'(s) (v(s) - u_-) = \big(A(v(s)) - \overline{\lambda}(s)\big) v'(s). \tag{1.13}$$

Differentiating (1.13) once more we obtain

$$\begin{aligned}
&\overline{\lambda}''(s) (v(s) - u_-) + 2 \overline{\lambda}'(s) v'(s) \\
&= \big(DA(v(s)) \cdot v'(s)\big) v'(s) + \big(A(v(s)) - \overline{\lambda}(s)\big) v''(s).
\end{aligned} \tag{1.14}$$

Letting first $s \to 0$ in (1.13) we find

$$\big(\overline{\lambda}(0) - A(u_-)\big) v'(0) = 0.$$

We already observed that $\bar{\lambda}(0) = \lambda_i(u_-)$, so the above relation shows that $v'(0)$ is a multiple of the eigenvector $r_i(u_-)$. Moreover, one can check that $v'(0) \neq 0$ so that, by modifying the parametrization if necessary (replacing s with $a\,s$ for some constant a) we obtain

$$v'(0) = r_i(u_-).$$

Next, letting $s \to 0$ in (1.14) we find

$$2\,\bar{\lambda}'(0)\,r_i(u_-) = \big(DA(u_-) \cdot r_i(u_-)\big)\,r_i(u_-) + \big(A(v(0)) - \lambda_i(u_-)\big)\,v''(0).$$

On the other hand, by differentiating the relation $A\,r_i = \lambda_i\,r_i$ we have

$$\big(DA \cdot r_i\big)r_i = -\big(A - \lambda_i\big)\,Dr_i\,r_i + \big(\nabla\lambda_i \cdot r_i\big)\,r_i,$$

so

$$\begin{aligned}
\big(2\,\bar{\lambda}'(0) - \nabla\lambda_i(u_-) \cdot r_i(u_-)\big)\,r_i(u_-) \\
= \big(A(v(0)) - \lambda_i(u_-)\big)\,\big(v''(0) - Dr_i(u_-)r_i(u_-)\big).
\end{aligned} \tag{1.15}$$

Multiplying (1.15) by the left-eigenvector $l_i(u_-)$, we deduce that

$$\bar{\lambda}'(0) = \frac{1}{2}\nabla\lambda_i(u_-) \cdot r_i(u_-).$$

Returning to (1.15) we see that $v''(0) - Dr_i(u_-)\,r_i(u_-)$ must be an eigenvector. In other words, for some scalar b we have

$$v''(0) = Dr_i(u_-)r_i(u_-) + b\,r_i(u_-).$$

By modifying again the parametrization if necessary (replacing s with $s + b\,s^2/2$) the term $b\,r_i(u_-)$ can be absorbed in the first term of the expansion (1.9).

Differentiating (1.14) once more we obtain

$$\begin{aligned}
3\bar{\lambda}''(0)\,r_i(u_-) + \frac{3}{2}\,(\nabla\lambda_i \cdot r_i)(u_-)\,(Dr_i\,r_i)(u_-) \\
= \big((D^2 A\,r_i)\,r_i\big)r_i(u_-) + \big(DA(Dr_i\,r_i)\big)\,r_i(u_-) \\
+ 2\,(DA\,r_i)(u_-)\,(Dr_i\,r_i)(u_-) + \big(A(u_-) - \lambda_i(u_-)\big)\,v'''(0).
\end{aligned}$$

On the other hand, by differentiating the relation $A\,r_i = \lambda_i\,r_i$ twice we have

$$\begin{aligned}
\big((D^2 A\,r_i)\,r_i\big)r_i(u_-) + \big(DA(Dr_i\,r_i)\big)\,r_i(u_-) + 2\,(DA\,r_i)(u_-)\,(Dr_i\,r_i)(u_-) \\
= -\big(A(u_-) - \lambda_i(u_-)\big)\,D(Dr_i\,r_i)\,r_i(u_-) + 2\,(\nabla\lambda_i \cdot r_i)(u_-)\,(Dr_i\,r_i)(u_-) \\
+ \big(\nabla(\nabla\lambda_i \cdot r_i) \cdot r_i\big)(u_-)\,r_i(u_-).
\end{aligned}$$

It follows that

$$\begin{aligned}
\Big(3\,\bar{\lambda}''(0) - \big(\nabla(\nabla\lambda_i \cdot r_i) \cdot r_i\big)(u_-)\Big)\,r_i(u_-) \\
= \frac{(\nabla\lambda_i \cdot r_i)(u_-)}{2}\,(Dr_i\,r_i)(u_-) + \big(A(u_-) - \lambda_i(u_-)\big)\,\Big(v'''(0) - D(Dr_i\,r_i)\,r_i(u_-)\Big).
\end{aligned}$$

It suffices to multiply this identity by the left-eigenvector l_i to complete the derivation of (1.10) and, therefore, the proof of Theorem 1.1. □

Rarefaction waves and integral curves.
We now turn to the discussion of the rarefaction waves. They are constructed from
smooth and self-similar solutions of the equation (1.3), i.e., by solving the equation

$$(-\xi + A(u))\, u' = 0. \tag{1.16}$$

If (1.16) holds and $u'(\xi) \neq 0$ there must exist $i \in \{1, \dots, N\}$ and a scalar $c(\xi)$ such
that (for all relevant values ξ)

$$u'(\xi) = c(\xi)\, r_i(u(\xi)), \quad \xi = \lambda_i(u(\xi)).$$

By differentiating the second relation above we obtain $1 = c(\xi)\, \nabla\lambda_i(u(\xi)) \cdot r_i(u(\xi))$,
which determines $c(\xi)$ when the genuine nonlinearity condition $\nabla\lambda_i(u) \cdot r_i(u) \neq 0$
holds. When the latter vanishes the corresponding coefficient $c(\xi)$ becomes infinite.

The range of a rarefaction wave describes the integral curve associated with the
vector field r_i. For each given u_- satisfying $\nabla\lambda_i(u_-) \cdot r_i(u_-) \neq 0$ let us denote by
$\xi \mapsto u(\xi; u_-)$ the solution of the following ordinary differential equation with prescribed data at $\xi_- := \lambda_i(u_-)$,

$$u' = \frac{r_i(u)}{\nabla\lambda_i(u) \cdot r_i(u)}, \quad u(\xi_-) = u_-. \tag{1.17}$$

The solution exists and is smooth, as long as the term $\nabla\lambda_i(u) \cdot r_i(u)$ does not vanish.
Importantly, the condition $\xi = \lambda_i(u(\xi))$ is a consequence of (1.17).

The i-rarefaction waves issuing from u_- are parametrized by their right-hand
speed ξ_+ and are defined as follows. For any $\xi_+ \geq \xi_-$, the i-**rarefaction wave
connecting** u_- **to** $u_+ = u(\xi_+; u_-)$ is defined by

$$u(x,t) = \begin{cases} u_-, & x \leq \xi_- t, \\ u(x/t; u_-), & \xi_- t \leq x \leq \xi_+ t, \\ u_+, & x \geq \xi_+ t, \end{cases} \tag{1.18}$$

which is a self-similar and Lipschitz continuous solution of (1.1). Since $\xi = \lambda_i(u(\xi))$
by construction, we have $\xi_+ = \lambda_i(u_+)$. We emphasize that ξ_+ must be greater than
ξ_-. Indeed, the part $\xi_+ < \xi_-$ of the integral curve cannot be used since it would lead
to a *multivalued* function in (1.18).

Often, we will be interested only in the range $\{u(\xi; u_-)\,/\,\xi_- < \xi < \xi_+\} \subset \mathbb{R}^N$,
for which the parametrization is irrelevant.

THEOREM 1.2. (Integral curves.) *There exist $\delta_1 < \delta_0$ and $\varepsilon > 0$ such that for each*
$u_- \in \mathcal{B}(\delta_1)$ *the following holds. For each $i \in \{1, \dots, N\}$ the **integral curve** $\mathcal{O}_i(u_-)$
defined by*

$$w_i' = r_i(w_i), \quad w_i(0; u_-) = u_-, \tag{1.19}$$

*is a curve $s \mapsto w_i(s; u_-)$ defined for $s \in (-\varepsilon, \varepsilon)$ and depending smoothly upon u_- and
s, which satisfies*

$$w_i(s; u_-) = u_- + s\, r_i(u_-) + \frac{s^2}{2}\, Dr_i(u_-)\, r_i(u_-) + O(s^3) \tag{1.20}$$

and

$$\lambda_i(w_i(s; u_-)) = \lambda_i(u_-) + s\, \nabla\lambda_i(u_-) \cdot r_i(u_-) + \frac{s^2}{2}\, \big(\nabla(\nabla\lambda_i \cdot r_i) \cdot r_i\big)(u_-) + O(s^3). \tag{1.21}$$

$\square$

Comparing (1.19) with (1.17), we note that the *singularity* in (1.17) is due to the choice of the parameter ξ imposed along the integral curve inside the rarefaction fan. Note also that, later on, it will be convenient to describe the (*range of the*) integral curve via a different parameter $\bar{s}$ and so to replace (1.19) with

$$w_i' = \alpha(\bar{s}) \, r_i(w_i), \quad w_i(0; u_-) = u_-, \tag{1.19'}$$

where $\alpha(\bar{s})$ is a smooth function bounded away from zero. (See also Remark 1.7 below.)

Finally, rarefaction waves of the form (1.18) are constructed as follows. Assuming that the genuine nonlinearity condition $\nabla \lambda_i(u_-) \cdot r_i(u_-) > 0$ holds at the point u_-, the wave speed $s \mapsto \lambda_i(w_i(s; u_-))$ is an increasing function, at least near $s = 0$. It follows that $s \mapsto \lambda_i(w_i(s; u_-))$ is a one-to-one mapping from $(-\varepsilon, \varepsilon)$ onto a neighborhood of $\lambda_i(u_-)$. So, the function $s \mapsto \xi = \lambda_i(w_i(s; u_-))$ admits an inverse denoted by $\xi \mapsto \sigma(\xi)$. Choosing any value $s_+ > 0$ and setting $u_+ = w_i(s_+; u_-)$ we define the corresponding **rarefaction wave connecting** u_- **to** u_+ by

$$u(x,t) = \begin{cases} u_-, & x \leq t\, \lambda_i(u_-), \\ w_i(\sigma(x/t); u_-), & t\, \lambda_i(u_-) \leq x \leq t\, \lambda_i(u_+), \\ u_+, & x \geq t\, \lambda_i(u_+). \end{cases} \tag{1.22}$$

Observe that, when $\nabla \lambda_i(u_-) \cdot r_i(u_-) > 0$, *only the part $s > 0$ of the integral curve can be used to construct rarefactions.* (The formula (1.22) would give a multivalued function for $s < 0$.)

Contact discontinuities.

Shock waves associated with a linearly degenerate field are called **contact discontinuities** and satisfy the following property.

THEOREM 1.3. (Contact discontinuities.) *There exist $\delta_1 < \delta_0$ and $\varepsilon > 0$ such that for each $u_- \in \mathcal{B}(\delta_1)$ the following holds. Suppose that the i-characteristic field is linearly degenerate, that is,*

$$\nabla \lambda_i \cdot r_i \equiv 0.$$

Then, the integral curve $\mathcal{O}_i(u_-)$ and the Hugoniot curve $\mathcal{H}_i(u_-)$ coincide. Moreover, the characteristic speed along the integral curve and the shock speed along the Hugoniot curve are constant and coincide.

PROOF. Note that along the integral curve $s \mapsto w(s) = w_i(s; u_-)$ the wave speed is constant since $\lambda_i(w(s))' = \nabla \lambda_i(w(s)) \cdot r_i(w(s)) = 0$. Then, consider

$$h(s) = -\lambda_i(w(s)) \, (w(s) - u_-) + f(w(s)) - f(u_-).$$

Using that $w'(s) = r_i(w(s))$ we obtain

$$h'(s) = -\lambda_i(w(s)) \, w'(s) + A(w(s)) \, w'(s) = 0.$$

Since $h(0) = 0$ we have $h(s) = 0$ for all relevant values of s. This proves that the Rankine-Hugoniot relation (1.8) holds along the integral curve and that the two curves under consideration do coincide with, furthermore, $\lambda_i(w(s)) = \bar{\lambda}_i(u_-, w(s))$. $\qquad\square$

Genuinely nonlinear and linearly degenerate fields.
We now assume that each field associated with (1.1) is either genuinely nonlinear, that is, $\nabla \lambda_k \cdot r_k > 0$ after normalization, or linearly degenerate, that is, $\nabla \lambda_k \cdot r_k \equiv 0$. Combining shock and rarefaction waves together, we are able to solve the Riemann problem (1.1) and (1.2). To ensure the uniqueness of the weak solution, we impose the entropy inequality (1.5), where (U, F) is a given entropy pair. For a shock wave of the form (1.7), the entropy inequality (see (1.6)) precisely imposes that the rate of entropy dissipation be non-positive, that is,

$$E(u_-, u_+) := -\lambda \left(U(u_+) - U(u_-) \right) + F(u_+) - F(u_-) \leq 0. \qquad (1.23)$$

The following lemma provides the sign of the entropy dissipation

$$\overline{E}(s; u_-) := E(u_-, v_i(s; u_-))$$

along the i-Hugoniot curve issuing from u_-. (For the notation, see Theorem 1.1.)

LEMMA 1.4. (Entropy dissipation.) *Along the i-Hugoniot curve issuing from u_-, we have*

$$\overline{E}(s; u_-) = \frac{s^3}{12} \left(r_i^T D^2 U \, r_i \right) (u_-) \, \nabla \lambda_i(u_-) \cdot r_i(u_-) + O(s^4). \qquad (1.24)$$

PROOF. We use the same notation as in the proof of Theorem 1.1. The entropy dissipation rate is expressed as a function of s:

$$\overline{E}(s; u_-) := -\overline{\lambda}(s) \left(U(v(s)) - U(u_-) \right) + F(v(s)) - F(u_-).$$

On one hand, we have immediately $\overline{E}(0; u_-) = \overline{E}'(0; u_-) = 0$ and, with some tedious calculation, one may also get $\overline{E}''(0; u_-) = 0$ and $\overline{E}'''(0; u_-) \neq 0$. The latter follows also conveniently from the following integral form of $\overline{E}(s; u_-)$:

$$\overline{E}(s; u_-) = \int_0^s \nabla U(v(m))^T \left(-\overline{\lambda}(s) + A(v(m)) \right) v'(m) \, dm$$

$$= -\int_0^s v'(m)^T D^2 U(v(m)) \left(-\overline{\lambda}(s)(v(m) - u_-) + f(v(m)) - f(u_-) \right) dm \qquad (1.25)$$

$$= -\int_0^s v'(m)^T D^2 U(v(m)) \left(\overline{\lambda}(m) - \overline{\lambda}(s) \right) (v(m) - u_-) \, dm,$$

where we used the compatibility condition on the entropy pair, i.e., $\nabla F^T = \nabla U^T Df$, and the Rankine-Hugoniot relation (1.8). In view of the Taylor expansions (1.9) and (1.10) in Theorem 1.1 we deduce from (1.25) that

$$\overline{E}(s; u_-) = -\frac{1}{2} \left(r_i^T D^2 U \, r_i \right) (u_-) \, (\nabla \lambda_i \cdot r_i) (u_-) \int_0^s m \, (m - s) \, dm + O(s^4).$$

$\square$

We now discuss another approach for the selection of shock waves. Consider again a shock wave connecting u_- to u_+ at the speed λ. **Lax shock inequalities**

$$\lambda_i(u_-) \geq \lambda \geq \lambda_i(u_+) \qquad (1.26)$$

can be regarded as a generalization to the system of conservation laws (1.1) of the inequalities discussed earlier for scalar equations. The following result shows that, based on (1.26), the part $s \leq 0$ of the Hugoniot curve again should be retained. The proof is straightforward in view of (1.10).

LEMMA 1.5. *Consider the i-Hugoniot curve $s \mapsto v_i(s; u_-)$ issuing from a state u_- and associated with a genuinely nonlinear family. Then, the shock speed $s \mapsto \bar{\lambda}_i(s; u_-)$ is an increasing function of s and, for all $s < 0$, the inequalities*

$$\lambda_i(u_-) > \bar{\lambda}_i(s; u_-) > \lambda_i(v_i(s; u_-)) \tag{1.27}$$

hold, while both inequalities are violated for $s > 0$.   □

In view of (1.24) and relying on the genuine nonlinearity assumption, the entropy inequality (1.23) holds along the Hugoniot curve if and only if $s \leq 0$. On the other hand, recall that this is consistent with the fact that only the part $s > 0$ of the integral curve can be used in that case. (See the discussion after Theorem 1.2.) By definition, the i-**shock curve** $\mathcal{S}_i(u_-)$ is the part $s \leq 0$ of the Hugoniot curve $\mathcal{H}_i(u_-)$. We call i-**rarefaction curve**, $\mathcal{R}_i(u_-)$, is the part $s \geq 0$ of the integral curve $\mathcal{O}_i(u_-)$. We refer to $\mathcal{W}_i(u_-) := \mathcal{S}_i(u_-) \cup \mathcal{R}_i(u_-)$ as the i-**wave curve** issuing from u_-. (See Figure VI-1.)

Combining Theorems 1.1 to 1.3 and Lemmas 1.4 and 1.5 we arrive at the following important existence result. For clarity in the presentation, we say that a function $u = u(x, t)$ belongs to the class **P** if it is self-similar (that is, $u = u(x/t)$), piecewise smooth, and made of constant states separated by shock waves or rarefaction fans. All the existence and uniqueness results in the present chapter are stated in this class, while a general uniqueness theory is postponed to Chapter X. (See Figure VI-2.)

THEOREM 1.6. (Riemann solver for genuinely nonlinear and linearly degenerate fields.) *Suppose that, in $\mathcal{B}(\delta_0)$, the system (1.1) is strictly hyperbolic and admits only genuinely nonlinear or linearly degenerate fields. Then, there exist $\delta_1 < \delta_0$ and $\varepsilon > 0$ with the following property. To any $u_- \in \mathcal{B}(\delta_1)$ and $i \in \{1, \dots, N\}$ we can associate the i-**wave curve** issuing from u_-*

$$\mathcal{W}_i(u_-) = \mathcal{S}_i(u_-) \cup \mathcal{R}_i(u_-) =: \{\psi_i(s; u_-) \, / \, s \in (-\varepsilon, \varepsilon)\},$$
$$\psi_i(s; u_-) = \begin{cases} v_i(s; u_-), & s \in (-\varepsilon, 0], \\ w_i(s; u_-), & s \in [0, \varepsilon). \end{cases} \tag{1.28}$$

The mapping $\psi_i : (-\varepsilon, \varepsilon) \times \mathcal{B}(\delta_1) \to \mathbb{R}^N$ admits continuous derivatives up to second-order and bounded third-order derivatives in s and u_-, and satisfies

$$\psi_i(s; u_-) = u_- + s \, r_i(u_-) + \frac{s^2}{2} \, Dr_i(u_-) \, r_i(u_-) + O(s^3). \tag{1.29}$$

*Given any u_l and $u_r \in \mathcal{B}(\delta_1)$ the Riemann problem (1.1) and (1.2) admits a unique self-similar solution (in the class **P**) made up of $N + 1$ constant states*

$$u_l = u^0, \, u^1, \dots, \, u^N = u_r$$

separated by elementary waves. The intermediate constant states satisfy

$$u^j \in \mathcal{W}_j(u^{j-1}), \quad u^j = \psi_j(s_j; u^{j-1}),$$

for some $s_j \in (-\varepsilon, \varepsilon)$. The states u^{j-1} and u^j are connected with either a contact discontinuity (if the j-field is linearly degenerate) or else, when $s_j \geq 0$, a rarefaction wave and, when $s_j < 0$, a shock wave satisfying the entropy inequality (1.23) and Lax shock inequalities (1.26).

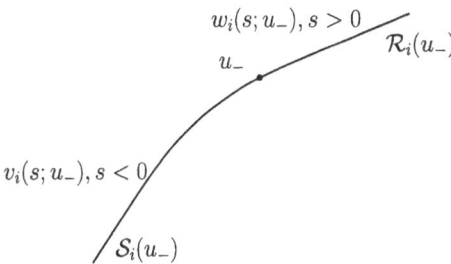

Figure VI-1 : The i-wave curve $\mathcal{W}_i(u_-)$.

PROOF. The mapping

$$s = (s_1, \ldots, s_N) \in (-\varepsilon, \varepsilon)^N \mapsto \Psi(s) = \psi_N(s_N) \circ \ldots \circ \psi_1(s_1)(u_l)$$

is locally invertible. This follows from the implicit function theorem since the differential at $s = 0$

$$D\Psi(0) = \big(r_k(u_l)\big)_{1 \le k \le N}$$

is an invertible $N \times N$ matrix. It follows from the monotonicity property of the shock speed and characteristic speed along the wave curves that a wave cannot follow another wave of the same family. This ensures that the Riemann problem cannot be solved with another combination of elementary waves. Each wave curve is smooth for $s < 0$ and $s > 0$. Furthermore, the first- and second-order derivatives (with respect to both variables s and u) of the shock and rarefaction curves coincide at $s = 0$, as follows from (1.9) and (1.20). □

REMARK 1.7. In Section 2 below, we will use a globally defined parameter $u \mapsto \mu_i(u)$ satisfying $\nabla \mu_i(u) \cdot r_i(u) \ne 0$, and we will re-parametrize all the wave curves accordingly, for instance the Hugoniot curve $m \mapsto v_i(m; u_-)$ (keeping here the same notation), in such a way that

$$\mu_i(v_i(m; u_-)) = m.$$

In this situation, by setting $\tilde{s} = m - \mu_i(u_-)$ the expansions (1.9) and (1.10) become

$$v_i(m; u_-) = u_- + \frac{\tilde{s}}{a(u_-)} r_i(u_-) + \frac{\tilde{s}^2}{2a(u_-)^2} \big((Dr_i \, r_i) + b \, r_i\big)(u_-) + O(\tilde{s}^3) \quad (1.30)$$

and

$$\begin{aligned}
\overline{\lambda}_i(m; u_-) = {} & \lambda_i(u_-) + \frac{\tilde{s}}{2a(u_-)} \big(\nabla \lambda_i \cdot r_i\big)(u_-) \\
& + \frac{\tilde{s}^2}{6a(u_-)^2} \Big(\big(\nabla(\nabla \lambda_i \cdot r_i) \cdot r_i\big) + c \nabla \lambda_i \cdot r_i\Big)(u_-) + O(\tilde{s}^3),
\end{aligned} \quad (1.31)$$

where a, b, and c are smooth and real-valued functions of u_- with, in particular, $a := \nabla \mu_i \cdot r_i \ne 0$. The formula (1.20) takes the same form as (1.30) while, with the same function a as above and for some function d, (1.21) becomes

$$\overline{\lambda}_i(w_i(m; u_-)) = \lambda_i(u_-) + \frac{\tilde{s}}{a(u_-)} \left(\nabla\lambda_i \cdot r_i\right)(u_-)$$

$$+ \frac{\tilde{s}^2}{2a(u_-)^2} \left(\left(\nabla(\nabla\lambda_i \cdot r_i) \cdot r_i\right) + d\,\nabla\lambda_i \cdot r_i\right)(u_-) + O(\tilde{s}^3).$$

(1.32)

$\square$

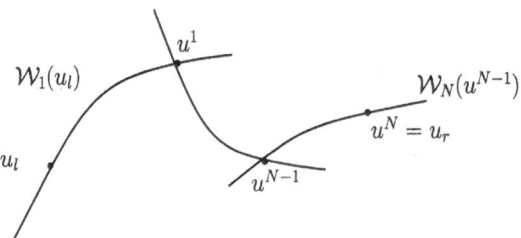

Figure VI-2 : The Riemann solution.

2. Classical Riemann solver

In this section we solve the Riemann problem (1.1) and (1.2) when the characteristic fields are not globally genuinely nonlinear. We restrict attention to systems for which the genuine nonlinearity condition may fail at one point (at most) along each wave curve. For scalar conservation laws Lax shock inequalities were found to be sufficiently discriminating to select a unique solution to the Riemann problem (Chapter II). This motivates us to proving here, for systems, that Lax shock inequalities (1.26) single out a unique Riemann solution, even for non-genuinely nonlinear fields. As observed already with scalar equation, a more discriminating condition (Liu entropy condition (2.6), below) is necessary only when the genuine nonlinearity fails at two or more points (along the wave curves). Note finally that for technical reasons we use a parameter (see (2.3) and (2.4), below) which is globally defined for all wave curves and does not coincide with the one in Section 1.

We assume that (1.1) is a strictly hyperbolic system of conservation laws and that there exists a partition $\{1,\dots,N\} = J_0 \cup J_1 \cup J_2$ such that

- $j \in J_0$ if the j-characteristic field is linearly degenerate,
- $j \in J_1$ if it is genuinely nonlinear,
- $j \in J_2$ if it is concave-convex or convex-concave, in a sense explained now.

In the latter case, the scalar-valued function

$$m_j(u) := \nabla\lambda_j(u) \cdot r_j(u)$$

does not keep a constant sign and we assume that

$$\mathcal{M}_j = \{u \in \mathcal{U} \,|\, m_j(u) = 0\}$$

(2.1)

is a smooth affine manifold with dimension $N - 1$ such that the vector field r_j is **transverse to the manifold** $\mathcal{M}_j$ and that one of the following two conditions holds:

$$
\begin{aligned}
\text{concave-convex field} &: \nabla m_j(u) \cdot r_j(u) > 0, \quad u \in \mathcal{U}, \\
\text{convex-concave field} &: \nabla m_j(u) \cdot r_j(u) < 0, \quad u \in \mathcal{U}.
\end{aligned}
\tag{2.2}
$$

This terminology generalizes the one introduced earlier for scalar equations in Chapter II. The notion is intrinsic, i.e., does not depend on the choice of the eigenvectors, since changing r_j into $-r_j$ does not change the sign of $\nabla m_j \cdot r_j$. For scalar equations, (2.2) constraints the third derivative of the flux-function f''' to be positive or negative, respectively, and we recover a stronger version of the notions of concave-convex and convex-concave functions introduced in Chapter II. Observe that for concave-convex fields $m_j(u)$ increases when u describes a wave curve in the direction of the right eigenvector r_j (while we have the opposite behavior in the convex-concave case). The transversality assumption (2.2) implies that for $j \in J_2$ the wave speed has exactly one critical point along each wave curve, that is, $\nabla \lambda_j(u) \cdot r_j(u) = 0$ if and only if $m_j(u) = 0$. In the concave-convex case the root of $m_j(u) = 0$ is associated with a *minimum value* of the wave speed (while in the convex-concave case it is associated with a *maximum value*.)

We are interested in solving the Riemann problem with data u_l and u_r in $\mathcal{B}(\delta_1)$ where δ_1 is small enough. Still, we assume that $\mathcal{M}_j \cap \mathcal{B}(\delta_1) \neq \emptyset$ for all $j \in J_2$ so that the problem does not reduce to the genuinely nonlinear case. Furthermore, to parametrize the wave curves it is convenient to have (for all j) a *globally defined parameter* $\mu_j(u) \in \mathbb{R}$ which should depend smoothly upon u and be strictly monotone along the wave curves. Specifically, we assume that a parameter μ_j is given such that

$$
\nabla \mu_j(u) \cdot r_j(u) \neq 0, \quad u \in \mathcal{U}
\tag{2.3}
$$

and for all $j \in J_2$

$$
\mu_j(u) = 0 \quad \text{if and only if} \quad m_j(u) = 0.
\tag{2.4}
$$

In view of the conditions (2.2), when $j \in J_2$ there is an obvious choice of μ_j which satisfies both requirements (2.3) and (2.4). When $j \in J_1$ a natural choice is the wave speed λ_j, while there is no canonical choice for $j \in J_0$. In summary, from now on, we assume that the wave curves are parametrized by ($u \in \mathcal{U}$)

$$
\mu_j(u) := \begin{cases} \lambda_j(u), & j \in J_1. \\ \nabla \lambda_j(u) \cdot r_j(u), & j \in J_2. \end{cases}
$$

Given $u_- \in \mathcal{U}$ and $j = 1, 2, \ldots, N$, we recall from Theorems 1.1 to 1.3 that the Hugoniot curve and the integral curve issuing from u_- are denoted by $\mathcal{H}_j(u_-) = \{v_j(m; u_-)\}$ and $\mathcal{O}_j(u_-) = \{w_j(m; u_-)\}$, respectively. The parameter m along these curves can be chosen to coincide with the parameter μ_j, that is,

$$
\mu_j\big(v_j(m; u_-)\big) = m, \quad \mu_j\big(w_j(m; u_-)\big) = m
\tag{2.5}
$$

for all relevant values m. Since μ_j is strictly monotone along the wave curves thanks to (2.2), the conditions (2.5) can be achieved by modifying the parameter s introduced earlier in Section 1. Recall that the local behavior of these curves is given by Remark 1.7.

Theorem 2.1 below establishes a generalization of Theorem 1.6, which was concerned with genuinely nonlinear or linearly degenerate fields only. In both the concave-convex and the convex-concave cases, we now prove that Lax shock inequalities (1.26) select a unique solution to the Riemann problem. For more general fields **Liu entropy criterion** is necessary and imposes that, along the Hugoniot curve $\mathcal{H}_j(u_-)$, the inequality

$$\overline{\lambda}_j(u_-, v_j(m; u_-)) \geq \overline{\lambda}_j(u_-, u_+) \qquad (2.6)$$

holds for all m between $\mu_j(u_-)$ and $\mu_j(u_+)$. In other words, the shock speed for m in the above range achieves its *minimum value* at the point u_+. This condition is a natural extension to systems of Oleinik entropy inequalities (see (II-1.6)).

We can now state the main existence result in this section.

THEOREM 2.1. (Classical Riemann solver.) *Suppose that* (1.1) *is strictly hyperbolic in* $\mathcal{B}(\delta_0)$ *and each* j-*field is either linearly degenerate* ($j \in J_0$), *genuinely nonlinear* ($j \in J_1$), *or else is concave-convex or convex-concave* ($j \in J_2$). *Then, there exist* $\delta_1 < \delta_0$ *and* $\varepsilon > 0$ *with the following property. For* $u_- \in \mathcal{B}(\delta_1)$ *and* $j \in J_2$ *the* j-**wave curve of right-hand states connected to** u_- *by a combination of* j-*elementary waves,*

$$\mathcal{W}_j^c(u_-) := \{\psi_j(m; u_-)\}$$

(m describing some open interval), is continuously differentiable with bounded second-order derivatives in m *and* u_-, *at least, and satisfies*

$$\psi_j(m; u_-) = u_- + \frac{m - \mu_j(u_-)}{(\nabla\mu_j \cdot r_j)(u_-)} r_j(u_-) + O\big(m - \mu_j(u_-)\big)^2. \qquad (2.7)$$

Given any u_l *and* $u_r \in \mathcal{B}(\delta_1)$, *the Riemann problem* (1.1) *and* (1.2) *admits a unique piecewise smooth solution (in the class* **P**) *made of* $N + 1$ *constant states*

$$u_l = u^0, u^1, \ldots, u^N = u_r$$

separated by j-*wave packets. The state* u^j *is connected to* u^{j-1} *by either a contact discontinuity (if* $j \in J_0$), *or a shock or rarefaction wave (if* $j \in J_1$), *or else (if* $j \in J_2$) *by (at most) two waves: either a shock from* u^{j-1} *to some intermediate state* $u^{j-1/2}$ *followed by a rarefaction connecting to* u^j, *or a rarefaction from* u^{j-1} *to some state* $u^{j-1/2}$ *followed by a shock connecting to* u^j. *Each shock wave satisfies Liu entropy criterion and Lax shock inequalities, any of these being sufficient to uniquely characterize the solution. Furthermore, the Riemann solution depends continuously upon its initial data in the sense that all of the states* $u^{j-1/2}$ *and* u^j *converge to* u_l *when* u_r *tends to* u_l.

The second-order derivatives of wave curves for non-genuinely nonlinear fields are *not continuous*, in general. In the course of the proof of Theorem 2.1 we will explicitly construct the wave curves $\mathcal{W}_j^c(u_-)$ for $j \in J_2$. From now on, we fix a left-hand state u_- satisfying (for definiteness) $\mu_j(u_-) > 0$. Concerning the integral curves $\mathcal{O}_j(u_-)$ it follows from the expansion (1.32) and the assumption (2.2) that:

LEMMA 2.2. (Characteristic speed along the integral curve.) *Let* $j \in J_2$ *and* u_- *be given with* $\mu_j(u_-) > 0$ *and consider the integral curve* $\mathcal{O}_j(u_-)$ *parametrized by the map* $m \mapsto w_j(m; u_-)$.

- *In the concave-convex case, the* j-**characteristic speed along the integral curve**

$$m \mapsto \lambda_j(w_j(m; u_-))$$

is a strictly convex function of m *achieving its minimum value at* $m = 0$. *In particular, it is decreasing for* $m < 0$ *and increasing for* $m > 0$.

- *In the convex-concave case, the characteristic speed $\lambda_j(w_j(m; u_-))$ is a strictly concave function of m achieving its maximum value at $m = 0$. In particular, it is increasing for $m < 0$ and decreasing for $m > 0$.*

PROOF. With (1.32) in Remark 1.7 we have at $m = \mu_j(u_-)$

$$\frac{\partial^2}{\partial m^2}\lambda_j(w_j(m; u_-))_{|m=\mu_j(u_-)} = \frac{1}{(\nabla\mu_j \cdot r_j)(u_-)}\left(1 + d(u_-)\frac{(\nabla\lambda_j \cdot r_j)(u_-)}{(\nabla\mu_j \cdot r_j)(u_-)}\right)$$

which is positive, since $\nabla\lambda_j \cdot r_j$ vanishes on the manifold $\mathcal{M}_j$ and

$$\nabla\mu_j \cdot r_j = \nabla(\nabla\lambda_j \cdot r_j) \cdot r_j$$

keeps a constant sign. $\square$

In view of Lemma 2.2 we can define the part of the curve $\mathcal{W}_j^c(u_-)$ associated with rarefaction waves, as follows. Given a point $u_+ \in \mathcal{O}_j(u_-)$, in order to construct a corresponding Lipschitz continuous, rarefaction wave $u = u(\xi)$ of the form (1.22) it is necessary that the wave speed $\xi \mapsto \lambda_j(u(\xi))$ be monotone increasing throughout the rarefaction fan. In the concave-convex case, this condition selects the part $m \geq \mu_j(u_-)$ of $\mathcal{O}_j(u_-)$. In the convex-concave case it is necessary that $m \leq \mu_j(u_-)$ but the desired monotonicity property is violated as one reaches $\mathcal{M}_j$ along the integral curve. Thus, this geometric restriction selects the part $0 \leq m \leq \mu_j(u_-)$ of the integral curve.

Now, to pursue the construction of the wave curve $\mathcal{W}_j^c(u_-)$ we rely on the Hugoniot curve $\mathcal{H}_j(u_-)$. The qualitative properties of the characteristic speed along $\mathcal{H}_j(u_-)$ are identical to those stated in Lemma 2.2 for $\mathcal{O}_j(u_-)$. The new features are concerned with the shock speed. (See Figure VI-3 for a graphical representation.)

LEMMA 2.3. (Characteristic speed and shock speed along the Hugoniot curve.) *Let $j \in J_2$ and u_- be given with $\mu_j(u_-) > 0$, and consider the Hugoniot curve $\mathcal{H}_j(u_-)$ parameterized by $m \mapsto v_j(m; u_-)$. In the concave-convex case the j-***characteristic speed along the Hugoniot curve***

$$m \mapsto \lambda_j(v_j(m; u_-))$$

*is a strictly convex function. On the other hand, the j-***shock speed along the Hugoniot curve***

$$m \mapsto \overline{\lambda}_j(m; u_-) := \overline{\lambda}_j(u_-, v_j(m; u_-))$$

is a strictly convex function, which either is globally increasing or else achieves a minimum value at a point

$$\mu_j^\natural(u_-).$$

Moreover, at this critical value the characteristic speed and the shock speed coincide:

$$\overline{\lambda}_j(m; u_-) = \lambda_j(v_j(m; u_-)) \quad at \ m = \mu_j^\natural(u_-) \tag{2.8}$$

and, for some smooth function $e = e(u_-) > 0$,

$$\partial_m v_j(\mu_j^\natural(u_-); u_-) = e(u_-)\, r_j(v_j(\mu_j^\natural(u_-); u_-)). \tag{2.9}$$

When $\mu_j(u_-) = 0$, the same properties hold with now $\mu_j^\natural(u_-) = 0$. In the convex-concave case, all the signs and monotonicity properties are reversed.

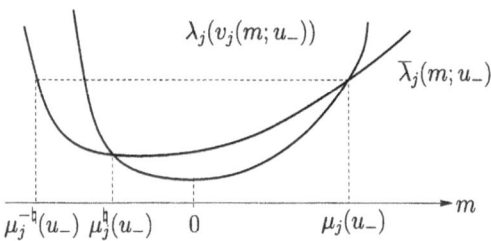

Figure VI-3 : Characteristic speed and shock speed.

In the following, we assume that the shock speed admits a minimum value. (The case of a globally increasing shock speed is simpler and can be easily deduced from the present discussion.) The proof of Lemma 2.3 (cf. the key formula (2.11), below) will imply that in the concave-convex case the characteristic speed and the shock speed satisfy

$$
\begin{aligned}
\overline{\lambda}_j(m; u_-) - \lambda_j(v_j(m; u_-)) &> 0, \quad m \in \left(\mu_j^\natural(u_-), \mu_j(u_-)\right), \\
\overline{\lambda}_j(m; u_-) - \lambda_j(v_j(m; u_-)) &< 0, \quad m < \mu_j^\natural(u_-) \text{ or } m > \mu_j(u_-).
\end{aligned}
\tag{2.10}
$$

PROOF. As in Lemma 2.2, the statement concerning the characteristic speed is a direct consequence of (2.2). With (1.32) (which also holds along the Hugoniot curve) in Remark 1.7 we have at $m = \mu_j(u_-)$

$$
\frac{\partial^2}{\partial m^2}\lambda_j(v_j(m; u_-))_{|m=\mu_j(u_-)} = \frac{1}{(\nabla\mu_j \cdot r_j)(u_-)}\left(1 + c(u_-)\frac{(\nabla\lambda_j \cdot r_j)(u_-)}{(\nabla\mu_j \cdot r_j)(u_-)}\right),
$$

which is positive since $\nabla\lambda_j \cdot r_j$ vanishes on the manifold $\mathcal{M}_j$ and $\nabla\mu_j \cdot r_j$ keeps a constant sign.

Similarly, using the notation $v(m) := v_j(m; u_-)$ and $\overline{\lambda}(m) := \overline{\lambda}_j(m; u_-)$, the formula (1.31) in Remark 1.7 gives

$$
\frac{\partial^2}{\partial m^2}\overline{\lambda}(m)_{|m=\mu_j(u_-)} = \frac{1}{(\nabla\mu_j \cdot r_j)(u_-)}\left(1/3 + c(u_-)\frac{(\nabla\lambda_j \cdot r_j)(u_-)}{(\nabla\mu_j \cdot r_j)(u_-)}\right),
$$

which gives the first statement for the shock speed. On the other hand, returning to (1.13) and multiplying this identity by $l_j(v(m))$ we obtain

$$
\overline{\lambda}'(m)\, l_j(v(m))\, (v(m) - u_-) = \left(\lambda_j(v(m)) - \overline{\lambda}(m)\right) l_j(v(m))\, v'(m).
$$

By relying on the expansion (1.9) it follows that

$$
m\, \overline{\lambda}'(m) = d(m)\left(\lambda_j(v(m)) - \overline{\lambda}(m)\right),
\tag{2.11}
$$

where $d = d(m)$ is a positive function bounded away from 0. Clearly, (2.8) follows from (2.11). Finally, (1.13) at $m = \mu_j^\natural(u_-)$ yields us

$$
0 = \left(A(v(\mu_j^\natural(u_-))) - \overline{\lambda}(\mu_j^\natural(u_-))\right) v'(\mu_j^\natural(u_-)),
$$

which yields (2.9). This completes the proof of Lemma 2.3.   □

PROOF OF THEOREM 2.1. Based on Lemmas 2.2 and 2.3 we can now construct the wave curves explicitly. We treat first the concave-convex case (represented in Figure VI-4). Given u_- with $\mu_j(u_-) > 0$, we already pointed out that the part $m > \mu_j(u_-)$ of the wave curve coincides with the integral curve $\mathcal{O}_j(u_-)$ and that the corresponding solutions are rarefaction waves. On the other hand, for m decreasing from $\mu_j(u_-)$, the wave curve coincides locally with the Hugoniot curve $\mathcal{H}_j(u_-)$ and the corresponding solutions are shock waves. This is correct as long as the entropy condition is satisfied. Relying on the property (2.10) which compares the characteristic speed and the shock speed along the Hugoniot curve, we see that Lax entropy inequalities (1.26) hold until we reach the value $m = \mu_j^\natural(u_-)$ only, at which point the equality holds in the right-hand side of (1.26). Furthermore, since the shock speed increases as m decreases from $\mu_j^\natural(u_-)$, Liu entropy criterion (2.6) is also violated exactly for $m < \mu_j^\natural(u_-)$. We conclude that the wave curve $\mathcal{W}_j^c(u_-)$ coincides with the part $m \in (\mu_j^\natural(u_-), \mu_j(u_-))$ of the Hugoniot curve, and no other point on this curve is admissible.

To extend the wave curve from the point $u_-^\natural := v_j(\mu_j^\natural(u_-); u_-)$, we consider the integral curve $\mathcal{O}_j(u_-^\natural)$. Any point $u_+ \in \mathcal{O}_j(u_-^\natural)$, with $\mu_j(u_+) < \mu_j^\natural(u_-)$, can be connected to u_- by a shock wave from u_- to $u_-^\natural$ followed by a rarefaction wave from $u_-^\natural$ to u_+. Note that the rarefaction is *attached to the shock* since the shock speed $\overline{\lambda}_j(\mu_j^\natural(u_-); u_-)$ *coincides* with the lowest speed of the rarefaction fan, $\lambda_j(u_-^\natural)$. (See (2.8).) Note also that the shock connecting u_- to $u_-^\natural$ is a **right-contact** in the sense that the propagation speed of the shock coincides with the characteristic speed of its right side. Finally, the wave curve is continuously differentiable at the point $u_-^\natural$ as follows from (2.9).

We now turn to the convex-concave case. For m increasing away from $\mu_j(u_-)$, the shock speed is decreasing and both Lax shock inequalities and Liu entropy criterion hold. Therefore, the wave curve contains the part $\mathcal{H}_j(u_-)$ for all $m > \mu_j(u_-)$. On the other hand, for m decreasing from $\mu_j(u_-)$ the wave curve coincides with the integral curve until m reaches the manifold, that is, $m = 0$. Values $m < 0$ cannot be attained using only a rarefaction wave since the characteristic speed is increasing for m decreasing from $m = 0$. This would violate the geometric requirement that the wave speed be increasing inside a rarefaction fan. To reach points $m < 0$ we proceed as follows. Take any point $u_+ \in \mathcal{O}_j(u_-)$ having $0 < \mu_j(u_+) < \mu_j(u_-)$ and define $u_+^\natural$ by the two conditions

$$u_+^\natural \in \mathcal{H}_j(u_+), \quad \mu_j^\natural(u_+^\natural) = \mu_j(u_+). \tag{2.12}$$

Then, we connect u_- to $u_+^\natural$ using first a rarefaction from u_- to u_+ followed by a shock wave connecting to $u_+^\natural$. By construction, the shock is *attached* to the rarefaction on the left and is called a **left-contact**.

This completes the construction of the waves curves. The existence and uniqueness of the Riemann solution follow from the implicit function theorem, as was discussed in the proof of Theorem 1.6. The proof of Theorem 2.1 is completed.   □

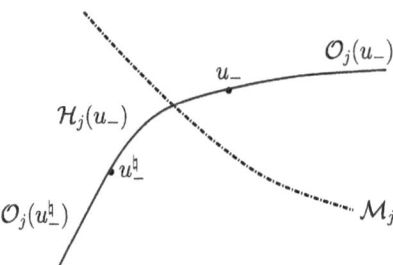

Figure VI-4 : The j-wave curve in the concave-convex case.

Finally, we prove some technical lemmas, say for concave-convex fields, which will be useful later on in this course.

LEMMA 2.4. (Regularity of the function $\mu_j^\natural$.) *The function $\mu_j^\natural$ depends smoothly upon its argument and*

$$\nabla \mu_j^\natural \cdot r_j \sim -\frac{1}{2} \nabla \mu_j \cdot r_j \quad \text{near the manifold } \mathcal{M}_j. \tag{2.13}$$

In particular, (2.13) implies that $\mu_j^\natural < 0$.

PROOF. The regularity property follows from the implicit function theorem applied to the mapping (see the condition (2.8))

$$H(m; u) = \frac{\overline{\lambda}_j(m; u) - \lambda_j(v_j(m; u))}{m - \mu_j(u)},$$

Indeed, $H(m; u)$ is smooth for $m \neq \mu_j(u)$ and, relying on the expansions in Remark 1.7, we see that $H(m; u)$ extends continuously at $m = \mu_j(u)$. We have also $H(\mu_j^\natural(u); u) = 0$ by definition and

$$H(m; u) = -\frac{(\nabla \lambda_j \cdot r_j)(u)}{2 \nabla (\nabla \lambda_j \cdot r_j) \cdot r_j(u)}$$
$$+ (m - \mu_j(u)) \left(-\frac{1}{3 \left(\nabla (\nabla \lambda_j \cdot r_j) \cdot r_j \right)(u)} + O(1) \left(\nabla \lambda_j \cdot r_j \right)(u) \right)$$
$$+ O\big(m - \mu_j(u)\big)^2.$$

Hence at $m = \mu_j(u)$ we find

$$\frac{\partial H}{\partial m}(\mu_j(u); u) = -\frac{1}{3 \left(\nabla (\nabla \lambda_j \cdot r_j) \cdot r_j \right)(u)} + O(1) \left(\nabla \lambda_j \cdot r_j \right)(u) \neq 0.$$

To derive (2.13) along the critical manifold we use the expansion of the shock speed (1.31) and the expansion of the characteristic speed (similar to (1.32)) along the Rankine-Hugoniot curve. The critical value $\tilde{s}(u) = \mu_j^{\natural}(u) - \mu_j(u)$ satisfies

$$\lambda_j(u) + \frac{\tilde{s}(u)}{2a(u)}\left(\nabla\lambda_j \cdot r_j\right)(u) + \frac{\tilde{s}(u)^2}{6a(u)^2}\left(\left(\nabla(\nabla\lambda_j \cdot r_j) \cdot r_j\right)(u) + \dots\right) + \dots$$

$$= \lambda_j(u) + \frac{\tilde{s}(u)}{a(u)}\left(\nabla\lambda_j \cdot r_j\right)(u) + \frac{\tilde{s}(u)^2}{2a(u)^2}\left(\left(\nabla(\nabla\lambda_j \cdot r_j) \cdot r_j\right)(u) + \dots\right) + \dots,$$

where we neglected high-order terms. Computing the second-order derivative in the direction r_j and letting the state u approach the manifold $\mathcal{M}_j$, we obtain

$$\frac{(\nabla\tilde{s} \cdot r_j)(u)}{2a(u)}\left(\nabla(\nabla\lambda_j \cdot r_j) \cdot r_j\right)(u) + \frac{(\nabla\tilde{s} \cdot r_j)(u)^2}{6a(u)^2}\left(\nabla(\nabla\lambda_j \cdot r_j) \cdot r_j\right)(u)$$

$$= \frac{(\nabla\tilde{s} \cdot r_j)(u)}{a(u)}\left(\nabla(\nabla\lambda_j \cdot r_j) \cdot r_j\right)(u) + \frac{(\nabla\tilde{s} \cdot r_j)(u)^2}{2a(u)^2}\left(\nabla(\nabla\lambda_j \cdot r_j) \cdot r_j\right)(u),$$

which yields

$$\left(\nabla\tilde{s} \cdot r_j\right)(u) = -(3/2)\left(\nabla\mu_j \cdot r_j\right)(u)$$

and, therefore, the identity (2.13) since $\nabla\tilde{s} \cdot r_j = \nabla\mu_j^{\natural} \cdot r_j - \nabla\mu_j \cdot r_j$. This completes the proof of Lemma 2.4. $\square$

To any two states u_- and u_+ with $u_+ = v_j(m; u_-) \in \mathcal{H}_j(u_-)$ we will associate a third state

$$\rho_j(u_-, u_+) = u_* := v_j(\mu_j^*; u_-) \in \mathcal{H}_j(u_-)$$

where the component $\mu_j^* = \mu_j^*(m; u_-)$ is determined so that the speed of the shock connecting u_- to u_* coincides with the speed of the shock connecting u_- to u_+, that is,

$$\overline{\lambda}_j(u_-, u_*) = \overline{\lambda}_j(u_-, u_+).$$

LEMMA 2.5. *The function μ_j^* depends smoothly upon its arguments and*

$$\mu_j^*(m; u) \sim -\mu_j(u) - m \quad \text{near the manifold } \mathcal{M}_j. \tag{2.14}$$

PROOF. The arguments are similar the the one in the proof of Lemma 2.4, so we only sketch the proof. Using the expansion of the shock speed we have

$$\lambda_j(u) + \frac{m - \mu_j(u)}{2\,a(u)}\left(\nabla\lambda_j \cdot r_j\right)(u) + \frac{(m - \mu_j(u))^2}{6\,a(u)^2}\left(\left(\nabla(\nabla\lambda_j \cdot r_j) \cdot r_j\right)(u) + \dots\right) + \dots$$

$$= \lambda_j(u) + \frac{\mu_j^*(m; u) - \mu_j(u)}{2\,a(u)}\left(\nabla\lambda_j \cdot r_j\right)(u)$$

$$+ \frac{(\mu_j^*(m; u) - \mu_j(u))^2}{6\,a(u)^2}\left(\left(\nabla(\nabla\lambda_j \cdot r_j) \cdot r_j\right)(u) + \dots\right) + \dots$$

which, since $\mu_{ij}(u) = \left(\nabla\lambda_j \cdot r_j\right)(u)$, yields

$$\frac{m - \mu_j^*(m; u)}{a(u)}\left((1/2)\,\mu_{ij}(u) + (1/6)\left(m - \mu_j(u) + \mu_j^*(m; u) - \mu_j(u)\right)\right) + \dots = 0.$$

$\square$

3. Entropy dissipation and wave sets

In the following two sections we study the Riemann problem (1.1) and (1.2) for non-genuinely nonlinear systems endowed with a strictly convex entropy pair (U, F). In order to encompass all possible diffusive-dispersive limits compatible with this entropy (see the discussion in Chapter I), we investigate the consequences of a *single entropy inequality* for the solutions of the Riemann problem. Not surprisingly, the class of admissible solutions will turn out to be larger than the one selected in Section 2 by Lax entropy inequalities or Liu entropy criterion. In particular, the solutions of the Riemann problem may now contain "nonclassical" shocks.

Nonclassical shocks and entropy dissipation.
Generalizing a notion introduced in Chapter II for scalar equations we set:

DEFINITION 3.1. A shock wave (1.7) is called a **nonclassical shock** if it satisfies the single entropy inequality (1.23) but not Lax shock inequalities (1.26). It is called a **classical shock** if (1.23) holds but (1.26) is violated.   □

For shocks with sufficiently small amplitude, the wave speeds associated with different wave families are totally separated. Therefore, our analysis can focus on each wave family separately. It will be useful to introduce the following terminology before imposing any entropy condition at this stage. A j-shock wave connecting a left-hand state u_- to a right-hand state u_+ can be:

- a **Lax shock**, satisfying

$$\lambda_j(u_-) \geq \overline{\lambda}_j(u_-, u_+) \geq \lambda_j(u_+), \tag{3.1}$$

- a **slow undercompressive shock** :

$$\overline{\lambda}_j(u_-, u_+) \leq \min(\lambda_j(u_-), \lambda_j(u_+)), \tag{3.2}$$

- a **fast undercompressive shock** :

$$\overline{\lambda}_j(u_-, u_+) \geq \max(\lambda_j(u_-), \lambda_j(u_+)), \tag{3.3}$$

- or a **rarefaction shock** :

$$\lambda_j(u_-) < \overline{\lambda}_j(u_-, u_+) < \lambda_j(u_+). \tag{3.4}$$

Let u_- such that $\mu_j(u_-) > 0$. Recalling Lemma 2.3 which describes the properties of the shock speed along the Hugoniot curve $\mathcal{H}_j(u_-)$, we denote by $\mu_j^{-\natural}(u_-)$ the point of $\mathcal{H}_j(u_-)$ such that

$$\overline{\lambda}_j(\mu_j^{-\natural}(u_-); u_-) = \lambda_j(u_-), \quad \mu_j^{-\natural}(u_-) < \mu_j^{\natural}(u_-), \tag{3.5}$$

whenever such a point exists. For simplicity in the presentation we assume that both points $\mu_j^{\natural}(u_-)$ and $\mu_j^{-\natural}(u_-)$ exist, since the present discussion would be much simpler otherwise. Note in passing that

$$
\begin{aligned}
&G_j^{\natural} \circ G_j^{-\natural} = G_j^{-\natural} \circ G_j^{\natural} = id, \\
&\text{where } G_j^{\natural}(u_-) := v_j(\mu_j^{\natural}(u_-); u_-), \quad G_j^{-\natural}(u_-) := v_j(\mu_j^{-\natural}(u_-); u_-).
\end{aligned}
\tag{3.6}
$$

This is easily checked using the (symmetric) form of the Rankine-Hugoniot relation. (With the notation introduced in Chapter II the property (3.6) represents a extension to systems of the property $\varphi^{\natural} \circ \varphi^{-\natural} = \varphi^{-\natural} \circ \varphi^{\natural} = id$ satisfied by scalar equations.)

Note also that the shock wave connecting u_- to $u_-^\natural$ is a *right-contact*, while the shock connecting u_- to $u_-^{-\natural}$ is a *left-contact*.

LEMMA 3.2. (Classification of shock waves.) *Let u_- be given with $\mu_j(u_-) \geq 0$ and consider a point u_+ on the Hugoniot curve $\mathcal{H}_j(u_-)$, say $u_+ = v_j(m; u_-)$ with $m = \mu_j(u_+)$.*
  - *In the concave-convex case, the shock connecting u_- to u_+ is*

      - *a rarefaction shock if $m > \mu_j(u_-)$ or $m < \mu_j^{-\natural}(u_-)$,*
      - *a Lax shock if $m \in \left[\mu_j^\natural(u_-), \mu_j(u_-)\right]$,* $\qquad$ (3.7i)
      - *an undercompressive shock if $m \in \left[\mu_j^{-\natural}(u_-), \mu_j^\natural(u_-)\right)$.*

    In the second instance, the shock also satisfies Liu entropy criterion.
  - *In the convex-concave case, the shock connecting u_- to u_+ is*

      - *a Lax shock if $m \geq \mu_j(u_-)$ or $m \leq \mu_j^{-\natural}(u_-)$,*
      - *a rarefaction shock if $m \in \left(\mu_j^\natural(u_-), \mu_j(u_-)\right)$,* $\qquad$ (3.7ii)
      - *an undercompressive shock if $m \in \left(\mu_j^{-\natural}(u_-), \mu_j^\natural(u_-)\right]$.*

    In the first instance, the shock also satisfies Liu entropy criterion.

PROOF. Consider for instance the concave-convex case, the other case being similar. According to Lemma 2.3 the function $m \mapsto \overline{\lambda}_j(m; u_-) - \lambda_j(v_j(m; u_-))$ is positive for $\mu_j^\natural(u_-) < m < \mu_j(u_-)$ and negative for $m < \mu_j^\natural(u_-)$ or $m > \mu_j(u_-)$. On the other hand the function $m \mapsto \overline{\lambda}_j(m; u_-) - \lambda_j(u_-)$ is positive for $m < \mu_j^{-\natural}(u_-)$ or $m > \mu_j(u_-)$, and negative for $m \in \left(\mu_j^{-\natural}(u_-), \mu_j(u_-)\right)$. The classification follows immediately from these two properties. $\qquad\square$

Next, we investigate the sign of the entropy dissipation E (see (1.23)) along the Hugoniot curve. (See Figure VI-5.)

LEMMA 3.3. (Entropy dissipation.) *Let u_- be given with $\mu_j(u_-) > 0$ and consider the Hugoniot curve $\mathcal{H}_j(u_-)$. Consider the concave-convex case (respectively the convex-concave case).*
  (i) *The entropy dissipation $s \mapsto \overline{E}(m; u_-) := E(u_-, v_j(m; u_-))$ vanishes at $\mu_j(u_-)$ and at a point*
  $$\mu_{j0}^\flat(u_-) \in \left(\mu_j^{-\natural}(u_-), \mu_j^\natural(u_-)\right).$$
  The entropy dissipation is decreasing (resp. increasing) for $m < \mu_j^\natural(u_-)$, increasing (resp. decreasing) for $m > \mu_j^\natural(u_-)$, and achieves a negative maximum value (resp. a positive maximum value) at the critical point of the wave speed $\mu_j^\natural(u_-)$.
  (ii) *A shock satisfying the entropy inequality (1.23) cannot be a rarefaction shock. A nonclassical shock is undercompressive and satisfies*
  $$m \in \left(\mu_{j0}^\flat(u_-), \mu_j^\natural(u_-)\right) \quad \text{(resp. } m \in \left(\mu_j^{-\natural}(u_-), \mu_{j0}^\flat(u_-)\right)\text{).}$$
  (iii) *Any shock satisfying Liu entropy criterion (2.6) also satisfies the entropy inequality (1.23).*

We refer to $u_- \mapsto \mu_{j0}^{\flat}(u_-)$ as the **zero-entropy dissipation function** associated with the j-characteristic field. In view of Lemma 2.3 we can also define the companion function $\mu_{j0}^{\sharp}$ by

$$\mu_{j0}^{\flat}(u_-) < \mu_j^{\natural}(u_-) < \mu_{j0}^{\sharp}(u_-), \quad \overline{\lambda}_j\left(\mu_{j0}^{\sharp}(u_-); u_-\right) = \overline{\lambda}_j\left(\mu_{j0}^{\flat}(u_-); u_-\right). \tag{3.8}$$

It can be checked using the implicit function theorem (along the same lines as in Remark II-4.4 and Lemma 2.4) that $\mu_{j0}^{\flat}$ and $\mu_{j0}^{\sharp}$ are smooth mappings.

PROOF. By (1.25) in the proof of Lemma 1.4 the entropy dissipation has the explicit form

$$\overline{E}(m; u_-) = \int_{\mu_j(u_-)}^{m} \left\{\overline{\lambda}_j(m; u_-) - \overline{\lambda}_j(t; u_-)\right\} g_j(t; u_-)\, dt, \tag{3.9}$$

where

$$g_j(t; u_-) := \frac{dv_j^T}{dt}(t; u_-)\, D^2U(v_j(t; u_-))\, (v_j(t; u_-) - u_-). \tag{3.10}$$

Using the expansion (1.30) along the Hugoniot curve and the strict convexity of U (implying that $r_j^T D^2U\, r_j > 0$), we see that

$$(t - \mu_j(u_-))\, g_j(t; u_-) > 0, \qquad t \neq \mu_j(u_-).$$

When Liu entropy criterion (2.6) holds, $\overline{\lambda}_j(m; u_-) - \overline{\lambda}_j(t; u_-)$ is non-positive and it follows that the entropy dissipation is non-positive. This proves the property (iii) of the lemma.

When, instead, the shock is a rarefaction shock (see (3.4)), the properties stated in Lemma 2.3 show that

$$\overline{\lambda}_j(m; u_-) - \overline{\lambda}_j(t; u_-) \geq 0, \quad t \in [\mu_j(u_-), m]. \tag{3.11}$$

Combining (3.9)–(3.11) shows that the entropy dissipation is positive for rarefaction shocks. This proves the property (ii).

Finally, we establish the property (i) by differentiating (3.9):

$$\frac{\partial}{\partial m}\overline{E}(m; u_-) = \int_{\mu_j(u_-)}^{m} \frac{\partial}{\partial m}\overline{\lambda}_j(m; u_-)\, g_j(t; u_-)\, dt,$$

which relates the entropy dissipation and the shock speed:

$$\frac{\partial}{\partial m}\overline{E}(m; u_-) = G(m; u_-)\, \frac{\partial}{\partial m}\overline{\lambda}_j(m; u_-), \quad G(m; u_-) := \int_{\mu_j(u_-)}^{m} g_j(t; u_-)\, dt, \tag{3.12}$$

with

$$C_1 \left|m - \mu_j(u_-)\right|^2 \leq G(m; u_-) \leq C_2 \left|m - \mu_j(u_-)\right|^2 \tag{3.13}$$

for some positive constants C_1 and C_2.

In view of (3.12), the entropy dissipation reaches a critical value when the shock speed has a critical point and at the point u_-. From the properties of the shock speed (Lemma 2.3), it follows that $\overline{E}(m; u_-)$ is decreasing for $m < \mu_j^{\natural}(u_-)$ and increasing for $m > \mu_j^{\natural}(u_-)$. On the other hand, from its definition it is clear that $\overline{E}(m; u_-)$ vanishes at $m = \mu_j(u_-)$, the values $m < \mu_j^{-\natural}(u_-)$ correspond to rarefaction shocks for which we already checked that the entropy dissipation is negative. Therefore there exists a unique value $\mu_{j0}^{\flat}(u_-)$ in the interval $\left(\mu_j^{-\natural}(u_-), \mu_j^{\natural}(u_-)\right)$ where the dissipation vanishes. This completes the proof of Lemma 3.3. $\square$

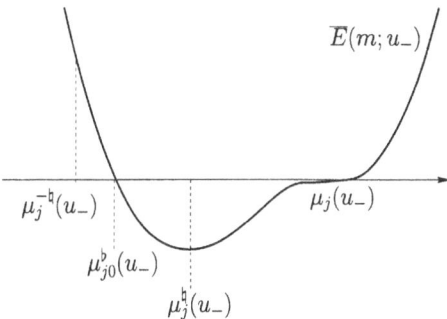

$E(m; u_-)$

$\mu_j^{-\natural}(u_-)$ $\mu_j(u_-)$

$\mu_{j0}^\flat(u_-)$

$\mu_j^\natural(u_-)$

Figure VI-5 : Entropy dissipation in the concave-convex case.

Nonclassical wave sets for general characteristic fields.

We now construct a multi-parameter family of solutions to the Riemann problem (1.1) and (1.2), based on the single entropy inequality (1.5). For each j-wave family we define here a "wave set" consisting of all states reachable from a given left-hand state using only admissible j-waves.

Consider a j-wave fan with left-hand state u_- and right-hand state u with $\mu_j(u_-) \geq 0$ and $j \in J_2$. (Recall that the wave curve was already constructed in Section 1 when $j \in J_0 \cup J_1$.) We consider first a concave-convex field. Recall (Section 2) that the j-wave fan using only classical waves contains

  (a) a rarefaction from u_- to $u \in \mathcal{O}_j(u_-)$ if $\mu_j(u) > \mu_j(u_-)$,

  (b) or a classical shock from u_- to $u \in \mathcal{H}_j(u_-)$ if $\mu_j(u) \in \big(\mu_j^\natural(u_-), \mu_j(u_-)\big)$,

  (c) or else a classical shock from u_- to $u_-^\natural := v_j(\mu_j^\natural(u_-); u_-)$ followed by an attached rarefaction connecting to $u \in \mathcal{O}_j(u_-^\natural)$ if $\mu_j(u) < \mu_j^\natural(u_-)$.

In the special case that $\mu_j(u_-) = 0$, the j-wave curve is the j-integral curve issuing from u_-. This completes the description of the **classical j-wave curve** $\mathcal{W}_j^c(u_-)$.

Given a left-hand state u_-, the set of all states that can be reached using only j-waves is called the **nonclassical j-wave set issuing from** u_- and denoted by $\mathcal{X}_j(u_-)$. (See Figure VI-6.)

THEOREM 3.4. (Nonclassical j-wave set – Concave-convex field.) *In addition to the classical one, the j-wave fan may contain a nonclassical j-shock connecting u_- to some intermediate state $u_+ \in \mathcal{H}_j(u_-)$ with $\mu_j(u_+) \in \big[\mu_{j0}^\flat(u_-), \mu_j^\natural(u_-)\big)$ followed by*

  (a) *either a non-attached rarefaction connecting u_+ to $u \in \mathcal{O}_j(u_+)$ if $\mu_j(u) < \mu_j(u_+)$,*

  (b) *or by a classical shock connecting u_+ to $u \in \mathcal{H}_j(u_+)$ if $\mu_j(u) > \mu_j(u_+)$.*

This defines a two-parameter family of right-hand states u which can be reached from u_- by nonclassical solutions.

We now consider a *convex-concave* characteristic field. Recall from Section 2 that the j-wave fan using only classical waves contains

(a) a classical shock connecting u_- to $u \in \mathcal{H}_j(u_-)$ if either $\mu_j(u) \geq \mu_j(u_-)$ or $\mu_j(u) \leq \mu_j^{-\natural}(u_-)$,

(b) or a rarefaction connecting u_- to $u \in \mathcal{O}_j(u_-)$ if $\mu_j(u) \in [0, \mu_j(u_-)]$,

(c) or a rarefaction wave connecting u_- to a point u_+, followed by an attached classical shock connecting to $u \in \mathcal{H}_j(u_+)$ with $\mu_j^\natural(u) = \mu_j(u_+)$, if $\mu_j(u) \in \left(\mu_j^{-\natural}(u_-), 0\right)$. (In this latter case, the set of u does not describe a rarefaction or shock curve.)

This completes the description of the classical j-wave curve $\mathcal{W}_j^c(u_-)$.

THEOREM 3.5. (Nonclassical j-wave set – Convex-concave field.) *The j-wave fan may also contain*

(a) *a rarefaction to $u_+ \in \mathcal{O}_j(u_-)$ with $\mu_j(u_+) \in (0, \mu_j(u_-))$, possibly followed by a non-attached nonclassical shock which connects the intermediate state u_+ to u, if $\mu_j(u) \in \left(\mu_j^{-\natural}(u_+), \mu_{j0}^\natural(u_+)\right]$; (in this case, the set of u does not describe a rarefaction or shock curve);*

(b) *or a classical shock to $u_+ \in \mathcal{H}_j(u_-)$ with $\mu_j(u_+) > \mu_j(u_-)$, followed by a nonclassical shock connecting to $u \in \mathcal{H}_j(u_+)$, if $\mu_j(u) \in \left(\mu_j^{-\natural}(u_+), \mu_{j0}^\flat(u_+)\right]$.*

This defines a two-parameter wave set $\mathcal{X}_j(u_-)$ of right-hand states u which can be reached from u_- by nonclassical solutions.

PROOF OF THEOREM 3.4. To construct the wave set $\mathcal{X}_j(u_-)$ for $u_- \in \mathcal{U}$ and $j \in J_2$ in the concave-convex case and we proceed as follows. Consider a point u_- away from the manifold with $\mu_j(u_-) > 0$. The construction of the wave curve will depend on the values $\mu_{j0}^\flat(u_-) < \mu_j^\natural(u_-) < \mu_{j0}^\natural(u_-)$ introduced earlier.

Recall first the construction of the classical part of the wave set. Considering first the region $\mu_j(u) > \mu_j(u_-)$, we see that the state u_- can be connected to any point on $\mathcal{O}_j(u_-)$ by a rarefaction, since the wave speed λ_j is increasing for $\mu_j(u)$ increasing (Lemma 2.2). Therefore, the wave set $\mathcal{X}_j(u_-)$ coincides with the rarefaction curve $\mathcal{O}_j(u_-)$ for $\mu_j(u) \geq \mu_j(u_-)$.

For $\mu_j(u)$ decreasing from $\mu_j(u_-)$ the shock speed is decreasing as long as $\mu_j(u)$ remains larger than the critical value $\mu_j^\natural(u_-)$ (Lemma 2.3). Therefore, all of the points in the Hugoniot curve $\mathcal{H}_j(u_-)$ with $\mu_j(u) \in [\mu_j^\natural(u_-), \mu_j(u_-)]$ can be reached from u_- by a classical shock satisfying Lax shock inequalities and Liu entropy criterion. According to Lemma 3.3, the entropy dissipation remains non-positive in the whole range $\mu_j(u) \in [\mu_{j0}^\flat(u_-), \mu_j(u_-)]$, thus the points of the Hugoniot curve $\mathcal{H}_j(u_-)$ with $\mu_j(u) \in [\mu_{j0}^\flat(u_-), \mu_j^\natural(u_-))$ can also be reached from u_- but, now, using a nonclassical shock. These are the only admissible solutions with a single j-wave issuing from u_-.

Consider now an admissible solution containing a single wave joining u_- to a state u_+. It is not difficult to see that if $\mu_j(u_+) > \mu_j^\natural(u_-)$ no further j-wave can be constructed from u_+ which would travel faster than the first wave. So, consider a nonclassical shock joining u_- to a state u_+ with $\mu_j(u_+) \in [\mu_{j0}^\flat(u_-), \mu_j^\natural(u_-))$. According to Lemma 2.3, the wave speed is increasing when μ_j decreases from $\mu_j(u_+) < 0$, so u_+ can be connected to any point $u = u_2$ in the rarefaction curve $\mathcal{O}_j(u_+)$ with $\mu_j(u_2) \leq \mu_j(u_+)$. Observe that the nonclassical shock is *not attached* to the rarefaction fan, i.e.,

$$\overline{\lambda}_j(u_-, u_+) < \lambda_j(u_+). \tag{3.14}$$

This describes all of the solutions containing a nonclassical shock followed by a rarefaction. It is not difficult to check that no further j-wave may follow the rarefaction.

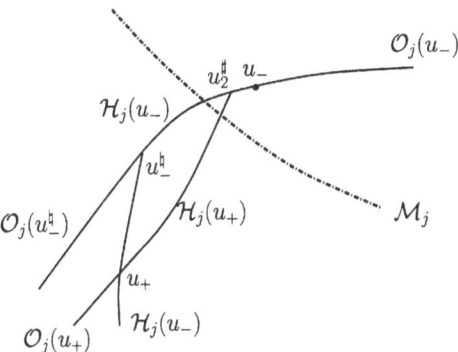

Figure VI-6 : Wave set in the concave-convex case.

Consider again a nonclassical shock joining u_- to u_+ with

$$\mu_j(u_+) \in \left[\mu_{j0}^\flat(u_-), \mu_j^\natural(u_-) \right).$$

The shocks with small strength issuing from u_+ have a larger speed than that of the nonclassical shock, i.e., $\overline{\lambda}_j(u_+, u_2) \approx \lambda_j(u_+) > \overline{\lambda}_j(u_-, u_+)$, for all states u_2 close to u_+. Hence, the speeds have the proper ordering and u_+ may be connected to any $u_2 \in \mathcal{H}_j(u_+)$, at least in the small. Such a shock is also admissible when μ_j increases (according to Lax shock inequalities and Liu entropy criterion) since the wave speed is decreasing when μ_j increases (Lemma 2.3). This construction can be continued, for u_+ given, until u_2 violates either of the two conditions:

$$\overline{\lambda}_j(u_+, u_2) > \overline{\lambda}_j(u_-, u_+), \qquad (3.15)$$

$$E(u_+, u_2) \leq 0. \qquad (3.16)$$

Actually, as $\mu_j(u_2)$ increases from $\mu_j(u_+)$ one reaches a maximum value, in which equality holds in (3.15) while the shock is still classical (and therefore (3.16) still holds). To check this property, consider the graphs of the two functions $h(m) := \overline{\lambda}_j(u_-, v_j(m; u_-))$ and $k(m) := \overline{\lambda}_j(u_+, v_j(m; u_+))$. By symmetry of the Rankine-Hugoniot relation one has $\overline{\lambda}_j(u_+, u_-) = \overline{\lambda}_j(u_-, u_+)$, so

$$\sigma := h(\mu_j(u_+)) = k(\mu_j(u_-)). \qquad (3.17)$$

In view of Lemma 2.3 and (2.10), the two graphs must intersect at exactly one point $m_2^\natural$ in the interval $\left(\mu_j(u_+), \mu_j(u_-) \right)$. We define $u_2^\natural$ by the conditions $\mu_j(u_2^\natural) = m_2^\natural$ and $u_2^\natural \in \mathcal{H}_j(u_+)$.

On the other hand, let us consider the point $u_3^\sharp \in \mathcal{H}_j(u_+)$ satisfying $\overline{\lambda}_j(u_+, u_3^\sharp) = \overline{\lambda}_j(u_-, u_+) = \sigma$. From the Rankine-Hugoniot relations

$$-\sigma \left(u_+ - u_- \right) + f(u_+) - f(u_-) = 0$$

and

$$-\sigma \left(u_3^\sharp - u_+ \right) + f(u_3^\sharp) - f(u_+) = 0,$$

we deduce that

$$-\sigma \left(u_3^\sharp - u_- \right) + f(u_3^\sharp) - f(u_-) = 0,$$

which proves that $\overline{\lambda}_j(u_-, u_3^\sharp) = \sigma$ and $u_3^\sharp \in \mathcal{H}_j(u_-)$.

From the above discussion we conclude that $u_2^\sharp = u_3^\sharp$ and

$$h(m_2^\sharp) = k(m_2^\sharp) = \sigma, \quad u_2^\sharp \in \mathcal{H}_j(u_-). \tag{3.18}$$

Then, it follows also that (3.15) holds for all $\mu_j(u_2) < m_2^\sharp$, and the equality holds in (3.15) at the critical value $u_2^\sharp$. Moreover, since $m_2^\sharp < \mu_j^\sharp(u_-)$, the shock speed is decreasing on the interval $\left(\mu_j(u_+), m_2^\sharp \right)$ and any shock from u_+ to u_2 (with $\mu_j(u_2) \leq m_2^\sharp$) satisfies Liu entropy criterion.

We have the inequalities $\mu_j(u_+) < \mu_j^\natural(u_-) < m_2^\sharp < \mu_j(u_-)$. As $\mu_j(u_+)$ increases, both $\mu_j(u_+)$ and $m_2^\sharp$ approach the limiting value $\mu_j^\natural(u_-)$. As $\mu_j(u_+)$ decreases, both $\mu_j(u_+)$ and $m_2^\sharp$ approach the limiting value $\mu_{j0}^\flat(u_-)$, while $m_2^\sharp$ approaches a limiting value which we denote by $\mu_{j0}^\sharp(u_-)$. Finally, one can also check from the properties of the wave speeds, that no third wave can follow a two-wave fan. This completes the proof of Theorem 3.4. $\square$

PROOF OF THEOREM 3.5. For $u_- \in \mathcal{M}_j$ it is not hard to see, using the properties (3.7ii), that $\mathcal{W}_j(u_-)$ coincides with the Hugoniot curve $\mathcal{H}_j(u_-)$. This is because the wave speed is decreasing when moving away from u_- in either direction. The construction is complete for $u_- \in \mathcal{M}_j$.

Consider the case $\mu_j(u_-) > 0$ and recall first the construction of the classical part of the wave set. For $\mu_j > \mu_j(u_-)$ the state u_- can be connected to any point on $\mathcal{H}_j(u_-)$ since the wave speed is decreasing for μ_j increasing. For $\mu_j < \mu_j(u_-)$, the wave speed is, locally, increasing for μ_j decreasing. So u_- can be connected to a point on $\mathcal{O}_j(u_-)$ by a rarefaction. This remains possible until μ_j reaches the value 0. It is also possible to connect any point $u_+ \in \mathcal{O}_j(u_-)$ satisfying $\mu_j(u_+) \in \left[0, \mu_j(u_-)\right]$ to a point $u_2 \in \mathcal{H}_j(u_+)$ provided

$$\overline{\lambda}_j(u_+, u_2) = \lambda_j(u_+). \tag{3.19}$$

This construction covers the range $\mu_j \in \left[\mu_j^{-\natural}(u_-), 0\right]$. It is also possible to connect u_- directly to a point $u \in \mathcal{H}_j(u_-)$ with $\mu_j(u) \leq \mu_j^{-\natural}(u_-)$, since the shock speed in this range satisfies Liu entropy criterion. This completes the construction of the classical wave curve $\mathcal{W}_j^c(u_-)$.

We now describe all nonclassical solutions with two j-waves. Consider an admissible one-wave solution from u_- to u_+. Suppose first that $\mu_j(u_+) \in \left(0, \mu_j(u_-)\right)$ and $u_+ \in \mathcal{O}_j(u_-)$. Then, one can connect u_+ to $u_2 \in \mathcal{H}_j(u_+)$ by a shock provided both conditions

$$\overline{\lambda}_j(u_+, u_2) \geq \lambda_j(u_+), \tag{3.20}$$

$$E(u_+, u_2) \leq 0 \tag{3.21}$$

hold. From the properties of the entropy dissipation (Lemma 3.3) we know that (3.21) is equivalent to

$$\mu_j(u_2) \leq \mu_{j0}^\flat(u_+) \quad (\text{or } \mu_j(u_2) \geq \mu_j(u_+)).$$

In view of the graph of the shock speed, (3.20) reads

$$\mu_j^{-\natural}(u_+) \leq \mu_j(u_2) \leq \mu_j(u_+).$$

Since we always have $\mu_{j0}^\flat(u_+) \in \left[\mu_j^{-\natural}(u_+), \mu_j^{\natural}(u_+)\right]$, it follows that the admissible interval in the case under consideration is $\mu_j(u_2) \in \left[\mu_j^{-\natural}(u_+), \mu_{j0}^\flat(u_+)\right]$. Moreover such a shock is classical only when $\mu_j(u_2) \leq \mu_j^{-\natural}(u_+)$, that is, only when $\mu_j(u_2) = \mu_j^{-\natural}(u_+)$.

Suppose now that $\mu_j(u_+) \geq \mu_j(u_-)$ with $u_+ \in \mathcal{H}_j(u_-)$. The, one can connect u_+ to a point $u_2 \in \mathcal{H}_j(u_+)$ provided

$$\overline{\lambda}_j(u_+, u_2) \geq \overline{\lambda}_j(u_-, u_+) \tag{3.22}$$

and

$$E(u_+, u_2) \leq 0. \tag{3.23}$$

The condition (3.23) is equivalent to saying that $\mu_j(u_2) \leq \mu_{j0}^\flat(u_+)$ (or $\mu_j(u_2) \geq \mu_j(u_+)$). As $\mu_j(u_2)$ decreases from $\mu_{j0}^\flat(u_+)$, the speed $\overline{\lambda}_j(u_+, u_2)$ satisfies (3.22) initially, decreases, and eventually reaches the value $\overline{\lambda}_j(u_-, u_+)$. Since $u_+ \in \mathcal{H}_j(u_-)$ and $u_2 \in \mathcal{H}_j(u_+)$ the same argument as in the concave-convex case shows that for that value of μ_j one has $u_2 \in \mathcal{H}_j(u_-)$.

This completes the description of the two wave patterns and, therefore, the proof of Theorem 3.5. □

4. Kinetic relations and nonclassical Riemann solver

In view of Theorems 3.4 and 3.5 the nonclassical wave set $\mathcal{X}_j(u_-)$ is a two-dimensional manifold when $j \in J_2$. It is our objective now to select the nonclassical wave curve $\mathcal{W}_j^{nc}(u_-)$ within the wave set $\mathcal{X}_j(u_-)$. In view of Theorems 3.4 and 3.5 one parameter should be prescribed for each non-genuinely nonlinear wave family. Generalizing the approach in Chapter II for scalar conservation laws, we postulate that for all u_- and $j \in J_2$ a *single* right-hand state u_+ can be reached from u_- with a nonclassical shock. As already pointed out for scalar conservation laws, the *kinetic function* to be introduced now is a given "constitutive function" which represents certain small-scale effects neglected at the hyperbolic level of modeling. For definiteness, we restrict now attention to *concave-convex fields*. (The results in this section extend to convex-concave fields by relying on Theorem 3.5 instead of Theorem 3.4).

DEFINITION 4.1. For each $j \in J_2$ a **kinetic function** for the j-characteristic field is a Lipschitz continuous mapping $\mu_j^\flat : \mathcal{B}(\delta_1) \to I\!R$ satisfying

$$\begin{aligned} \mu_{j0}^\flat(u) < \mu_j^\flat(u) \leq \mu_j^\natural(u), \quad \mu_j(u) > 0, \\ \mu_j^\natural(u) \leq \mu_j^\flat(u) < \mu_{j0}^\flat(u), \quad \mu_j(u) < 0, \end{aligned} \tag{4.1}$$

We shall say that a solution $u = u(x, t)$ (in the class **P**) satisfies the **kinetic relation** associated with the kinetic function $\mu_j^\flat$ if for every j-nonclassical shock the right-hand state u_+ is determined from the left-hand state u_- by

$$u_+ = v_j(m; u_-) \quad \text{with } m = \mu_j^\flat(u_-). \tag{4.2}$$

$\square$

To the kinetic function we shall associate its companion function $\mu_j^\sharp(u)$ by

$$\overline{\lambda}_j(u_-, u_+^\sharp) = \overline{\lambda}_j(u_-, u_+^\flat),$$
$$\text{where } u_+^\flat := v_j(\mu_j^\flat(u_-); u_-), \ u_+^\sharp := v_j(\mu_j^\sharp(u_-); u_-). \tag{4.3}$$

It can be checked, by the implicit function theorem for Lipschitz continuous mappings, that $\mu_j^\sharp(u)$ exists and depends Lipschitz continuously upon its argument. From the discussion in Chapter II we know that an additional constraint is needed to avoid selecting the classical solution.

DEFINITION 4.2. We shall say that a weak solution u satisfies the **nucleation criterion** associated with the kinetic function $\mu_j^\flat$ if for every classical shock connecting u_- to u_+ we have

$$\mu_j^\sharp(u_-) \le \mu_j(u_+) \le \mu_j(u_-) \quad \text{when } \mu_j(u_-) \ge 0,$$
$$\mu_j(u_-) \le \mu_j(u_+) \le \mu_j^\sharp(u_-) \quad \text{when } \mu_j(u_-) \le 0. \tag{4.4}$$

$\square$

THEOREM 4.3. (Nonclassical Riemann solver.) *Suppose the system (1.1) admits linearly degenerate, genuinely nonlinear, and concave-convex fields.*

(a) *For $j \in J_2$ consider a (Lipschitz continuous) kinetic function $\mu_j^\flat$ (satisfying (4.1)). Then, for each $u_- \in \mathcal{B}(\delta_1)$ the kinetic relation (4.2) and the nucleation criterion (4.4) select a unique **nonclassical j-wave curve** $\mathcal{W}_j^{nc}(u_-)$ within the wave set $\mathcal{X}_j(u_-)$. When $\mu_j(u_-) > 0$ it is composed of the following four pieces:*

$$\mathcal{W}_j^{nc}(u_-) = \begin{cases} \mathcal{O}_j(u_-), & \mu_j(u) \ge \mu_j(u_-), \\ \mathcal{H}_j(u_-), & \mu_j^\sharp(u_-) \le \mu_j(u) \le \mu_j(u_-), \\ \mathcal{H}_j(u_+^\flat), & \mu_j^\flat(u_-) \le \mu_j(u) < \mu_j^\sharp(u_-), \\ \mathcal{O}_j(u_+^\flat), & \mu_j(u) \le \mu_j^\flat(u_-), \end{cases}$$

where $u_+^\flat := v_j(\mu_j^\flat(u_-); u_-)$. The Riemann solution is a single rarefaction shock, or a single classical shock, or a nonclassical shock followed by a classical shock, or finally a nonclassical shock followed by a rarefaction, respectively. The curve $\mathcal{W}_j^{nc}(u_-)$ is continuous and monotone in the parameter $m = \mu_j(u)$. It is of continuously differentiable with bounded second-order derivatives for all $m \ne \mu_j^\sharp(u_-)$ and Lipschitz continuous (at least) at $m = \mu_j^\sharp(u_-)$.

(b) *For all u_l and u_r in $\mathcal{B}(\delta_1)$ the Riemann problem (1.1) and (1.2) admits a unique solution determined by combining together the (classical) wave curves $\mathcal{W}_j(u_-)$ for $j \in J_0 \cup J_1$ and the (nonclassical) wave curves $\mathcal{W}_j^{nc}(u_-)$ for $j \in J_2$.*

PROOF. Let $u_- \in \mathcal{U}$ and $j \in J_2$ be given. In view of the assumption (4.1) the criterion (4.2) selects a unique nonclassical shock along the Hugoniot curve $\mathcal{H}_j(u_-)$, which we denote by $u_+^b := v_j(\mu_j^b(u_-); u_-)$. Once this state is selected the construction in Theorem 3.4 determines a unique wave curve $\mathcal{W}_j^{nc}(u_-)$ having the form described in the theorem. Furthermore, without the nucleation criterion (4.4) the classical wave curve $\mathcal{W}_j^c(u_-)$ is admissible, since the kinetic relation does not prevent one from solving the Riemann problem by using classical waves only. The nucleation criterion precisely excludes the single shock solution when the nonclassical construction is available.

The nonclassical wave curve is continuous in the parameter μ_j which, by construction, is monotone increasing along it. Finally, having constructed the (possibly only Lipschitz continuous) wave curves $\mathcal{W}_j^{nc}$ for $j \in J_2$ and the (smooth) wave curves $\mathcal{W}_j^c$ for $j \notin J_2$, and using the condition that $\{r_k\}$ is a basis of $\mathbb{R}^N$, we can solve the Riemann problem with data in $\mathcal{B}(\delta_1)$ by combining together the wave curves and relying on the implicit function theorem for Lipschitz continuous mappings. $\square$

In Theorem 4.3, for $j \in J_2$ we can also recover the classical wave curve $\mathcal{W}_j^c(u_-)$ with the trivial choice $\mu_j^b(u) = \mu_j^\sharp(u)$ for all u. With this choice, the nonclassical shock have the maximal negative entropy dissipation while another particular choice, $\mu_j^b(u) = \mu_{j0}^b(u)$, leads to nonclassical shocks with vanishing entropy dissipation.

As was already pointed out for scalar conservation laws (in Section II-4), the nonclassical Riemann solution depends continuously in the L^1 norm upon its initial data, but not in a pointwise sense. In the classical solution, the value of the intermediate state (if any) in the Riemann solution varies continuously as $u_+ \in \mathcal{W}_j^c(u_-)$ describes the wave curve; the solution in the (x, t) plane varies continuously in the L^1 norm and its total variation is a continuous function of the end points.

Along a nonclassical wave curve, the speeds of the (rarefaction or shock) waves change continuously. We simply observe that at the point $\mu_j^\sharp(u_-)$ one has to compare, on one hand, the shock speed of the nonclassical shock and, on the other hand, the shock speeds of the nonclassical and the classical shocks. All three terms coincide at $\mu_j^\sharp(u_-)$, i.e.,

$$\lim_{\substack{m \to \mu_j^\sharp(u_-) \\ m > \mu_j^\sharp(u_-)}} \overline{\lambda}_j(u_-, v_j(m; u_-)) = \lim_{\substack{m \to \mu_j^\sharp(u_-) \\ m < \mu_j^\sharp(u_-)}} \overline{\lambda}_j(u_+^b, v_j(m; u_+^b)) = \overline{\lambda}_j(u_-, u_+^b).$$

(4.5)

The continuous dependence of the wave speeds implies the L^1 continuous dependence of the solution. For the nonclassical wave curve the wave speeds (only) are continuous and the total variation of the nonclassical Riemann solution is *not* a continuous function of its end points. This lack of continuity makes it delicate to control the strengths of waves at interactions; see Chapter VIII which is concerned with the Cauchy problem.

REMARK 4.4.
- We may also constrain the entropy dissipation $E(u_-, u_+)$ of nonclassical shocks through a kinetic relation of the general form

$$E(u_-, u_+) = \phi_j(u_-, u_+).$$

(4.6)

- A special class of such kinetic functions of particular interest is based on the *entropy dissipation function* and depends solely on the shock speed $\overline{\lambda}_j(u_-, u_+)$, i.e.,

$$E(u_-, u_+) = \varphi\big(\overline{\lambda}_j(u_-, u_+)\big). \tag{4.7}$$

A left-hand state u_- being fixed, one observes that the entropy dissipation along the Hugoniot curve (when re-written as a function of the shock speed) is increasing from its maximal negative value

$$E_j^\flat(u_-) = \min_{u_+ \in \mathcal{H}_j(u_-)} E(u_-, u_+),$$

achieved at $\lambda^\flat := \overline{\lambda}_j\big(u_-, w_j(\mu_j^\flat(u_-); u_-)\big)$, to the value 0 which is achieved at the speed $\lambda = \overline{\lambda}_j\big(u_-, w_j(\mu_{j0}^\flat(u_-); u_-)\big)$. Provided the function $\varphi(s)$ is decreasing and that $\varphi\big(\overline{\lambda}_j(u_-, u_+)\big)$ lies in the interval

$$E_j^\flat(u_-) \leq \varphi\big(\overline{\lambda}_j(u_-, u_+)\big) \leq 0,$$

there exists a unique point $m = \mu_j^\flat(u_-)$ such that the kinetic relation (4.7) is satisfied.

- In the applications concerning scalar conservation laws and the 2×2 system of nonlinear elastodynamics, it turns out that the kinetic function can always be expressed as a function of the shock speed, i.e., in the form (4.7). In many physical systems, the entropy dissipation is related to the mechanical energy and regarded as **a force driving the propagation** of discontinuities. The kinetic relation (4.7) imposes a one-to-one relationship between the propagation speed and the driving force.

$\square$

CHAPTER VII

CLASSICAL ENTROPY SOLUTIONS
OF
THE CAUCHY PROBLEM

In this chapter we establish the existence of a *classical* entropy solution to the Cauchy problem associated with a strictly hyperbolic system of conservation laws when the initial data have small total variation. We cover here the general class of systems whose each characteristic field is either genuinely nonlinear or concave-convex. With minor changes, the results in this chapter extend to linearly degenerate and convex-concave fields. In Section 1 we discuss fundamental properties of (exact and approximate) classical entropy solutions to the Riemann problem, studied earlier in Sections VI-1 and VI-2. The key property is given by the *interactions estimates* in Theorem 1.1: at each interaction, the wave strengths may increase by an amount which is bounded by the product of the strengths of the two incoming waves. In Section 2 we describe the *approximation scheme* which generalizes the one given in Section IV-2 for scalar conservation laws, and we state the main *existence result;* see Theorem 2.1. Technical aspects of the proof are postponed to Section 3. Finally, in Section 4 we briefly discuss *pointwise regularity* properties of the solutions.

1. Glimm interaction estimates

Consider the system

$$\partial_t u + \partial_x f(u) = 0, \quad u = u(x,t) \in \mathcal{U}, \ x \in \mathbb{R}, \ t > 0. \tag{1.1}$$

The set $\mathcal{U} := \mathcal{B}(\delta_0)$ is a ball with center $0 \in \mathbb{R}^N$ and radius δ_0, and the flux $f : \mathcal{U} \to \mathbb{R}^N$ is assumed to be strictly hyperbolic. For each u we denote by $\lambda_1(u) < \ldots < \lambda_N(u)$ the eigenvalues of the matrix $Df(u)$ and by $l_j(u)$ and $r_j(u)$, $1 \leq j \leq N$ corresponding basis of left- and right-eigenvectors. For δ_0 sufficiently small the *averaging matrix*

$$\overline{A}(u_-, u_+) := \int_0^1 Df\big(\theta\, u_- + (1-\theta)\, u_+\big)\, d\theta$$

is also strictly hyperbolic for all $u_-, u_+ \in \mathcal{B}(\delta_0)$. We denote by $\overline{\lambda}_i(u_-, u_+), \overline{l}_i(u_-, u_+)$, and $\overline{r}_i(u_-, u_+)$ its eigenvalues and left- and right-eigenvectors, respectively, normalized so that

$$\overline{l}_i(u_-, u_+)\, \overline{r}_i(u_-, u_+) = 1.$$

Exact Riemann solver.
The Riemann problem associated with (1.1) and

$$u(x,0) = \begin{cases} u_l, & x < 0, \\ u_r, & x > 0, \end{cases} \tag{1.2}$$

where u_l and $u_r \in \mathcal{B}(\delta_2)$ and $\delta_2 < \delta_0$, was solved in Sections VI-1 and VI-2, in the class of classical entropy solutions. Let us summarize the main results as follows. It is convenient to introduce, for each wave family, a *global parameter* $\mu_i = \mu_i(u)$ such that $\nabla \mu_i \cdot r_i \neq 0$. For $\delta_1 < \delta_0$ sufficiently small and for all $u_- \in \mathcal{B}(\delta_1)$ and $1 \leq i \leq N$ the *i-wave curve* $\mathcal{W}_i(u_-)$ issuing from u_- is parameterized by a mapping $m \mapsto \psi_i(m; u_-)$ with

$$\mu_i(\psi_i(m; u_-)) = m$$

for all m varying in some open and bounded interval containing $\mu_i(u_-)$. Each state $\psi_i(m; u_-)$ can be connected to u_- on the right by elementary i-waves. For genuinely nonlinear fields one obtains a i-*rarefaction wave* when $m \geq \mu_i(u_-)$ and a i-*shock wave* when $m < \mu_i(u_-)$. (See Section VI-1.) For concave-convex fields the wave curves are made of three different parts: a single rarefaction wave, a single shock wave, or else *a right-contact plus a rarefaction wave*. (See Section VI-2.) Moreover, the mapping ψ_i has bounded second-order derivatives in $(m; u_-)$ with

$$\psi_i(m; u_-) = u_- + \frac{m - \mu_i(u_-)}{(\nabla \mu_i \cdot r_i)(u_-)} \, r_i(u_-) + O(m - \mu_i(u_-))^2. \tag{1.3}$$

The solution of the Riemann problem contains (at most) N wave fans associated with each of the characteristic families. Since $\{r_i\}$ is a basis of $\mathbb{R}^N$ and in view of (1.3), for any fixed $u_l \in \mathcal{B}(\delta_2)$ (with $\delta_2 < \delta_1$ sufficiently small) the mapping

$$(s_1, \dots, s_N) \mapsto \Psi(s_1, \dots, s_N; u_l) = u_N,$$
$$u_0 := u_l, \quad u_i := \psi_i(\mu_i(u_{i-1}) + s_i; u_{i-1}), \quad 1 \leq i \leq N,$$

is one-to-one from a neighborhood of 0 in $\mathbb{R}^N$ onto a subset of $\mathcal{B}(\delta_0)$ containing $\mathcal{B}(\delta_2)$. For u_l and u_r in $\mathcal{B}(\delta_2)$ the **wave strengths** $\sigma_i = \sigma_i(u_l, u_r)$ of the Riemann solution of (1.1) and (1.2) are defined implicitly by

$$\Psi(\sigma_1, \dots, \sigma_N; u_l) = u_r.$$

They have second-order bounded derivatives, and they are equivalent to the usual distance in $\mathbb{R}^N$ in the sense that for some constant $C \geq 1$

$$\frac{1}{C} \, |u_r - u_l| \leq \sum_{i=1}^{N} |\sigma_i(u_l, u_r)| \leq C \, |u_r - u_l|. \tag{1.4}$$

It will be convenient to introduce the general notation

$$\sigma_k(u_l, u_r) = \sigma_k(u_l, u_r)^S + \sigma_k(u_l, u_r)^R, \quad 1 \leq k \leq N,$$

where $\sigma_k(u_l, u_r)^S$ and $\sigma_k(u_l, u_r)^R$ represent the strengths of the k-shock wave and of the k-rarefaction wave in the corresponding Riemann solution, respectively. Observe in passing that $\sigma_k(u_l, u_r)^S$ and $\sigma_k(u_l, u_r)^R$ always have the same sign so that, for instance,

$$\max \big(|\sigma_k(u_l, u_r)^S|, |\sigma_k(u_l, u_r)^R| \big) \leq |\sigma_k(u_l, u_r)|.$$

In this section our main objective is to derive "wave interaction estimates". That is, we consider the solution of the Cauchy problem associated with (1.1) when the initial data take *three constant values*:

$$u(x,0) = \begin{cases} u_l, & x < -1, \\ u_m, & -1 < x < 1, \\ u_r, & x > 1, \end{cases} \tag{1.5}$$

where u_l, u_m, and $u_r \in \mathcal{B}(\delta_2)$. By combining the two Riemann solutions associated with the left- and right-hand data u_l, u_m and u_m, u_r, respectively, it is easy to construct the solution of (1.1) and (1.5) for *small* time. Waves originating from the initial discontinuities located at $x = -1$ and $x = 1$ propagate until their trajectories meet eventually. After all possible interactions have taken place, the solution has reached an asymptotic state which is determined by the Riemann solution connecting u_l to u_r. The **wave interaction estimates** relate the wave strengths of the two **incoming** Riemann solutions with the ones of the **outgoing** Riemann solution. For instance, if the flux f were linear we would simply write

$$\sigma_k(u_l, u_r) = \sigma_k(u_l, u_m) + \sigma_k(u_m, u_r), \quad 1 \le k \le N. \tag{1.6}$$

The formula (1.6) extends to nonlinear flux-functions *up to a quadratic error term*. Since this is sufficient for our purpose we assume that each incoming Riemann solutions contains a *single* wave fan.

THEOREM 1.1. (Glimm interaction estimates – Exact Riemann solver.) *For all u_l, u_m, and $u_r \in \mathcal{B}(\delta_2)$ we have the following property. Suppose that u_l is connected to u_m by an i-wave fan and that u_m is connected to u_r by a j-wave fan $(1 \le i, j \le N)$. Then, the wave strengths $\sigma_k(u_l, u_r)$ of the outgoing Riemann solution satisfy*

$$\sigma_k(u_l, u_r) = \sigma_k(u_l, u_m) + \sigma_k(u_m, u_r) + O(1)\, Q_{\mathrm{in}}(u_l, u_m, u_r), \quad 1 \le k \le N,$$

$$= \begin{cases} \sigma_i(u_l, u_m) + O(1)\, Q_{\mathrm{in}}(u_l, u_m, u_r), & k = i \ne j, \\ \sigma_j(u_m, u_r) + O(1)\, Q_{\mathrm{in}}(u_l, u_m, u_r), & k = j \ne i, \\ \sigma_i(u_l, u_m) + \sigma_j(u_m, u_r) + O(1)\, Q_{\mathrm{in}}(u_l, u_m, u_r), & k = j = i, \\ O(1)\, Q_{\mathrm{in}}(u_l, u_m, u_r), & otherwise, \end{cases} \tag{1.7}$$

where the **interaction potential** *between the two* incoming *waves is defined as*

$$Q_{\mathrm{in}}(u_l, u_m, u_r) = |\sigma_i(u_l, u_m)\, \sigma_j(u_m, u_r)|$$

and the symbol $O(1)$ denotes some uniformly bounded functions.

With the notation of the theorem note that, when either $i < j$ or else $i = j$ and both incoming waves are rarefaction waves, the waves do not truly interact and the formula (1.6) holds (without error term).

PROOF. We can describe the set of solutions under consideration by fixing the left-hand state u_l and using the wave strengths

$$s_l := \sigma_i(u_l, u_m), \quad s_r := \sigma_j(u_m, u_r)$$

as parameters. The state u_r is regarded as a function of s_l and s_r, that is,

$$u_r := \psi_j(\mu_j(u_m) + s_r; u_m), \quad u_m := \psi_i(\mu_i(u_l) + s_l; u_l).$$

We set

$$H_k(s_l, s_r) := \sigma_k(u_l, u_r) - \begin{cases} s_l, & k = i \neq j, \\ s_r, & k = j \neq i, \\ s_l + s_r, & k = j = i, \\ 0, & \text{otherwise,} \end{cases}$$

and

$$K(s_l, s_r) = |s_l \, s_r|.$$

Obviously, if $K(s_l, s_r) = 0$, then either $s_l = 0$ or $s_r = 0$ and one of the incoming wave is trivial. In both cases we have $H_k(s_l, s_r) = 0$. This motivates us to show that

$$\left| H_k(s_l, s_r) \right| \leq C \, K(s_l, s_r) \tag{1.8}$$

for all relevant values s_l, s_r and for some constant $C > 0$. Indeed, we have

$$H_k(s_l, s_r) = H_k(s_l, 0) + \int_0^{s_r} \frac{\partial H_k}{\partial s_r}(s_l, \sigma'') \, d\sigma''$$

$$= H_k(s_l, 0) + \int_0^{s_r} \left(\frac{\partial H_k}{\partial s_r}(0, \sigma'') + \int_0^{s_l} \frac{\partial^2 H_k}{\partial s_l \partial s_r}(\sigma', \sigma'') \, d\sigma' \right) d\sigma'',$$

which gives (1.8) with

$$C := \sup \left| \frac{\partial^2 H_k}{\partial s_l \partial s_r} \right|,$$

since $H_k(s_l, 0) = H_k(0, s_r) = 0$ and the functions σ_k and ψ_k have bounded second-order derivatives.   □

Approximate interaction solvers.

We now generalize Theorem 1.1 to "approximate" wave fronts, since for techni-cal reasons we will need to solve the Riemann problem approximately. Fix some (small) parameter $h > 0$. By definition, a (classical) **approximate i-wave front** $(1 \leq i \leq N + 1)$ is a *propagating discontinuity* connecting two constant states u_- and u_+ at some speed λ, with

    (a) either $i < N + 1$, $\lambda = \overline{\lambda}_i(u_-, u_+) + O(h)$, and $u_+ \in \mathcal{W}_i(u_-)$;

    (b) or $i = N + 1$, the states u_- and u_+ are arbitrary, and the speed $\lambda := \lambda_{N+1}$ is
        a fixed constant satisfying

$$\lambda_{N+1} > \sup_{u \in \mathcal{B}(\delta_0)} \lambda_N(u).$$

In Case (a) the **strength** of the wave is the usual length $\sigma_i(u_-, u_+)$ measured along the wave curve $\mathcal{W}_i(u_-)$. The propagating jump is a (classical approximate) **shock front** if u_+ belongs to the Hugoniot curve starting from u_-, or an (approximate) **rarefaction front** if u_+ belongs to the integral curve starting from u_-. (We will not use the remaining part of concave-convex i-wave curves involving two-wave patterns.) In Case (b) we refer to the front as an **artificial front** or $(N + 1)$-**wave front** and its **strength** is defined by

$$\sigma_{N+1}(u_-, u_+) := |u_+ - u_-|. \tag{1.9}$$

Observe that a shock front need not propagate with the Rankine-Hugoniot speed as an error of order $O(h)$ is allowed (and specified later in Section 2) provided Lax shock inequalities are kept. Similarly, a rarefaction front travels with an averaged of the associated speed of the rarefaction fan, up to an error of order $O(h)$.

Relying on the above terminology, we consider an approximate i-wave front connecting u_l to u_m, followed with an approximate j-wave front connecting u_m to u_r. We suppose that they collide at some point, which implies that

$$i \geq j, \quad 1 \leq i \leq N + 1, \quad 1 \leq j \leq N.$$

In particular, two $(N+1)$-fronts cannot meet since they travel at the same (constant) speed. To extend the solution beyond the interaction time we will introduce suitable approximations of the Riemann solution connecting u_l to u_r. More precisely, we distinguish between an "accurate" wave interaction solver (to be used in Section 2 for waves with "large" strength) and a "rough" solver (for waves with "small" strength). Here, we call **wave interaction solver** a mapping which, to the incoming fronts connecting u_l to u_m and u_m to u_r, respectively, associates a (piecewise constant) approximate solution to the Riemann problem with data u_l and u_r. Note that the interaction solvers introduced now depend on the *middle state* u_m, as well as on the given parameter $h > 0$.

The **accurate interaction solver** is defined when $i, j < N + 1$. We consider the Riemann solution associated with u_l and u_r, and we decompose any existing rarefaction fan into several propagating jumps with small strength $\leq h$.

More precisely, suppose that the Riemann solution contains a k-rarefaction fan connecting a state u_- to a state $u_+ = \psi_k(m; u_-)$ for some m. Let p be the largest integer such that

$$ph \leq |m - \mu_k(u_-)| \qquad (1.10)$$

and set $\varepsilon = \mathrm{sgn}(m - \mu_k(u_-))$. Then, we replace the rarefaction fan by a k-rarefaction front connecting u_- to

$$v_1 := \psi_k(\mu_k(u_-) + \varepsilon\, h; u_-)$$

and propagating at the speed $\overline{\lambda}_k(u_-, v_1)$, followed by another k-rarefaction connecting v_1 to

$$v_2 := \psi_k(\mu_k(v_1) + \varepsilon\, h; v_1)$$

and propagating at the speed $\overline{\lambda}_k(v_1, v_2)$, etc., and finally followed by k-rarefaction front connecting v_p to

$$\psi_k(\mu_k(v_p) + m - \mu_k(u_-) - \varepsilon p h; v_p) = u_+$$

and propagating at the speed $\overline{\lambda}_k(v_p, u_+)$. However, we have also the freedom of changing the above wave speeds by adding small terms of order $O(h)$ (while always keeping the ordering of wave fronts).

The following terminology will be useful. All outgoing k-waves with $k \neq i, j$ are called **secondary waves**. When $i \neq j$, all i- and j-waves are called **primary waves**. If $i = j$, then the shock and the "first" (left-hand) rarefaction front –if any– are called **primary waves**, while all other i-waves are called **secondary waves**. With this terminology, the interaction estimates (1.7) (see also (1.13), below) ensure that *all secondary waves are quadratic in the incoming wave strengths*.

On the other hand, in the **rough interaction solver** we neglect the nonlinear interaction between incoming waves and we treat them as linear waves. This is done at the expense of introducing an artificial wave front.

If $i \leq N$ and $i > j$, then the rough Riemann solution contains a j-wave fan with strength $\tilde{s} := \sigma_j(u_m, u_r)$ connecting to $\tilde{u} := \psi_j(\mu_j(u_l) + \tilde{s}; u_l)$, followed by an

i-wave fan with strength $\tilde{\tilde{s}} = \sigma_i(u_l, u_m)$ connecting to $\tilde{\tilde{u}} := \psi_i(\mu_i(\tilde{u}) + \tilde{\tilde{s}}; \tilde{u})$, plus an artificial front connecting to the right-hand state u_r. Here, we also decompose each of the two wave fans into a shock front plus rarefaction fronts with strength $\leq h$, as already explained above. All of the i- and j-waves are called primary waves, while the artificial front is called a secondary wave.

Finally, if $i \leq N$ and $i = j$, then the Riemann solution contains an i-shock front with strength $\tilde{s}$ connecting to $\tilde{u} := \psi_i(\mu_i(u_l) + \tilde{s}; u_l)$, an i-rarefaction front with strength $\tilde{\tilde{s}}$ connecting to $\tilde{\tilde{u}} := \psi_i(\mu_i(\tilde{u}) + \tilde{\tilde{s}}; u_l)$, followed by an artificial front connecting to u_r. The strengths $\tilde{s}$ and $\tilde{\tilde{s}}$ are determined so that

$$\tilde{s} + \tilde{\tilde{s}} = \sigma_i(u_l, u_m) + \sigma_i(u_m, u_r) + O(1) |\sigma_i(u_l, u_m) \, \sigma_i(u_m, u_r)|,$$

and $|\tilde{\tilde{s}}| \leq h$. This is indeed possible, thanks to the estimate (1.13) in Theorem 1.3 below which shows that the strength of a rarefaction essentially *diminishes* at interactions and since incoming rarefactions are kept of strength $\leq h$ throughout our construction. The i-wave fronts are called primary waves, while the artificial front is called a secondary wave. Again, all secondary waves are quadratic in the incoming wave strengths.

This completes the description of the interaction solvers. It remains to define the wave strengths of the approximate Riemann solutions. This is obvious for the accurate interaction solver, since we are always using states lying on some wave curves and we can therefore measure the strengths from the parametrization given along the wave curves. The same is true for the rough solver, except for the artificial fronts for which we simply use the definition (1.9). The wave strengths of the accurate interaction solver are identical with the wave strengths $\sigma_k(u_l, u_r)$ $(1 \leq k \leq N)$ of the exact Riemann solver. With some abuse of notation, the wave strengths of the rough interaction solver will still be denoted by $\sigma_k(u_l, u_r)$ $(1 \leq k \leq N+1)$. For the accurate solver it is convenient to set $\sigma_{N+1}(u_l, u_r) = 0$. It is easy to extend Theorem 1.1 to approximate wave fronts, as follows.

THEOREM 1.2. (Wave interaction estimates – Approximate interaction solvers.) *For all u_l, u_m, and $u_r \in \mathcal{B}(\delta_1)$, we have the following property. Suppose that u_l is connected to u_m by an approximate i-wave front $(1 \leq i \leq N+1)$ and that u_m is connected to u_r by an approximate j-wave front $(1 \leq j \leq N)$. Then, the wave strengths of the accurate interaction solver satisfy the estimates (1.7) and*

$$\sigma_{N+1}(u_l, u_r) = \sigma_{N+1}(u_l, u_m) + O(1) \, Q_{\text{in}}(u_l, u_m, u_r). \tag{1.11}$$

In the rest of this section we will estimate the outgoing **interaction potential** defined by

$$Q_{\text{out}}(u_l, u_m, u_r) = \sum |\sigma \, \sigma'|,$$

in terms of the incoming interaction potential $Q_{\text{in}}(u_l, u_m, u_r)$. Here, the summation is over all pairs (σ, σ') of waves of the *same* family within the corresponding accurate or rough interaction solvers.

THEOREM 1.3. (Refined interaction estimates.) *Considering u_l, u_m, and $u_r \in \mathcal{B}(\delta_1)$, suppose that u_l is connected to u_m by an approximate i-wave front, that u_m is connected to u_r by an approximate j-wave front $(1 \leq j \leq i \leq N)$, and that the left-hand front travels faster than the right-hand front. If $i > j$, then for some $C > 0$ we have*

$$Q_{\text{out}}(u_l, u_m, u_r) \leq C \, Q_{\text{in}}(u_l, u_m, u_r). \tag{1.12}$$

If $i = j$, then for some constants $C > 0$ and $c \in (0,1)$ we have

$$|\sigma_i(u_l, u_r)^R| \leq \max(|\sigma_i(u_l, u_m)^R|, |\sigma_i(u_m, u_r)^R|) + C\, Q_{\text{in}}(u_l, u_m, u_r) \qquad (1.13)$$

and

$$Q_{\text{out}}(u_l, u_m, u_r) \leq (1 - c)\, Q_{\text{in}}(u_l, u_m, u_r). \qquad (1.14)$$

PROOF. For genuinely nonlinear fields the estimate (1.13) follows immediately from the standard interaction estimates (1.7). A wave is either a shock or a rarefaction, therefore (1.12) and (1.14) are trivial in this case.

Consider the case $i > j$ and a concave-convex characteristic field. The crossing of two waves of different families corresponds to a "shifting" of the waves in the phase space: for instance, roughly speaking, the wave connecting u_l to u_m is shifted by the distance $\sigma_j(u_m, u_r)$. Consider the outgoing i-wave fan together with decomposition

$$\sigma_i(u_l, u_r) = \sigma_i(u_l, u_r)^S + \sigma_i(u_l, u_r)^R.$$

If the incoming i-wave is a shock, then the outgoing rarefaction part $\sigma_i(u_l, u_r)^R$ depends at most linearly upon the incoming strength. Precisely, since $\sigma_i(u_l, u_r)^R$ depends (at least) Lipschitz continuously upon $\sigma_i(u_l, u_m)$ and $\sigma_j(u_m, u_r)$ and vanishes when one of the latter vanishes, it follows that

$$|\sigma_i(u_l, u_r)^R| \leq C\, \min(|\sigma_i(u_l, u_m)|, |\sigma_j(u_m, u_r)|).$$

Of course, in the accurate and rough solvers the new rarefaction fan may need to be decomposed into small rarefaction fans with strength $\leq h$. The "self" interaction potential between these waves is at most $|\sigma_i(u_l, u_r)^R|^2$. In turn, we can estimate the terms in Q_{out} concerned with i-waves, say Q_{out}^i, as follows:

$$\begin{aligned} Q_{\text{out}}^i(u_l, u_m, u_r) &\leq |\sigma_i(u_l, u_r)^S\, \sigma_i(u_l, u_r)^R| + |\sigma_i(u_l, u_r)^R|^2 \\ &\leq C\, Q_{\text{in}}(u_l, u_m, u_r). \end{aligned}$$

Next, if the incoming i-wave is a rarefaction, then both the outgoing shock strength $\sigma_i(u_l, u_r)^S$ and the change in the rarefaction strength depend at most linearly upon the incoming strengths, since they depend (at least) Lipschitz continuously upon $\sigma_i(u_l, u_m)$ and $\sigma_j(u_m, u_r)$ and vanish when one of the latter vanishes:

$$|\sigma_i(u_l, u_r)^S| + |\sigma_i(u_l, u_r)^R - \sigma_i(u_l, u_m)| \leq C\, \min(|\sigma_i(u_l, u_m)|, |\sigma_j(u_m, u_r)|).$$

Here, by construction we always have $|\sigma_i(u_l, u_m)| \leq h$ and we only may need to decompose the part $\sigma_i(u_l, u_r)^R - \sigma_i(u_l, u_m)$ of the rarefaction fan. We estimate Q_{out}^i as follows:

$$\begin{aligned} &Q_{\text{out}}^i(u_l, u_m, u_r) \\ &\leq \big(|\sigma_i(u_l, u_r)^S| + |\sigma_i(u_l, u_r)^R - \sigma_i(u_l, u_m)|\big)\, |\sigma_i(u_l, u_r)^R| \\ &\quad + |\sigma_i(u_l, u_r)^S|\, |\sigma_i(u_l, u_r)^R - \sigma_i(u_l, u_m)| + |\sigma_i(u_l, u_r)^R - \sigma_i(u_l, u_m)|^2 \\ &\leq C\, Q_{\text{in}}(u_l, u_m, u_r). \end{aligned}$$

This establishes the estimate (1.12).

Consider the case $i = j$ of a concave-convex characteristic field. We use here the notation introduced in Chapter VI where we constructed the i-wave curve. Following the general classification given in the proof of Theorem IV-4.1, we distinguish between several interaction patterns depending on the relative positions of $\mu_i(u_l)$, $\mu_i(u_m)$, and $\mu_i(u_r)$. We restrict attention to the cases with $\mu_i(u_l) > 0$, the other cases being completely similar. The argument below strongly uses the monotonicity properties of the shock speed and characteristic speed determined in Section VI-3, which imply, for instance, that two rarefaction fronts cannot meet. We emphasize that all the calculations and inequalities below should include *error terms of quadratic order* in the incoming wave strengths which, for simplicity in the presentation, we neglect in the rest of this discussion. (Strictly speaking, there are more interaction cases for systems than for scalar conservation laws, but the "new" cases can be regarded as quadratic perturbations of the cases listed below.) We use the notation

$$[R] := |\sigma_i(u_l, u_r)^R| - \max(|\sigma_i(u_l, u_m)^R|, |\sigma_i(u_m, u_r)^R|).$$

and

$$[Q] := |\sigma_i(u_l, u_r)^S| \, |\sigma_i(u_l, u_r)^R| - |\sigma_i(u_l, u_m)| \, |\sigma_i(u_m, u_r)|.$$

Recall from Lemma VI-2.4 that $\nabla\mu_i^\natural \cdot r_i \sim -(1/2)\nabla\mu_i \cdot r_i$ near the critical manifold $\mathcal{M}_i$ which implies that, for states u and u'' taken along any i-wave curve and for some constant $\theta \in (0,1)$, we have

$$|\mu_i^\natural(u) - \mu_i^\natural(u')| \le \theta \, |\mu_i(u) - \mu_i(u')| \tag{1.15a}$$

and

$$|\mu_i(u) - \mu_i^\natural(u')| \le \theta \, |\mu_i^{-\natural}(u) - \mu_i(u')|. \tag{1.15b}$$

Case RC-1 : That is, (R_+C)–(C') when $0 < \mu_i(u_l) < \mu_i(u_m)$ and $\mu_i^\natural(u_l) \le \mu_i(u_r) < \mu_i(u_l)$ (up to a quadratic error $O(1)\,|\sigma_i(u_l, u_m)|\,|\sigma_i(u_m, u_r)|$). There is only one outgoing i-wave, and the incoming pattern is non-monotone in the variable μ_i. We find

$$[R] = -|\mu_i(u_l) - \mu_i(u_m)| \le 0.$$

Case RC-2 : That is, $(R_+C_\pm)$–$(C'_\pm R'_-)$ when $0 < \mu_i(u_l) < \mu_i(u_m)$ and $\mu_i^\natural(u_m) \le \mu_i(u_r) < \mu_i^\natural(u_l)$. The exact outgoing pattern contains a shock wave and a rarefaction wave. We have

$$
\begin{aligned}
[R] &= |\mu_i(u_r) - \mu_i^\natural(u_l)| - |\mu_i(u_m) - \mu_i(u_l)| \\
&\le \theta \, |\mu_i^{-\natural}(u_r) - \mu_i(u_l)| - |\mu_i(u_m) - \mu_i(u_l)| \\
&\le -(1 - \theta) \, |\mu_i(u_m) - \mu_i(u_l)|.
\end{aligned}
$$

Using $\mu_i(u_l) < \mu_i^{-\natural}(u_r) \le \mu_i(u_m)$ and (1.16), we find that for every $\kappa \in (0,1)$

$$
\begin{aligned}
[Q] &= |\mu_i^\natural(u_l) - \mu_i(u_l)| \, |\mu_i(u_r) - \mu_i^\natural(u_l)| - |\mu_i(u_m) - \mu_i(u_l)| \, |\mu_i(u_r) - \mu_i(u_m)| \\
&\le -(1 - \kappa) \, |\mu_i(u_m) - \mu_i(u_l)| \, |\mu_i(u_r) - \mu_i(u_m)| \\
&\quad + \theta \, |\mu_i^\natural(u_l) - \mu_i(u_l)| \, |\mu_i^{-\natural}(u_r) - \mu_i(u_l)| - \kappa \, |\mu_i^{-\natural}(u_r) - \mu_i(u_l)| \, |\mu_i(u_r) - \mu_i^{-\natural}(u_r)| \\
&\le -(1 - \kappa) \, |\mu_i(u_m) - \mu_i(u_l)| \, |\mu_i(u_r) - \mu_i(u_m)| \\
&\quad - (\kappa - \theta) \, |\mu_i^\natural(u_l) - \mu_i(u_l)| \, |\mu_i^{-\natural}(u_r) - \mu_i(u_l)| \\
&\le -(1 - \kappa) \, |\mu_i(u_m) - \mu_i(u_l)| \, |\mu_i(u_r) - \mu_i(u_m)| \le 0,
\end{aligned}
$$

since $\mu_i(u_r) < \mu_i^\natural(u_l) < \mu_i(u_l) < \mu_i^{-\natural}(u_r)$ and provided we choose κ such that $1 > \kappa > \theta$.

Case CR-1 : That is, $(C_\pm R_-)$–$(C'_\pm)$ when $\mu_i^\natural(u_l) \le \mu_i(u_r) < \mu_i(u_m) < 0$. There is only one outgoing wave and the incoming solution is monotone. We find here

$$[R] = -|\mu_i^{-\natural}(u_r) - \mu_i^{-\natural}(u_m)| \le 0.$$

Case CR-2 : That is, (C_+R_+)–(C'_+) when $0 < \mu_i(u_m) < \mu_i(u_r) < \mu_i(u_l)$. There is only one outgoing wave and some cancellation is taking place. This case is similar to Case RC-1.

Case CR-3 : That is, $(C_\pm R_-)$–$(C'_\pm R'_-)$ when $\mu_i^\star(u_l, u_m) < \mu_i(u_r) < \mu_i^\natural(u_l) \le \mu_i(u_m) < 0$. Here the value $\mu = \mu_i^\star(u_l, u_m)$ is defined by the conditions (v_i denoting the parametrization of the Hugoniot curve)

$$u_* = v_i(\mu; u_l), \quad \overline{\lambda}_i(u_l, u_*) = \overline{\lambda}_i(u_l, u_r).$$

Let us first observe that, possibly using a larger value $\theta \in (0, 1)$ if necessary, it follows from Lemma VI-2.5 (see also the proof of Theorem IV-4.2 for scalar equation) that

$$
\begin{aligned}
& |\mu_i^\natural(u_l) - \mu_i(u_l)|\,|\mu_i(u_r) - \mu_i^\natural(u_l)| \\
& \le \theta \min\big(|\mu_i(u_l)|\,|\mu_i(u_r)|, |\mu_i(u_l) - \mu_i^\star(u_l, u_r)|\,|\mu_i(u_r) - \mu_i^\star(u_l, u_r)|\big).
\end{aligned}
\tag{1.16}
$$

The outgoing pattern contains two waves and the incoming solution is monotone. We have

$$
\begin{aligned}
[R] &= |\mu_i(u_r) - \mu_i^\natural(u_l)| - |\mu_i(u_r) - \mu_i(u_m)| \\
&= -|\mu_i^\natural(u_l) - \mu_i(u_m)| \le 0,
\end{aligned}
$$

and for every $\kappa \in (0, 1)$

$$
\begin{aligned}
[Q] =& |\mu_i^\natural(u_l) - \mu_i(u_l)|\,|\mu_i(u_r) - \mu_i^\natural(u_l)| - |\mu_i(u_m) - \mu_i(u_l)|\,|\mu_i(u_r) - \mu_i(u_m)| \\
=& |\mu_i^\natural(u_l) - \mu_i(u_l)|\,|\mu_i(u_r) - \mu_i^\natural(u_l)| - \kappa\,|\mu_i(u_m) - \mu_i(u_l)|\,|\mu_i(u_r) - \mu_i(u_m)| \\
& - (1 - \kappa)\,|\mu_i(u_m) - \mu_i(u_l)|\,|\mu_i(u_r) - \mu_i(u_m)|.
\end{aligned}
$$

The polynomial function $\mu_i(u_m) \mapsto |\mu_i(u_m) - \mu_i(u_l)|\,|\mu_i(u_r) - \mu_i(u_m)|$ over the interval determined by $\mu_i^\star(u_l, u_r) \le \mu_i(u_m) \le 0$ satisfies the inequality

$$
\begin{aligned}
& |\mu_i(u_m) - \mu_i(u_l)|\,|\mu_i(u_r) - \mu_i(u_m)| \\
& \ge \min\big(|\mu_i(u_l)|\,|\mu_i(u_r)|, |\mu_i(u_l) - \mu_i^\star(u_l, u_r)|\,|\mu_i(u_r) - \mu_i^\star(u_l, u_r)|\big).
\end{aligned}
$$

Therefore, by (1.16) we conclude that

$$
\begin{aligned}
[Q] \le& |\mu_i^\natural(u_l) - \mu_i(u_l)|\,|\mu_i(u_r) - \mu_i^\natural(u_l)| \\
& - \kappa \min\big(|\mu_i(u_l)|\,|\mu_i(u_r)|, |\mu_i(u_l) - \mu_i^\star(u_l, u_r)|\,|\mu_i(u_r) - \mu_i^\star(u_l, u_r)|\big) \\
& - (1 - \kappa)\,|\mu_i(u_m) - \mu_i(u_l)|\,|\mu_i(u_r) - \mu_i(u_m)| \\
\le& - (1 - \kappa)\,|\mu_i(u_m) - \mu_i(u_l)|\,|\mu_i(u_r) - \mu_i(u_m)|,
\end{aligned}
$$

provided κ is chosen such that $\theta \le \kappa < 1$.

Case CC-1 : That is, (C_+C)–(C') when $0 < \mu_i(u_m) < \mu_i(u_l)$ and $\mu_i^\natural(u_m) \le \mu_i(u_r) \le \mu_i(u_m)$. This case is similar to Case CR-1.

Case CC-2 : That is, $(C_\pm C)$–(C') when $\mu_i^\natural(u_l) \le \mu_i(u_m) < 0$ and $\mu_i(u_m) < \mu_i(u_r) \le \mu_i^\natural(u_m)$. This case is similar to Case RC-1.

This completes the proof of (1.13) and (1.14) and, therefore, the proof of Theorem 1.3. $\qquad\Box$

2. Existence theory

We will now construct a sequence of piecewise constant approximate solutions u^h : $\mathbb{R} \times \mathbb{R}_+ \to \mathcal{U}$ of the Cauchy problem

$$\partial_t u + \partial_x f(u) = 0, \quad u = u(x,t) \in \mathcal{U}, \ x \in \mathbb{R}, \ t \geq 0, \qquad (2.1)$$

$$u(x,0) = u_0(x), \quad x \in \mathbb{R}. \qquad (2.2)$$

The solutions will be made of a large number of approximate wave fronts of the type introduced in the previous section. We assume that the given function $u_0 : \mathbb{R} \to \mathcal{U}$ in (2.2) has bounded total variation, denoted by $TV(u_0)$. Given a sequence $h \to 0$, it is easy to construct a piecewise constant approximation $u_0^h : \mathbb{R} \to \mathcal{U}$ with compact support which has only finitely many jump discontinuities (say, $1/h$ at most) and satisfies

$$u_0^h \to u_0 \quad \text{almost everywhere,} \quad TV(u_0^h) \leq TV(u_0). \qquad (2.3)$$

Then, the Cauchy problem associated with the initial data u_0^h can be solved explicitly for small time t. One simply solves a Riemann problem at each jump discontinuity of u_0^h, each problem being treated independently from each other. Each (approximate) Riemann solution may contain one or several (approximate) wave fronts. When two of these wave fronts collide, we can again solve a Riemann problem and continue the solution beyond the interaction time. If the algorithm does not break down (we return to this issue shortly), we continue this construction globally in time by resolving all interactions one by one.

Our main objective is to show that the approximate solutions are globally defined in time and to derive the uniform bound on the total variation of the approximate solutions,

$$TV(u^h(t)) \leq C\, TV(u_0), \quad t \geq 0. \qquad (2.4)$$

By Helly's compactness theorem (see the appendix) (2.4) implies that the limit $u = \lim_{h \to 0} u^h$ exists almost everywhere and, in turn, satisfies (2.1) and (2.2). In contrast with scalar conservation laws (in Chapter IV), the total variation $TV(u^h(t))$ may well increase in time, and to establish (2.4) it will be necessary to rely on the interactions estimates derived in Section 1.

Several important obstacles must be overcome in order to implement the above strategy:

- The Riemann problem is known to be solvable for data in $\mathcal{B}(\delta_2)$, only, and therefore we must ensure that the values $u^h(x,t)$ remain in this neighborhood of 0.
- A Riemann solution may contain centered rarefaction fans, and we will replace them with several propagating jumps with small strength h (with the minor drawback that such fronts violate the entropy condition).
- When two wave fronts interact, the *number of outgoing fronts* may be greater than the number of incoming ones so that the number of waves could become infinite in finite time. To prevent this from happening, we will neglect waves with "small" strength.
- Additionally, we will check that the *number of interactions* between these waves is finite so that the scheme does not break down in finite time.
- Finally, in order to implement this strategy *artificial wave fronts* will be needed and will propagate "very small" error terms away, with the fixed large speed λ_{N+1}.

By slightly changing the wave speeds and introducing an error term of order $O(h)$ if necessary, one can assume that, at each time, *at most one interaction* is taking place and that this interaction involves *exactly two wave fronts*.

The **wave front tracking approximations** u^h are defined as follows, based on the two approximate interaction solvers proposed in Section 1 and on a threshold function $\mathcal{E} = \mathcal{E}(h)$ satisfying

$$\lim_{h \to 0} \frac{\mathcal{E}(h)}{h^n} = 0, \quad n = 0, 1, 2, \ldots \tag{2.5}$$

First of all, at the initial time $t = 0$, one solves a Riemann problem at each jump discontinuity of u_0^h using the *accurate* interaction solver. (All rarefaction fans at time $t = 0+$ are replaced with several propagating jumps with strength less than h.) Next, at each interaction involving two incoming waves of families i and j and with strengths s_l and s_r, respectively, we proceed as follows:

1. If $|s_l\, s_r| > \mathcal{E}(h)$ and $i \le N$, we resolve the interaction by using the *accurate* interaction solver.
2. If $|s_l\, s_r| \le \mathcal{E}(h)$ or if $i = N + 1$, we use the *rough* interaction solver.

The main result in this chapter is the following one.

THEOREM 2.1. (Existence of classical entropy solutions.) *Consider the system of conservation laws* (2.1) *with a smooth flux f defined in the ball $\mathcal{U} = \mathcal{B}(\delta_0)$. Assume that each characteristic field of* (2.1) *is either genuinely nonlinear or concave-convex. Then, there exist $\delta_3 < \delta_0$, $c_*, C_* > 0$ such that the following property holds for all initial data $u_0 : \mathbb{R} \to \mathcal{B}(\delta_3)$ satisfying $TV(u_0) \le c_*$.*

Consider a sequence of approximate initial data $u_0^h : \mathbb{R} \to \mathcal{B}(\delta_3)$ satisfying (2.3) *and a threshold function $\mathcal{E}$ satisfying* (2.5). *Then, the above algorithm generates a sequence of approximate solutions u^h which are globally defined in time, contain finitely many lines of discontinuity and finitely many points of interaction, and satisfy the uniform estimates*

$$
\begin{aligned}
& u^h(x,t) \in \mathcal{B}(\delta_0), \quad x \in \mathbb{R}, \, t \ge 0, \\
& TV(u^h(t)) \le C_*\, TV(u_0), \quad t \ge 0, \\
& \|u^h(t_2) - u^h(t_1)\|_{L^1(\mathbb{R})} \le \big(o(1) + C_*\, \Lambda\, TV(u_0)\big)\, |t_2 - t_1|, \quad t_1, t_2 \ge 0,
\end{aligned}
\tag{2.6}
$$

where $o(1) \to 0$ as $h \to 0$, and

$$\Lambda := \sup_{\substack{u \in \mathcal{B}(\delta_1) \\ 1 \le i \le N}} |\lambda_i(u)|.$$

After extracting a subsequence if necessary, u^h converges (almost everywhere in (x,t) and in L^1_{loc} in x for all t) to a weak solution u of the Cauchy problem (2.1) *and* (2.2) *with*

$$
\begin{aligned}
& u(x,t) \in \mathcal{B}(\delta_0), \quad x \in \mathbb{R}, \, t \ge 0, \\
& TV(u(t)) \le C_*\, TV(u_0), \quad t \ge 0, \\
& \|u(t_2) - u(t_1)\|_{L^1(\mathbb{R})} \le C_*\, \Lambda\, TV(u_0)\, |t_2 - t_1|, \quad t_1, t_2 \ge 0.
\end{aligned}
\tag{2.7}
$$

If the system (2.1) *admits a convex entropy pair (U, F), then the limit u satisfies the entropy inequality*

$$\partial_t U(u) + \partial_x F(u) \le 0 \tag{2.8}$$

in the weak sense.

Note that the solution satisfies also Lax shock inequalities; see Section 4 below. The rest of this section is devoted to a proof of Theorem 2.1, assuming the uniform bounds (2.6) together with an estimate for the strength of artificial waves (Lemma 2.2 below). Proving these two technical estimates is postponed to Section 3.

We will use the following notation. For each time t which is not an interaction tile, we denote by $\mathcal{J}(u^h(t))$ the set of the points of jump of the piecewise constant function $x \mapsto u^h(x,t)$. At each point $x \in \mathcal{J}(u^h(t))$ the approximate solution contains a wave with strength $\sigma^h(x,t)$ of the family $i^h(x,t) \leq N+1$, propagating at the speed $\lambda^h(x,t)$. Recall that, if the front is associated with a wave family $i^h(x,t) \leq N$, the (signed) strength is measured along the wave curve:

$$u_+^h(x,t) = \psi_{i^h(x,t)}\big(\mu_{i^h(x,t)}(u_-^h(x,t)) + \sigma^h(x,t); u_-^h(x,t)\big).$$

If this is an artificial front, we have $i^h(x,t) = N+1$ and the (non-negative) strength is

$$\sigma^h(x,t) = \big|u_+^h(x,t) - u_-^h(x,t)\big|.$$

Additionally, when $i^h(x,t) \leq N$ it will be convenient to use the notation

$$\sigma^h(x,t) = \sigma^h(x,t)^S + \sigma^h(x,t)^R, \qquad (2.9)$$

where, for instance,

$$\sigma^{hR}(x,t) = \sigma_{i^h(x,t)}\big(u_-^h(x,t), u_+^h(x,t)\big)^R.$$

The total sum of wave strengths in u^h is controlled by the total variation estimate in (2.6). Indeed thanks to (1.4) and (2.3) we have

$$\sum_{x \in \mathcal{J}^h(t)} |\sigma^h(x,t)| \leq C_2 \sum_{x \in \mathcal{J}^h(t)} \big|u_+^h(x,t) - u_-^h(x,t)\big|$$

$$= C_2\, TV(u^h(t)) \leq C_2\, C_*\, TV(u_0) < +\infty.$$

LEMMA 2.2. (Strengths of rarefaction and artificial fronts.) *Under the assumptions of Theorem 2.1 we have*

$$|\sigma^h(x,t)^R| \leq h, \quad 1 \leq i^h(x,t) \leq N, x \in \mathcal{J}^h(t), \qquad (2.10)$$

and

$$\lim_{h \to 0}\Big(\sup_{t \geq 0} \sum_{\substack{x \in \mathcal{J}^h(t) \\ i^h(x,t)=N+1}} |\sigma^h(x,t)|\Big) = 0. \qquad (2.11)$$

Of course, (2.10) is obvious by construction since all rarefaction fronts have strength $\leq h$ at the initial time, and that the accurate and rough interaction solvers do not increase the size of existing rarefactions and create only new ones with strength $\leq h$.

PROOF OF THEOREM 2.1. In view of the uniform estimates (2.6) we can apply Helly's compactness theorem as explained in the appendix. For each time $t > 0$ there exists a converging subsequence $u^h(t)$ and a limiting function $u(t)$. By a standard diagonal argument we can find a subsequence so that for all rational t

$$u^h(x,t) \to u(x,t) \text{ for almost every } x. \qquad (2.12)$$

The uniform Lipschitz bound in (2.6) then implies that $u(t)$ is well defined for all t and that, in fact, $u^h(t)$ converges to $u(t)$ in L^1_{loc} for all t. The inequalities (2.7) follow immediately from (2.6) by using the lower semi-continuity property of the L^1 norm and total variation.

We claim that the limit u is a weak solution. Given a smooth scalar-valued function θ with compact support in $\mathbb{R} \times [0, +\infty)$, we consider

$$E^h(\theta) := \int_{\mathbb{R}} \theta(0)\, u_0^h\, dx + \iint_{\mathbb{R} \times (0, +\infty)} \left(u^h\, \partial_t \theta + f(u^h)\, \partial_x \theta \right) dx dt.$$

Since $u^h \to u$ almost everywhere,

$$E^h(\theta) \to \int_{\mathbb{R}} \theta(0)\, u_0\, dx + \iint_{\mathbb{R} \times (0, +\infty)} \left(u\, \partial_t \theta + f(u)\, \partial_x \theta \right) dx dt,$$

thus we simply have to prove $E^h(\theta) \to 0$.

Recall that u^h is piecewise constant. Defining

$$[u^h](x, t) := u_+^h(x, t) - u_-^h(x, t),$$

and similarly for $[f(u^h)](x, t)$ and using Green's formula, we find

$$E^h(\theta) = \int_{\mathbb{R}_+} \sum_{x \in \mathcal{J}^h(t)} \left(\lambda^h(x, t)\, [u^h](x, t) - [f(u^h)](x, t) \right) \theta(x, t)\, dt.$$

By construction if the front located at x is a shock, the left- and right-hand states satisfy the Rankine-Hugoniot relation but an error of order $O(h)$ is allowed on the speed, therefore the Rankine-Hugoniot relation holds approximately:

$$\left| \lambda^h(x, t)\, [u^h](x, t) - [f(u^h)](x, t) \right| \le C\, h\, |\sigma^h(x, t)|. \tag{2.13}$$

If the front is a rarefaction, its speed is of the form

$$\lambda^h(x, t) = \overline{\lambda}_{i^h(x, t)}(u_-^h(x, t), u_+^h(x, t)) + O(h).$$

And, since $u_+^h(x, t)$ lies on the integral curve issuing from $u_-^h(x, t)$, it is easy to check that

$$\left| \lambda^h(x, t)\, [u^h](x, t) - [f(u^h)](x, t) \right| \le C\, |\sigma^h(x, t)|^2 + C\, h\, |\sigma^h(x, t)|. \tag{2.14}$$

Finally, if the front is an artificial wave, we use the simple estimate

$$\left| \lambda^h(x, t)\, [u^h](x, t) - [f(u^h)](x, t) \right| \le C\, |\sigma^h(x, t)|. \tag{2.15}$$

By assumption, the support of the function θ is contained in $\mathbb{R} \times [0, T]$ for some $T > 0$. Combining (2.13)–(2.15) we obtain

$$|E^h(\theta)| \le C \int_0^T \left(\sum_{1 \le i^h(x, t) \le N} h\, |\sigma^h(x, t)| + |\sigma^h(x, t)^R|^2 + \sum_{i^h(x, t) = N+1} |\sigma^h(x, t)| \right) dt.$$

In view of Lemma 2.2 we deduce that

$$|E^h(\theta)| \le T\, C\, C_3\, h \sup_{t \in [0, T]} \sum_x |\sigma^h(x, t)| + C \sup_{t \in [0, T]} \sum_{i^h(x, t) = N+1} |\sigma^h(x, t)|$$

$$\le C'\, h \sup_{t \in [0, T]} TV(u^h(t)) + C \sup_{t \in [0, T]} \sum_{i^h(x, t) = N+1} |\sigma^h(x, t)| \longrightarrow 0.$$

Hence, the limiting function u is a weak solution of (2.1) and (2.2).

Since the underlying exact Riemann solution satisfies the entropy inequality (2.8), the accurate and rough Riemann solutions satisfy (2.8) up to error terms which are completely analogous to those studied above. So, proving that u satisfies the entropy inequality is similar, provided equalities are replaced with inequalities throughout. This completes the proof of Theorem 2.1 (when the uniform estimates (2.6) and (2.11) hold).   □

3. Uniform estimates

This section provides a proof of the technical estimates in Theorem 2.1 and Lemma 2.2. We use the notation given after the statement of Theorem 1.1. To simplify the notation we often suppress the explicit dependence in h and in t.

Total variation bound.
The total strength of waves is controlled by the **linear functional**

$$V(t) = \sum_{x \in \mathcal{J}^h(t)} |\sigma^h(x,t)|,$$

while the potential increase due to wave interactions will be measured by the **quadratic functional**

$$Q(t) = M \sum_{(x,y) \in \mathcal{A}_d^h(t)} |\sigma^h(x,t)\,\sigma^h(y,t)| + \sum_{(x,y) \in \mathcal{A}_s^h(t)} |\sigma^h(x,t)\,\sigma^h(y,t)|,$$

where $M > 1$ is a sufficiently large constant. We count in $Q(t)$ all the quadratic products of strengths between: (i) all waves of different families, provided the wave on the left-hand side is faster than the wave on the right-hand side, (ii) and all waves of the same family except artificial waves. In other words, the set $\mathcal{A}_d^h(t)$ ("different families") contain pairs (x,y) of **approaching waves** having $x < y$ and $1 \le i^h(y,t) < i^h(x,t) \le N+1$, while the set $\mathcal{A}_s^h(t)$ ("same family") is defined as $x < y$ and $i^h(x,t) = i^h(y,t) < N+1$.

Recall (from (1.4)) that

$$\frac{1}{C}\left|u_+^h(x,t) - u_-^h(x,t)\right| \le |\sigma^h(x,t)| \le C\left|u_+^h(x,t) - u_-^h(x,t)\right|, \qquad (3.1)$$

so that the functional V is equivalent to the usual total variation functional,

$$\frac{1}{C} TV(u^h(t)) \le V(t) \le C\,TV(u^h(t)), \quad t \ge 0. \qquad (3.2)$$

On the other hand, the interaction potential is dominated by the linear functional:

$$Q(t) \le M \left(V(t)\right)^2.$$

Estimating V and Q is based on the wave interaction estimates derived in Section 1. On one hand, the wave strengths are increased by a small quadratic term at interactions, at most. On the other hand, the function Q decreases at interactions by the same quadratic amount. To take advantage of these facts, consider the functional

$$V(t) + C_4\,Q(t).$$

By choosing a sufficiently large constant C_4, the increase of $V(t)$ can be compensated by the decrease of $Q(t)$.

LEMMA 3.1. (Decreasing functional.) *For $C_4 > 0$ sufficiently large the (piecewise constant) function*

$$t \mapsto V(t) + C_4 \, Q(t)$$

decreases at each interaction time.

PROOF. Let t be an interaction time and consider two waves meeting at some point x: an i^α-wave with strength s^α located on the left-hand side of an i^β-wave with strength s^β. Let s^γ be the strengths of the outgoing waves, where γ describes a finite set of indices.

First of all, since V is the sum of all wave strengths which possibly increase at the time t but by (at most) the product of the strengths of the two incoming waves (Theorem 1.2), we have

$$\sum_\gamma |s^\gamma| \le |s^\alpha| + |s^\beta| + \frac{C}{2} \, |s^\alpha \, s^\beta|.$$

Thus the total increase is

$$[V(t)] := V(t+) - V(t-) \le \frac{C}{2} \, |s^\alpha \, s^\beta|. \tag{3.3}$$

Consider an interaction between waves of different families. The term $M \, |s^\alpha \, s^\beta|$ is counted in $Q(t-)$, but no longer in $Q(t+)$ since the two waves are no longer approaching after the interaction. (See the definition of the set $\mathcal{A}_d^h(t)$ above.) Additionally, the estimate (1.12) in Theorem 1.3 shows that the self-interaction between outgoing waves of the *same* family is less than $C \, |s^\alpha \, s^\beta|$. Hence, by choosing M (arising in the expression of $Q(t)$) sufficiently large, we see that the latter is dominated by the former. Moreover, by Theorem 1.2 the waves in the other families are of quadratic order.

Consider next an interactions involving two waves of the same family. The estimate (1.14) derived in Theorem 1.3 shows that the self-interaction between outgoing waves of the same family is at most $(1 - c) \, |s^\alpha \, s^\beta|$ for some $c \in (0, 1)$, while by Theorem 1.2 the waves in the other families are of quadratic order.

Therefore, in all cases we find for some $c \in (0, 1)$

$$Q(t+) \le Q(t-) - c \, |s^\alpha \, s^\beta| + \frac{C}{2} \, V(t-) \, |s^\alpha \, s^\beta|.$$

If we *assume* that $V(t-) \le c/C$ it follows that

$$[Q(t)] := Q(t+) - Q(t-) \le -\frac{c}{2} \, |s^\alpha \, s^\beta| \tag{3.4}$$

and, then, by summation we have for any $C_4 \ge C/c$

$$[V(t)] + C_4 \, [Q(t)] \le 0. \tag{3.5}$$

It remains to observe that

$$
\begin{aligned}
V(t+) &\le V(t-) + C_4 \, Q(t-) \le V(0) + C_4 \, Q(0) \\
&\le C_2 \, TV(u_0) + C_4 C_2^2 \, TV(u_0)^2 \\
&\le \frac{c}{C}.
\end{aligned}
\tag{3.6}
$$

The latter inequality holds as soon as $TV(u_0)$ is less than a numerical constant c_* which depends upon the constants C_2 and C_4, only. (Take $c_* := 1/(2\,C_2 C_4)$, for instance.)

By induction, we conclude that $V(t-) \leq c/C$ and that the functional is decreasing. This completes the proof that the total variation of $u^h(t)$ is uniformly bounded, that is, the proof of the second property in (2.6). $\square$

Moreover, since the approximations have compact support and the scheme satisfies the property of propagation with finite speed, the approximate solutions are bounded in amplitude:

$$\sup |u^h(t+)| \leq \sup |u_0^h| + TV(u^h(t)) \leq \sup |u_0^h| + C_*\,TV(u_0). \qquad (3.7)$$

The Lipschitz estimate is a consequence of the total variation estimate and the property of propagation at finite speed. Indeed, in any interval $[t_1, t_2]$ containing no interaction, the speed $\lambda^h(x,t)$ of each wave front $x = x^h(t)$ is constant and we can write

$$\|u^h(t_2) - u^h(t_1)\|_{L^1(I\!R)} \leq \sum |u_+^h(x^h(t_1), t_1) - u_-^h(x^h(t_1), t_1)| \, |x^h(t_2) - x^h(t_1)|.$$

For non-artificial fronts we have

$$|x^h(t_2) - x^h(t_1)| = |\lambda^h| \, |t_2 - t_1| \leq \left(O(h) + \sup_{\substack{B(\delta_0) \\ 1 \leq i \leq N}} |\lambda_i(.)|\right) |t_2 - t_1|.$$

On the other hand, for artificial fronts we have

$$|x^h(t_2) - x^h(t_1)| = \lambda_{N+1} |t_2 - t_1|,$$

but the total strength tends to zero with h, by Lemma 2.2. This completes the proof of the estimates (2.6).

Number of wave fronts.
In view of the estimate (3.4) the number of interactions having $|s^\alpha s^\beta| > \mathcal{E}(h)$ must be *finite* since the non-negative function $Q(t)$ decreases by the amount $c\,\mathcal{E}(h)$ (at least) across any such interaction. Then, disregarding first the artificial waves we observe that:

- In the case $|s^\alpha s^\beta| \leq \mathcal{E}(h)$ the rough solver is used and generate two outgoing waves for any two incoming waves of the same family.
- Two waves of different families may cross at most once.

Therefore, the number of non-artificial waves is also *finite* for all time. It follows also that the number of artificial waves is finite since new artificial waves are created by interactions between non-artificial waves. This establishes that *the total number of waves is finite*.

We just observed that the number of interactions involving the accurate solver is finite. On the other hand, when the rough solver is used, the system being strictly hyperbolic, two waves of different families can cross each other at most once. So, we may restrict attention to the solution u^h *after* a sufficiently large time and we may assume that only interactions between waves of the same family are taking place. Consider a concave-convex i-characteristic field (since the result is obvious for genuinely nonlinear fields). In view of the interaction cases listed in the proof of Theorem 1.3 and without taking into account artificial waves, we see that the interaction of two waves of the same family leads either to a two-wave pattern, one of them being a right-contact, or to a one-wave pattern. Moreover, a right-contact may not interact with waves on its right-hand side. When it interacts with a wave on its left-hand side the outgoing pattern contains a single wave. In consequence, for each concave-convex i-field and for all t sufficiently large, the functional

$$F_i^h(t) := \left(G_i^h(t) + 1\right) G_i^h(t) + \sum_{C_i \cup R_i} G_i^h(t) + \sum_{C_i^\natural} G_{C_i^\natural,\text{left}}^h(t),$$

is strictly decreasing at each interaction, where the sums are over all i-shock fronts C_i and rarefaction fronts R_i and over all right-contacts, respectively, and $G_i^h(t)$ is the total number of waves at time t while $G_{C_i^\natural,\text{left}}^h(t)$ is the total number of waves located on the right hand-side of the right-contact $C_i^\natural$. This completes the proof that the *number of interaction times* is finite.

We now want to derive an estimate on the number of waves. To each wave we associate a **generation order** $r = 1, 2, \ldots$. This number keeps track of the number of interactions that were necessary to generate that wave. All the waves generated at time $t = 0+$ have by definition the order $r = 1$. At each interaction the "old" waves keep the same order, while we assign a higher order to the "new" waves. More precisely, consider an interaction involving an i-wave of order r^α and strength s^α and a j-wave of order r^β and strength s^β.

1. If $i, j \leq N$ and $|s^\alpha s^\beta| > \mathcal{E}(h)$, then we used the accurate interaction solver. We choose the order of the secondary outgoing waves to be $\max(r^\alpha, r^\beta) + 1$. When $i \neq j$, the order of the primary outgoing i- and j-waves is r^α and r^β, respectively. When $i = j$, the order of two primary outgoing i-waves is defined to be $\min(r^\alpha, r^\beta)$.
2. If $i, j \leq N$ and $|s^\alpha s^\beta| \leq \mathcal{E}(h)$, we used the rough interaction solver. When $i \neq j$, the outgoing i- and j-waves keep their orders r^α and r^β, respectively. When $i = j$, the order of the i-waves is defined to be $\min(r^\alpha, r^\beta)$. The artificial wave is assigned the order $\max(r^\alpha, r^\beta) + 1$.
3. If $i = N + 1$ and $j \leq N$, we used the rough solver. The solution contains a j-wave and an artificial wave and the outgoing waves keep their orders r^α and r^β, respectively.

We checked earlier that the total number of fronts is finite for each fixed h. So there exists a **maximal generation order**, say $r_{\max}$ (depending on h).

LEMMA 3.2. (Number of wave fronts.) *The number M_r of fronts with order r is at most polynomial in $1/h$,*

$$M_r \leq \frac{C(r)}{h^{m(r)}}, \quad r = 1, 2, \ldots, r_{\max} \tag{3.8}$$

for some constants $C(r) > 0$ and some integer exponents $m(r)$.

PROOF. By construction, the initial data contain at most $1/h$ jumps. Then, each initial jump may generate N waves, possibly decomposed into $O(1/h)$ small jumps (when there are rarefaction fans). Therefore we obtain

$$M_1 = O(\frac{1}{h^2}). \tag{3.9}$$

There are less than M_1^2 points of interaction between waves of the first generation order and at most $C\,N/h$ outgoing waves from each interaction. Thus the number

M_2 of wave fronts of the second generation satisfies

$$M_2 \leq O\left(\frac{1}{h}\right) M_1^2 \leq O\left(\frac{1}{h^5}\right).$$

More generally, the waves of order $r - 1$ can produce waves of order r by interacting with waves of order $\leq r - 1$. Thus the number of waves of order r is found to be

$$M_r \leq N \frac{C}{h} (M_1 + M_2 + \cdots + M_{r-1}) M_{r-1},$$

which implies (3.8) by induction from (3.9).   □

Strength of artificial waves.
Denote by $p^h(x,t)$ the order of the wave located at $x \in \mathcal{J}^h(t)$. Away from interaction times t and for each integer r, let $V_r(t)$ be the sum of the strengths of all waves of order $\geq r$,

$$V_r(t) = \sum_{\substack{x \in \mathcal{J}^h(t) \\ p^h(x,t) \geq r}} |\sigma^h(x,t)|$$

and define also $Q_r(t)$ by

$$Q_r(t) = M \sum_{\substack{(x,y) \in \mathcal{A}_d^h(t) \\ \max\left(p^h(x,t), p^h(y,t)\right) \geq r}} |\sigma^h(x,t)\,\sigma^h(y,t)| + \sum_{\substack{(x,y) \in \mathcal{A}_a^h(t) \\ \max\left(p^h(x,t), p^h(y,t)\right) \geq r}} |\sigma^h(x,t)\,\sigma^h(y,t)|.$$

Denote by I_r the set of interaction times where two incoming waves of order $p^h(x,t)$ and $p^h(y,t)$ interact with $\max\left(p^h(x,t), p^h(y,t)\right) = r$.

Similarly as done above in the proof of Lemma 3.1, one can check the following **precised estimates:**

- The strengths of waves of order $\geq r$ do not change when two waves of order $\leq r - 2$ interact:

$$[V_r(t)] = 0, \quad t \in I_1 \cup \ldots \cup I_{r-2}. \tag{3.10i}$$

- The change in the strength of waves of order $\geq r$ is compensated by the interaction potential between waves of order $\geq r - 1$:

$$[V_r(t)] + C_4\,[Q_{r-1}(t)] \leq 0, \quad t \in I_{r-1} \cup I_r \cup \ldots \tag{3.10ii}$$

Similarly for the interaction potentials, we have:

- At interaction times involving low-order waves, the possible increase in the potential Q_r is controlled by the decrease of the potential Q:

$$[Q_r(t)] + C_4\,V_r(t-)\,[Q(t)] \leq 0, \quad t \in I_1 \cup \ldots \cup I_{r-2}, \tag{3.11i}$$

- At interaction times involving a wave of order $r-1$ and a wave of order $\leq r-1$, the possible increase in Q_r is controlled by (the decrease of) Q_{r-1}:

$$[Q_r(t)] + C_4\,V(t-)\,[Q_{r-1}(t)] \leq 0, \quad t \in I_{r-1}, \tag{3.11ii}$$

- At the remaining interactions, the potential of interaction is non-increasing

$$[Q_r(t)] \leq 0, \quad t \in I_r \cup I_{r+1} \cup \ldots \tag{3.11iii}$$

On the other hand, observe also that

$$V_1(t) = V(t), \quad Q_1(t) = Q(t),$$
$$V_r(0+) = Q_r(0+) = 0, \quad r \geq 2.$$

(3.12)

We now claim that:

LEMMA 3.3. (Total strength of waves of a given order.) *There exist constants $C_5 > 0$ and $\eta \in (0,1)$ such that*

$$V_r(t) \leq C_5 \, \eta^r, \quad t \geq 0, r = 1, 2, \ldots, r_{\max}.$$

PROOF. The estimates (3.10) and (3.182) yield ($r \geq 2$)

$$V_r(t) = \sum_{0 < \tau \leq t} [V_r(\tau)] + V_r(0+) \leq -C_4 \sum_{0 < \tau \leq t} [Q_{r-1}(\tau)],$$

thus

$$V_r(t) \leq C_4 \sum_{0 < \tau \leq t} [Q_{r-1}(\tau)]_-.$$

(3.13)

On the other hand, for the interaction potentials we find ($r \geq 2$)

$$0 \leq Q_r(t) \leq Q_r(0+) + \sum_{0 < \tau \leq t} [Q_r(\tau)]_+$$
$$\leq C_4 \sum_{0 < \tau \leq t} [Q_r(\tau)]_- \sup_{t'} V_r(t') + C_4 \sum_{0 < \tau \leq t} [Q_{r-1}(\tau)]_- \sup_{t'} V(t'),$$

(3.14)

where we used (3.11) and (3.12).

Note that, since Q_r is non-negative and $Q_r(0+) = 0$ for $r \geq 2$,

$$0 \leq Q_r(t) = \sum_{0 < \tau \leq t} [Q_r(\tau)]_+ - \sum_{0 < \tau \leq t} [Q_r(\tau)]_-,$$

thus

$$\sum_{0 < \tau \leq t} [Q_r(\tau)]_- \leq \sum_{0 < \tau \leq t} [Q_r(\tau)]_+.$$

(3.15)

With the uniform total variation estimate we have also

$$V_r(t) \leq V(t) \leq C_* C_2 \, TV(u_0),$$

$$\sum_{0 < \tau < +\infty} [Q_r(\tau)]_- \leq \sup Q_r \leq \sup Q \leq Q(0+) \leq C_*^2 C_2^2 \, TV(u_0)^2.$$

Define now

$$\tilde{V}_r := \sup_{t>0} V_r(t), \quad \tilde{Q}_r = \sum_{0 < \tau < +\infty} [Q_r(\tau)]_+.$$

Therefore, we obtain from (3.13) and (3.14)

$$\tilde{V}_r \leq C_4 \, \tilde{Q}_{r-1},$$

$$\tilde{Q}_r \leq C' \, TV(u_0)^2 \tilde{V}_r + C'' \, TV(u_0) \, \tilde{Q}_{r-1}.$$

Thus we have

$$\tilde{Q}_r \leq \left(C' C_4^2 \, TV(u_0)^2 + C'' \, TV(u_0) \right) \tilde{Q}_{r-1} \leq \eta \tilde{Q}_{r-1}$$

with $\eta \in (0,1)$, provided again that the total variation of u_0 is sufficiently small.

Then, we have

$$\tilde{Q}_r \leq \eta^r \, \tilde{Q}_1 \leq \eta^r \, C_*^2 C_2^2 \, TV(u_0)^2,$$

and for all $t > 0$

$$V_r(t) \leq \tilde{V}_r \leq C \, \eta^{r-1},$$

which yields the desired inequality on V_r. $\square$

To complete the proof of (2.11) in Lemma 2.2 we rely on Lemma 3.3 which shows that the total amount of waves with large order is "small". On the other hand, by construction, artificial waves –when they are generated– have a small strength less than $C_4 \, |s^\alpha \, s^\beta| \leq C_4 \, \mathcal{E}(h)$ (where s^α and s^β denoted the strengths of the incoming waves).

Given $\varepsilon > 0$, choose an integer r_* such that

$$C_5 \sum_{r \geq r_*} \eta^r \leq \frac{\varepsilon}{2}.$$

Consider the total strength of artificial waves

$$E(t) := \sum_{i^h(x,t)=N+1} |\sigma^h(x,t)|.$$

For waves with orders $p^h \geq r_*$ we take advantage of the estimate in Lemma 3.3, while for waves with low orders $p^h < r_*$ we rely on Lemma 3.2, as follows:

$$\begin{aligned} E(t) &\leq \sum_{r \leq r_*} \frac{C(r)}{h^{m(r)}} \, C_4 \mathcal{E}(h) + C_5 \sum_{r \geq r_*} \eta^r \\ &\leq \frac{\tilde{C}(r_*)}{h^{\bar{m}(r_*)}} \, \mathcal{E}(h) + \frac{\varepsilon}{2}, \end{aligned}$$

by keeping the worst constant and exponent among $C(r)$ and $h^{m(r)}$. Finally, in view of the assumption (2.5) on $\mathcal{E}$, we can choose h sufficiently small so that the first term in the right-hand side above is less than $\varepsilon/2$, hence

$$E(t) \leq \varepsilon \quad \text{for all sufficiently small } h.$$

Since ε is arbitrary this establishes (2.11) and completes the proof of Lemma 2.2. $\square$

4. Pointwise regularity properties

In this last section we state without proof some regularity properties of the solution constructed in Theorem 2.1. Solutions to conservation laws turn out to be much more regular than arbitrary functions of bounded variation. (Compare with the statement in Theorem A.5 of the appendix.)

THEOREM 4.1. (Structure of shock curves). *Let $u = u(x,t)$ be a solution of (1.1) given by Theorem 2.1 and let $\varepsilon > 0$ be given. Then, there exist finitely many Lipschitz continuous curves, $x = y_k(t)$ for $t \in [\underline{T}_k, \overline{T}_k]$, $k = 1, \ldots, \bar{k}$, such that the following holds.*

For each m and all (but countably many) times $t_0 \in [\underline{T}_k, \overline{T}_k]$ the derivative $\dot{y}_k(t_0)$ and the left- and right-hand limits

$$u_- := \lim_{\substack{(x,t) \to (y_k(t_0), t_0) \\ x < y_k(t_0)}} u(x,t), \qquad u_+ := \lim_{\substack{(x,t) \to (y_k(t_0), t_0) \\ x > y_k(t_0)}} u(x,t) \qquad (4.1)$$

exist. The states u_- and u_+ determine a shock wave with strength $|\sigma_k(t_0)| \geq \varepsilon/2$, satisfying the Rankine-Hugoniot relations and Lax shock inequalities. The total variation of the mapping $t \mapsto \sigma_k(t)$ and of $t \mapsto \dot{y}_k(t)$ are bounded. At each point (x_0, t_0) outside the set

$$\mathcal{J}_\varepsilon(u) := \left\{ (y_k(t), t) \,/\, t \in [\underline{T}_k, \overline{T}_k],\ k = 1, \dots, \bar{k} \right\}$$

and outside a finite set $\mathcal{I}_\varepsilon(u)$, the function u has small oscillation:

$$\limsup_{(x,t) \to (x_0, t_0)} \big| u(x,t) - u(x_0, t_0) \big| \leq 2\,\varepsilon. \tag{4.2}$$

Using a countable sequence $\varepsilon \to 0$ we arrive at:

THEOREM 4.2. (Regularity of solutions). *Let u be a solution of (1.1) given by Theorem 2.1. Then, there exists a countable set $\mathcal{I}(u)$ of* **interaction points** *and a countable family of Lipschitz continuous* **shock curves**

$$\mathcal{J}(u) := \left\{ (y_k(t), t) \,/\, t \in [\underline{T}_k, \overline{T}_k],\ k = 1, 2, \dots \right\}$$

(both being possibly empty) such that the following holds. For each k and each $t \in [\underline{T}_k, \overline{T}_k]$ such that $(y_k(t), t) \notin \mathcal{I}(u)$, the left- and right-hand limits in (4.1) exist at $(y_k(t), t)$; the shock speed $\dot{y}_k(t)$ also exists and satisfies the Rankine-Hugoniot relations and Lax shock inequalities. Moreover, u is continuous at each point outside the set $\mathcal{J}(u) \cup \mathcal{I}(u)$. □

CHAPTER VIII

NONCLASSICAL ENTROPY SOLUTIONS OF THE CAUCHY PROBLEM

In this chapter we give a general existence result for *nonclassical* entropy solutions to the Cauchy problem associated with a system of conservation laws whose characteristic fields are genuinely nonlinear or concave-convex. (The result can be extended to linearly degenerate and convex-concave fields as well.) The proof is based on a generalization of the algorithm described in Chapter VII. Here, we use the nonclassical Riemann solver based on a given kinetic function for each concave-convex as was described in Section VI-3. Motivated by the examples arising in the applications (see Chapter III) we can assume that the kinetic functions satisfy the following *threshold condition*: any shock wave with strength less than some critical value is classical. In Section 1 we introduce a *generalized total variation* functional which is non-increasing for nonclassical solutions (Theorem 1.4) and whose decay rate can be estimated (Theorem 1.5). In Section 2 we introduce a *generalized interaction potential* and we extend Theorem IV-4.3 to nonclassical solutions; see Theorem 2.1. Section 3 and 4 are concerned with the *existence and regularity* theory for systems; see Theorems 3.1 and 4.2 respectively.

1. A generalized total variation functional

Consider the Cauchy problem for a scalar conservation law

$$\partial_t u + \partial_x f(u) = 0, \quad u = u(x,t) \in I\!\!R,$$
$$u(x,0) = u_0(x), \quad x \in I\!\!R, \tag{1.1}$$

when the flux $f : I\!\!R \to I\!\!R$ is assumed to be a concave-convex function (in the sense (II-2.5)) and the data $u_0 : I\!\!R \to I\!\!R$ are integrable functions with bounded variation. We consider the piecewise constant approximations $u^h = u^h(x,t)$ associated with (1.1) defined earlier in Section IV-3. To control the total variation of approximate solutions we introduce a "generalized total variation" functional $\tilde{V}(u^h(t))$ which is non-increasing in time and reduces to the standard total variation functional in the classical regime. This functional will be sufficiently robust to work for systems of equations (Section 3, below).

We will use the same notation as in Section II-4. The monotone decreasing function $\varphi^\natural : I\!\!R \to I\!\!R$ is characterized by

$$\frac{f(u) - f(\varphi^\natural(u))}{u - \varphi^\natural(u)} = f'(\varphi^\natural(u)), \quad u \neq 0,$$

and that $\varphi^{-\natural}$ denotes the inverse of the function $\varphi^\natural$. Recall that $\varphi^{\natural'}(u)$ tends to $-1/2$ when u tends to 0. We assume here that

$$-1 \leq \varphi^{\natural'}(u) < 0, \quad u \in I\!\!R,$$

but, alternatively, we could restrict attention to a bounded range of values u.

Given a **threshold coefficient** $\beta > 0$, consider a *kinetic function* $\varphi^\flat : I\!\!R \to I\!\!R$, which is smooth everywhere but possibly only Lipschitz continuous at $u = \pm\beta$ and satisfies the inequalities

$$\varphi^{-\natural}(u) < \varphi^\flat(u) \leq \varphi^\natural(u), \quad u > 0,$$
$$\varphi^\natural(u) \leq \varphi^\flat(u) < \varphi^{-\natural}(u), \quad u < 0.$$

Define also $\varphi^\sharp : I\!\!R \to I\!\!R$ by the inequalities

$$\varphi^\natural(u) \leq \varphi^\sharp(u) < u, \quad u > 0,$$
$$u < \varphi^\sharp(u) \leq \varphi^\natural(u), \quad u < 0,$$

and the condition

$$\frac{f(u) - f(\varphi^\flat(u))}{u - \varphi^\flat(u)} = \frac{f(u) - f(\varphi^\sharp(u))}{u - \varphi^\sharp(u)}, \quad u \neq 0.$$

Our main assumptions on the kinetic function are the following ones. For some constants $\beta > 0$ and

$$c_1 \in (0, 1], \quad c_2, c_3 \in [0, 1), \quad c_2 \leq c_3 < 1 - c_1 + c_2,$$
$$\text{or else } c_1 = 1, \ c_2 = c_3 = 0, \tag{1.2}$$

we impose
- the *monotonicity of $\varphi^\flat$*, more precisely

$$-c_1 \leq \varphi^{\flat'}(u) < 0, \quad u \in I\!\!R, \tag{1.3}$$

- the *threshold condition*

$$\varphi^\flat(u) = \varphi^\natural(u), \quad |u| \leq \beta, \tag{1.4}$$

- and the *monotonicity of $\varphi^\sharp$*, more precisely

$$-c_3 \leq \varphi^{\sharp'}(u) \leq -c_2, \quad |u| \geq \beta. \tag{1.5}$$

REMARK 1.1.
- Our assumptions imply for instance

$$\varphi^\natural(\beta) - c_1 (u - \beta) \leq \varphi^\flat(u) \leq \varphi^\natural(\beta), \quad u \geq \beta,$$
$$\varphi^\natural(-\beta) \leq \varphi^\flat(u) \leq \varphi^\natural(-\beta) - c_1 (u + \beta), \quad u \leq -\beta,$$

  and

$$\varphi^\natural(\beta) - c_3 (u - \beta) \leq \varphi^\sharp(u) \leq \varphi^\natural(\beta) - c_2 (u - \beta), \quad u \geq \beta,$$
$$\varphi^\natural(-\beta) - c_2 (u + \beta) \leq \varphi^\sharp(u) \leq \varphi^\natural(-\beta) - c_3 (u + \beta), \quad u \leq -\beta.$$

- A typical example of interest is given by $f(u) = u^3$,

$$\varphi^\flat(u) = \begin{cases} \beta/2 - c_1 (u + \beta), & u \leq -\beta, \\ -u/2, & -\beta \leq u \leq \beta, \\ -\beta/2 - c_1 (u - \beta), & u \geq \beta, \end{cases} \tag{1.6a}$$

and

$$\varphi^{\natural}(u) = \begin{cases} \beta/2 - c_2\,(u + \beta), & u \le -\beta, \\ -u/2, & -\beta \le u \le \beta, \\ -\beta/2 - c_2\,(u - \beta), & u \ge \beta, \end{cases} \tag{1.6b}$$

where the constants c_1 and c_2 satisfy

$$1/2 \le c_1 \le 1, \quad c_1 + c_2 = 1.$$

The kinetic function derived in Section III-2 from a dispersive-diffusive regularization of (1.1) has the form (1.6) with $c_1 = 1$ and $c_2 = 0$. This last case is covered by the second line in (1.2).

• More generally, the assumptions (1.3)–(1.5) are satisfied by any concave-convex flux-function and any kinetic function generated by nonlinear diffusion-dispersion, at least as far as values near the origin are concerned. (This is, of course, the situation of interest in the application to systems, in Section 3 below.) The conditions (1.2) will be motivated in Remark 1.6, below.

$\square$

We will work with a **generalized total variation functional** $\tilde{V}(u)$, defined as follows. If $u : \mathbb{R} \mapsto \mathbb{R}$ is a piecewise constant function, then we set

$$\tilde{V}(u) := \sum_x \tilde{\sigma}(u_-(x), u_+(x)),$$

where the summation is over all points of discontinuity of u and $\tilde{\sigma}(u_-, u_+)$ denotes the **generalized strength** of the wave connecting the left- and right-hand traces $u_- := u_-(x)$ and $u_+ := u_+(x)$ and is defined as follows. Note first that, if $\tilde{\sigma}(u_-, u_+) = |u_+ - u_-|$, then $\tilde{V}(u)$ coincides with the standard total variation $TV(u)$. Instead, to handle nonclassical solutions we choose (see Figure VIII-1)

$$\tilde{\sigma}(u_-, u_+) := \begin{cases} |u_+ - u_-|, & u_- u_+ \ge 0, \\ |u_- - (1 - K(u_-))\,u_+|, & u_- u_+ \le 0,\ |u_+| \le |\varphi^{\natural}(u_-)|, \\ |u_- + \varphi^{\flat}(u_-) - (2 - K(u_-))\,\varphi^{\natural}(u_-)|, & u_+ = \varphi^{\flat}(u_-), \end{cases} \tag{1.7}$$

where K is a Lipschitz continuous function satisfying

$$\begin{aligned} &K(u_-) \in [0, 2], \\ &K(u_-) = 0 \quad \text{when } |u_-| \le \beta. \end{aligned} \tag{1.8}$$

Observe that when $|u_-| \le \beta$ we have $\varphi^{\natural}(u_-) = \varphi^{\flat}(u_-) = \varphi^{\natural}(u_-)$ and, therefore, $\tilde{\sigma}$ coincides with the standard wave strength, that is, $\tilde{\sigma}(u_-, u_+) = |u_+ - u_-|$ for all u_+. Note that the definition (1.7) does not specify $\tilde{\sigma}(u_-, u_+)$ when u_+ belongs to the range $u_- u_+ < 0$, $|u_+| > |\varphi^{\natural}(u_-)|$, and $|u_+| \ne |\varphi^{\flat}(u_-)|$, since the generalized strength will not be of real use within this range. For convenience in the presentation we can define $\tilde{\sigma}(u_-, u_+)$ for arbitrary u_- and u_+ as being the sum of the generalized strengths of the two waves in the associated Riemann solution. (See the dashed lines on Figure VIII-1.)

Observe also that

$$\tilde{\sigma}(u_-, u_+) > 0 \quad \text{for all } u_+ \ne u_-,$$

and, more precisely, for every $M > 0$ we have

$$\frac{1 - c_1'}{1 + c_1'} |u_+ - u_-| \leq \tilde{\sigma}(u_-, u_+) \leq |u_+ - u_-|, \quad u_-, u_+ \in [-M, M], \qquad (1.9)$$

where

$$c_1' := \sup\{|\varphi^\flat(u)/u| \, / \, u \in [-M, M]\} < 1. \qquad (1.10)$$

(When $c_1 < 1$, c_1' can be replaced with c_1 and, then, (1.9) holds for arbitrary u_-, u_+.) Hence, the generalized strength and the generalized total variation are *equivalent* to the standard ones.

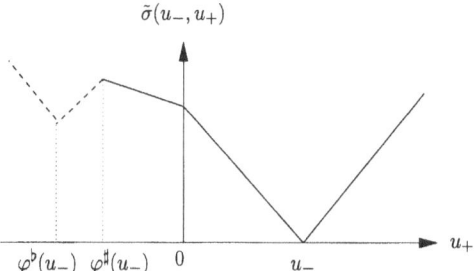

Figure VIII-1 : Generalized strength.

REMARK 1.2. The form of the generalized total variation is motivated as follows:
- The standard total variation cannot be used here since, when a nonclassical shock is generated at some interaction, the standard total variation *increases* by some (large) amount of the order of the strength of the incoming waves.
- From considering the interaction cases listed in Section IV-3 one can show that, to obtain a non-increasing functional, it is necessary that the strengths of nonclassical and of crossing classical shocks be weighted less than the standard strengths. (Note that a crossing classical shock may be transformed into a nonclassical shock through interactions.)
- Additionally, for the generalized total variation of the nonclassical Riemann solution to depend *continuously* upon the left- and the right-hand states, the strength of the classical shock from u_- to $\varphi^\sharp(u_-)$ must coincide with the strength of the nonclassical shock from u_- to $\varphi^\flat(u_-)$ plus the strength of the classical shock from $\varphi^\flat(u_-)$ to $\varphi^\sharp(u_-)$. The definition (1.7) satisfies this property as is clear from

$$\tilde{\sigma}(u_-, \varphi^\flat(u_-)) = |u_- - (1 - K(u_-)) \varphi^\sharp(u_-)| - |\varphi^\flat(u_-) - \varphi^\sharp(u_-)|$$
$$= \tilde{\sigma}(u_-, \varphi^\sharp(u_-)) - \tilde{\sigma}(\varphi^\flat(u_-), \varphi^\sharp(u_-)).$$

$\square$

The generalized strength depends on the function K, which should satisfy some constraints for the total variation of approximate solutions of (1.1) to be non-increasing.

THEOREM 1.3. (Diminishing generalized total variation.) *Under the assumptions listed above, suppose that the function K satisfies (1.8) and the following differential inequalities for all $|u| > \beta$*

$$\max\left(-(2+K\,\varphi^{\sharp\prime}), (2-K)\,\varphi^{\sharp\prime} - 2\left(1+\varphi^{b\prime}\right)\right) \leq \varphi^{\sharp}\,K' \leq (2-K)\,\varphi^{\sharp\prime}. \qquad (1.11)$$

Consider the piecewise constant approximations $u^h = u^h(x,t)$ defined in Section IV-3 and based on the nonclassical Riemann solver. Then, the function

$$t \mapsto \tilde{V}(u^h(t))$$

is non-increasing.

It is easy to check that by our assumptions (1.3)–(1.5) the intervals involved in (1.11) are not empty. The actual existence of a function K is established in Theorem 1.4 below.

PROOF. Observe first that (1.11) implies that the functions K, $2\,u+\varphi^{\sharp}(u)$, $2\,(u+\varphi^{b})-(2-K)\,\varphi^{\sharp}$, and $(2-K)\,\varphi^{\sharp}$ are non-decreasing for $u > \beta$. The first three functions are also non-decreasing in the region $0 \leq u \leq \beta$ (region in which we simply have $K \equiv 0$). We consider a front connecting a left-hand state u_l to a right-hand state u_m and interacting at some time t_0 with some other front connecting u_m to some state u_r. For definiteness we always assume that $u_l > 0$, the other cases being completely similar. We use the notation and classification given in Section IV-3.

Case RC-1 : Recall that $u_l \leq u_m$ and that $\varphi^{\sharp}(u_l) \leq u_r$. We distinguish between two subcases. When $u_r \geq 0$, the result is trivial and

$$[\tilde{V}(u^h(t_0))] = -2\,|u_m - u_l| \leq 0.$$

When $u_r < 0$ we have

$$\begin{aligned}
[\tilde{V}(u^h(t_0))] &= \left(u_l - (1 - K(u_l))\,u_r\right) - (u_m - u_l) - \left(u_m - (1 - K(u_m))\,u_r\right) \\
&= 2\,(u_l - u_m) + u_r\left(K(u_l) - K(u_m)\right) \\
&= \left(2\,(u_l - u_m) + \varphi^{\sharp}(u_l)\left(K(u_l) - K(u_m)\right)\right) + \left(u_r - \varphi^{\sharp}(u_l)\right)\left(K(u_l) - K(u_m)\right) \\
&\leq \left(2\,u_l + K(u_l)\,\varphi^{\sharp}(u_l)\right) - \left(2\,u_m + K(u_m)\,\varphi^{\sharp}(u_m)\right),
\end{aligned}$$

since $u_r - \varphi^{\sharp}(u_l) \geq 0$ and K is non-decreasing by (1.11). Furthermore, by (1.11) we have that $2\,u + \varphi^{\sharp}\,K$ is non-decreasing, so that $[\tilde{V}(u^h(t_0))] \leq 0$.

Case RC-2 : We have $u_l \leq u_m$ and $\varphi^{\sharp}(u_m) \leq u_r \leq \varphi^{b}(u_l)$.

$$\begin{aligned}
[\tilde{V}(u^h(t_0))] &= \left(u_l + \varphi^{b}(u_l) - (2 - K(u_l))\,\varphi^{\sharp}(u_l)\right) + \left(\varphi^{b}(u_l) - u_r\right) \\
&\quad - (u_m - u_l) - \left(u_m - (1 - K(u_m))\,u_r\right) \\
&= 2\,(u_l - u_m) + 2\,\varphi^{b}(u_l) - (2 - K(u_l))\,\varphi^{\sharp}(u_l) - K(u_m)\,u_r \\
&= -2\,|\varphi^{\sharp}(u_l) - \varphi^{b}(u_l)| - K(u_m)\,|u_r - \varphi^{\sharp}(u_m)| \\
&\quad + \left(2\,u_l + K(u_l)\,\varphi^{\sharp}(u_l)\right) - \left(2\,u_m + K(u_m)\,\varphi^{\sharp}(u_m)\right).
\end{aligned}$$

The first two terms are non-positive and may vanish. For the last two terms we use the assumption made in (1.11) that $2\,u + \varphi^{\sharp}(u)\,K(u)$ is non-decreasing.

Case RC-3 : We have $\max\bigl(\varphi^{\flat}(u_l), \varphi^{\sharp}(u_m)\bigr) \le u_r \le \varphi^{\sharp}(u_l) \le 0 \le u_l < u_m$, thus

$$
\begin{aligned}
\bigl[&\tilde{V}(u^h(t_0))\bigr] \\
&= u_l + \varphi^{\flat}(u_l) - (2 - K(u_l))\,\varphi^{\sharp}(u_l) + \bigl(u_r - \varphi^{\flat}(u_l)\bigr) \\
&\quad - (u_m - u_l) - \bigl(u_m - (1 - K(u_m))\,u_r\bigr) \\
&= 2\,(u_l - u_m) - (2 - K(u_l))\,\varphi^{\sharp}(u_l) + (2 - K(u_m))\,u_r \\
&= -|2 - K(u_m)|\,|\varphi^{\sharp}(u_l) - u_r| + \Bigl(2\,(u_l - u_m) + \varphi^{\sharp}(u_l)\,\bigl(K(u_l) - K(u_m)\bigr)\Bigr),
\end{aligned}
$$

and we conclude as in Case RC-1.

Case RN : We have $0 < u_l < u_m$ and $u_r = \varphi^{\flat}(u_m)$, thus

$$
\begin{aligned}
\bigl[\tilde{V}(u^h(t_0))\bigr] &= \bigl(u_l + \varphi^{\flat}(u_l) - (2 - K(u_l))\,\varphi^{\sharp}(u_l)\bigr) + \bigl(\varphi^{\flat}(u_l) - \varphi^{\flat}(u_m)\bigr) - (u_m - u_l) \\
&\quad - \bigl(u_m + \varphi^{\flat}(u_m) - (2 - K(u_m))\,\varphi^{\sharp}(u_m)\bigr) \\
&= \Bigl(2\,u_l + 2\,\varphi^{\flat}(u_l) - (2 - K(u_l))\,\varphi^{\sharp}(u_l)\Bigr) \\
&\quad - \Bigl(2\,u_m + 2\,\varphi^{\flat}(u_m) - (2 - K(u_m))\,\varphi^{\sharp}(u_m)\Bigr),
\end{aligned}
$$

which is non-positive since the function $2\,u + 2\,\varphi^{\flat} - (2 - K)\,\varphi^{\sharp}$ is non-decreasing by (1.11).

Case CR-1 : We have $\varphi^{\sharp}(u_l) \le u_r < u_m \le 0 < u_l$, thus

$$
\begin{aligned}
\bigl[\tilde{V}(u^h(t_0))\bigr] &= \bigl(u_l - (1 - K(u_l))\,u_r\bigr) - \bigl(u_l - (1 - K(u_l))\,u_m\bigr) - (u_m - u_r) \\
&= -K(u_l)\,|u_m - u_r|.
\end{aligned}
$$

Case CR-2 : This case is trivial and

$$
\bigl[\tilde{V}(u^h(t_0))\bigr] = -2\,|u_r - u_m|.
$$

Case CR-3 : We have $u_r \le \varphi^{\flat}(u_l) \le \varphi^{\sharp}(u_l) \le u_m \le 0 < u_l$, thus

$$
\begin{aligned}
\bigl[\tilde{V}(u^h(t_0))\bigr] &= \bigl(u_l + \varphi^{\flat}(u_l) - (2 - K(u_l))\,\varphi^{\sharp}(u_l)\bigr) + \bigl(\varphi^{\flat}(u_l) - u_r\bigr) \\
&\quad - \bigl(u_l - (1 - K(u_l))\,u_m\bigr) - (u_m - u_r) \\
&= -K(u_l)\,|u_m - \varphi^{\sharp}(u_l)| - 2\,|\varphi^{\sharp}(u_l) - \varphi^{\flat}(u_l)|.
\end{aligned}
$$

Case CR-4 : We have $\varphi^{\flat}(u_l) \le u_r \le \varphi^{\sharp}(u_l) \le u_m \le 0 < u_l$, thus

$$
\begin{aligned}
\bigl[\tilde{V}(u^h(t_0))\bigr] &= \bigl(u_l + \varphi^{\flat}(u_l) - (2 - K(u_l))\,\varphi^{\sharp}(u_l)\bigr) + \bigl(u_r - \varphi^{\flat}(u_l)\bigr) \\
&\quad - \bigl(u_l - (1 - K(u_l))\,u_m\bigr) - (u_m - u_r) \\
&= -K(u_l)\,|u_m - \varphi^{\sharp}(u_l)| - 2\,|\varphi^{\sharp}(u_l) - u_r|.
\end{aligned}
$$

Case CC-1 : We have $\varphi^{\sharp}(u_m) \le u_r < u_m < u_l$. When $u_r \ge 0$ the result is trivial and

$$
\bigl[\tilde{V}(u^h(t_0))\bigr] = 0.
$$

When $u_r < 0$ we have

$$\left[\tilde{V}(u^h(t_0))\right] = \left(u_l - (1 - K(u_l))\,u_r\right) - (u_l - u_m) - \left(u_m - (1 - K(u_m))\,u_r\right)$$
$$= -|K(u_l) - K(u_m)|\,|u_r|,$$

which is non-positive since K is non-decreasing.

Case CC-2 : We have $\varphi^\sharp(u_l) \leq u_m < u_r \leq \varphi^\sharp(u_m) \leq u_l$. When $u_r \leq 0$ we find

$$\left[\tilde{V}(u^h(t_0))\right] = \left(u_l - (1 - K(u_l))\,u_r\right) - \left(u_l - (1 - K(u_l))\,u_m\right) - (u_r - u_m)$$
$$= -|2 - K(u_l)|\,|u_r - u_m|.$$

When $u_r > 0$ we find

$$\left[\tilde{V}(u^h(t_0))\right] = (u_l - u_r) - \left(u_l - (1 - K(u_l))\,u_m\right) - \left((1 - K(u_m))\,u_r - u_m\right)$$
$$= -|2 - K(u_l)|\,|u_m| - |2 - K(u_m)|\,|u_r|,$$

since $u_m \leq 0$.

Case CC-3 : This case does not occur here since the function $\varphi^\sharp$ is non-increasing.

Case CN-1 : We have $0 \leq u_m < u_l$ and $\varphi^\sharp(u_l) \leq u_r = \varphi^\flat(u_m)$, thus

$$\left[\tilde{V}(u^h(t_0))\right] = \left(u_l - (1 - K(u_l))\,\varphi^\flat(u_m)\right) - (u_l - u_m)$$
$$- \left(u_m + \varphi^\flat(u_m) - (2 - K(u_m))\,\varphi^\sharp(u_m)\right)$$
$$= (2 - K(u_m))\left(\varphi^\sharp(u_m) - \varphi^\flat(u_m)\right) + \left(K(u_l) - K(u_m)\right)\varphi^\flat(u_m).$$

When $u_m \leq \beta$ the result is immediate (same formula as in Case CC-1):

$$\left[\tilde{V}(u^h(t_0))\right] = -K(u_l)\,|\varphi^\sharp(u_m)|.$$

When $u_m > \beta$ we write

$$\left[\tilde{V}(u^h(t_0))\right] = - (2 - K(u_l))\left|\varphi^\flat(u_m) - \varphi^\sharp(u_l)\right|$$
$$- \left((2 - K(u_l))\,\varphi^\sharp(u_l) - (2 - K(u_m))\,\varphi^\sharp(u_m)\right) \leq 0$$

since $(2 - K(u))\,\varphi^\sharp(u)$ is non-decreasing for $u \geq \beta$ by the second inequality in (1.11).

Case CN-2 : We have $\varphi^\sharp(u_l) \leq u_m \leq 0$ and $u_r = \varphi^\flat(u_m)$, thus

$$\left[\tilde{V}(u^h(t_0))\right] = \left(u_l - \varphi^\flat(u_m)\right) - \left(u_l - (1 - K(u_l))\,u_m\right)$$
$$- \left(-u_m - \varphi^\flat(u_m) + (2 - K(u_m))\,\varphi^\sharp(u_m)\right)$$
$$= -|2 - K(u_l)|\,|u_m| - |2 - K(u_m)|\,|\varphi^\sharp(u_m)|,$$

since $u_m \leq 0$ and $\varphi^\sharp(u_m) \geq 0$.

Case CN-3 : We have $0 \leq u_m < u_l$ and $u_r = \varphi^\flat(u_m) \leq \varphi^\sharp(u_l)$, thus

$$\left[\tilde{V}(u^h(t_0))\right] = \left(u_l + \varphi^\flat(u_l) - (2 - K(u_l))\,\varphi^\sharp(u_l)\right) + \left(\varphi^\flat(u_m) - \varphi^\flat(u_l)\right) - (u_l - u_m)$$
$$- \left(u_m + \varphi^\flat(u_m) - (2 - K(u_m))\,\varphi^\sharp(u_m)\right)$$
$$= (K(u_l) - 2)\,\varphi^\sharp(u_l) - (K(u_m) - 2)\,\varphi^\sharp(u_m),$$

which is non-positive since $(K - 2)\,\varphi^\sharp$ is non-increasing by (1.11). Note that as in Case CN-1 above, the second inequality in (1.11) is required in the region $|u| \geq \beta$, only.

Case NC : We have $\varphi^\sharp(u_l) \leq u_r \leq \varphi^\sharp(u_m) \leq u_l$ and $u_m = \varphi^\flat(u_l)$. When $u_r \leq 0$ we find

$$\left[\tilde{V}(u^h(t_0))\right]$$
$$= \left(u_l - (1 - K(u_l))\,u_r\right) - \left(u_l + \varphi^\flat(u_l) - (2 - K(u_l))\,\varphi^\sharp(u_l)\right) - \left(u_r - \varphi^\flat(u_l)\right)$$
$$= -|2 - K(u_l)|\,|u_r - \varphi^\sharp(u_l)|.$$

When $u_r \geq 0$ we find

$$\left[\tilde{V}(u^h(t_0))\right]$$
$$= (u_l - u_r) - \left(u_l + \varphi^\flat(u_l) - (2 - K(u_l))\,\varphi^\sharp(u_l)\right) - \left((1 - K(u_m))\,u_r - u_m\right)$$
$$= -|2 - K(u_l)|\,|\varphi^\sharp(u_l)| - |2 - K(u_m)|\,|u_r|.$$

Case NN : We have $u_m = \varphi^\flat(u_l)$ and $u_r = \varphi^\flat(u_m)$, thus

$$\left[\tilde{V}(u^h(t_0))\right] = \left(u_l - \varphi^\flat(u_m)\right) - \left(u_l + \varphi^\flat(u_l) - (2 - K(u_l))\,\varphi^\sharp(u_l)\right)$$
$$- \left(-\varphi^\flat(u_l) - \varphi^\flat(u_m) + (2 - K(u_m))\,\varphi^\sharp(u_m)\right)$$
$$= -|2 - K(u_l)|\,|\varphi^\sharp(u_l)| - |2 - K(u_m)|\,|\varphi^\sharp(u_m)|.$$

This completes the proof of Theorem 1.3.   □

We now establish the existence of a function K satisfying (1.11). We also estimate the rate of decay of the generalized total variation.

THEOREM 1.4. (Existence of a function K.) *Consider a concave-convex flux-function f, a kinetic-functions $\varphi^\flat$, and constants β, c_1, c_2, c_3 satisfying the assumptions (1.2)–(1.5) on an interval*

$$I_{\beta,\kappa} := \left[-(1 + \kappa)\,\beta, (1 + \kappa)\,\beta\right] \tag{1.12}$$

for some $\kappa > 0$. Then, the function K defined by

$$K(u) := \begin{cases} -K_* \,(u + \beta), & u \leq -\beta, \\ 0, & |u| \leq \beta, \\ K_* \,(u - \beta), & u \geq \beta, \end{cases} \tag{1.13}$$

satisfies the differential inequalities (1.11), provided the constant $K_* \geq 0$ satisfies

$$A_0 \leq K_* \leq \min(A_1, A_2, A_3) \tag{1.14}$$

where

$$A_0 := \frac{2\,c_3}{|\varphi^\natural(\beta)|}, \quad A_1 := \frac{2}{\kappa\,\beta}, \quad A_2 := \frac{2}{|\varphi^\natural(\beta)| + 2c_3\,\kappa\,\beta},$$
$$A_3 := \frac{2\,(1 - c_1 + c_2)}{|\varphi^\natural(\beta)| + (c_2 + c_3)\,\kappa\,\beta}, \tag{1.15}$$

(as well as analogous inequalities with $|\varphi^\natural(\beta)|$ replaced with $|\varphi^\natural(-\beta)|$).

THEOREM 1.5. (Decay rate for generalized total variation.) *Under the assumption made in Theorem 1.4, consider the piecewise constant approximations $u^h = u^h(x,t)$ of the Cauchy problem (1.1), defined in Section IV-3, based on the nonclassical Riemann solver associated with the kinetic function $\varphi^\flat$. Suppose that the range of u^h is included in the interval $I_{\beta,\kappa}$. If in (1.14) we have also the strict inequality*

$$K_* < \min(A_1, A_2),$$

then at each interaction time t_0 involving three constant states u_l, u_m, and u_r, we have the decay rate

$$\left[\tilde{V}(u^h(t_0))\right] \leq -c\,\Theta^h(t_0) \tag{1.16}$$

for some uniform constant $c > 0$, where the **cancelled strength** $\Theta^h(t_0)$ *is defined using the classification in Section IV-3 by*

$$\Theta^h(t_0) := \begin{cases} |u_m - u_l|, & \text{Cases RC-1, RC-2, RC-3,} \\ |u_r - u_m|, & \text{Cases CR-2, CC-2, CN-2,} \\ |u_r - \varphi^\flat(u_l)|, & \text{Cases CN-1 (when } |u_m| \geq |\beta|), \text{ NC,} \\ |\varphi^\flat(u_m) - \varphi^\flat(u_l)|, & \text{Case NN,} \\ 0, & \text{other cases.} \end{cases} \tag{1.17}$$

REMARK 1.6.
- When $f(u) = u^3$ we find $\varphi^\flat(u) = -u/2$. Therefore the inequalities (1.14) on K_* become

$$c_3 \leq \frac{\beta}{4}\,K_* \leq \min\left(\frac{1}{2\kappa}, \frac{1}{1 + 2\,c_3\,\kappa}, \frac{1 - c_1 + c_2}{1 + 2\,(c_2 + c_3)\,\kappa}\right),$$

  which, under the assumption made in (1.2), always determines a *non-empty* interval of values K_*, if κ is sufficiently small at least. (Either $c_3 < 1 - c_1 + c_2$ and we can find κ sufficiently small satisfying these inequalities, or else $c_3 = K_* = 1 - c_1 + c_2 = 0$ and there is no constraint on κ.)
- More generally, for general concave-convex flux-functions we observed earlier (Chapter II) that $\varphi^\flat(u) \sim -u/2$ near the origin, so that (1.14) always determines a *non-empty* interval of K_* if attention is restricted to a small neighborhood of the origin. Observe that when c_1, c_2, and c_3 are close to $1/2$ we can take κ close to $1/2$.
- In the special case (1.6) with $c_1 = 1$ and $c_2 = c_3 = 0$, we find $K \equiv 0$ and

$$\tilde{\sigma}(u_-, u_+) := \begin{cases} 3\,|\varphi^\flat(u_-)| & \text{if } u_+ = \varphi^\flat(u_-), \\ |u_+ - u_-| & \text{in other cases.} \end{cases}$$

  We have also

$$A_0 = 0, \quad A_1 = \frac{2}{\kappa\,\beta}, \quad A_2 = \frac{4}{\beta}, \quad A_3 = 0,$$

  and, concerning the constants ω_i defined below, $\omega_3 = \omega_4 = 0$.
- In view of the condition $A_0 \leq K_*$ in (1.14) (assuming that $c_3 \neq 0$), we see that K_* tends to infinity when $\beta \to 0$, so that the approach in this chapter is limited to the (large) class of kinetic functions admitting a threshold parameter. $\square$

Theorem 1.4 gives us a new proof for the uniform estimate of the total variation established earlier in Theorem IV-3.2. We can restate here this result as follows.

THEOREM 1.7. (Existence result for the Cauchy problem.) *Given a concave-convex flux-function f, a kinetic function φ^b, and constants β, c_1, c_2, and c_3 satisfying (1.2)– (1.5) on some interval $I_{\beta,\kappa}$, then provided β and κ are sufficiently small we have the following property. For all initial data u_0 in $L^1 \cap BV$ with range included in the interval $I_{\beta,\kappa}$, consider the sequence of piecewise constant approximations u^h based on the associated nonclassical Riemann solver and defined in Section IV-3. Then, the total variation of u^h remains uniformly bounded, and the sequence u^h converges almost everywhere to a weak solution of the Cauchy problem (1.1).*

Observe that if the range of the initial data is included in the interval $I_{\beta,\kappa}$, then by our assumptions the same is true for the (approximate and exact) solution at time t. Theorem 1.7 covers general concave-convex functions in a neighborhood of the origin and a large class of kinetic functions satisfying a threshold condition. Still, the assumptions made here on the kinetic function are stronger than the one required in the analysis of Section IV-3. But, the result here is also stronger since we determine the decay rate of the generalized total variation at each interaction. The interest of the present approach lies in the fact that it can be generalized to systems, as we will see in Section 3.

PROOF OF THEOREMS 1.4 AND 1.5. Observe that the inequalities (1.11) are trivially satisfied in the region $|u| \leq \beta$ since $K \equiv 0$. We will determine $K_* > 0$ so that (1.8) and (1.11) hold in the range $u \in \big(\beta, (1+\kappa)\beta\big)$, say. Dealing with the interval $\big(-(1+\kappa)\beta, -\beta\big)$ is completely similar.

To guarantee $K \leq 2$ we need that $K\big((1+\kappa)\beta\big) \leq 2$, that is,

$$K_* \leq \frac{2}{\kappa\beta}.$$

To guarantee that $|\varphi^\sharp|\, K' \leq 2 - |\varphi^{\sharp'}|\, K$ we need

$$\big(|\varphi^\natural(\beta)| + c_3\,(u-\beta)\big)\, K_* \leq 2 - c_3\, K_*\,(u-\beta),$$

that is,

$$K_* \leq \frac{2}{|\varphi^\natural(\beta)| + 2c_3\,\kappa\,\beta}.$$

To guarantee that $(K-2)\,\varphi^{\sharp'} \leq |\varphi^\sharp|\, K'$ we need

$$\big(2 - K_*\,(u-\beta)\big)\, c_3 \leq \big(|\varphi^\natural(\beta)| + c_2\,(u-\beta)\big)\, K_*, \quad u \in \big[\beta, (1+\kappa)\beta\big],$$

that is,

$$K_* \geq \frac{2\,c_3}{|\varphi^\natural(\beta)|}.$$

Finally, to guarantee that $|\varphi^\sharp|\, K' \leq 2 + 2\varphi^{b'} + (K-2)\,\varphi^{\sharp'}$ we need

$$\big(|\varphi^\natural(\beta)| + c_3\,(u-\beta)\big)\, K_* \leq 2 - 2\,c_1 + \big(2 - K_*\,(u-\beta)\big)\, c_2,$$

that is,

$$K_* \leq \frac{2\,(1 - c_1 + c_2)}{|\varphi^\natural(\beta)| + (c_2 + c_3)\,\kappa\,\beta}.$$

This completes the derivation of the inequalities (1.16) and (1.17).

We now estimate the rate of decay when the constants

$$\omega_1 := \inf_{0 \le u \le (1+\kappa)\beta} (2 - K(u)) = 2 - K_* \, \kappa \, \beta > 0$$

and

$$\omega_2 := \inf_{0 \le u \le (1+\kappa)\beta} \left(2 + K'(u) \, \varphi^\sharp(u) + K(u) \, \varphi^{\sharp'}(u)\right)$$

$$\ge 2 - \left(|\varphi^\natural(\beta)| + 2 \, c_3 \, \kappa \, \beta\right) K_* > 0$$

are positive. For completeness in the calculation, we also use the notation

$$\omega_3 := \inf_{\beta \le u \le (1+\kappa)\beta} \left((2 - K(u)) \, \varphi^{\sharp'}(u) - K'(u) \, \varphi^\sharp(u)\right)$$

$$\ge |\varphi^\natural(\beta)| \, K_* - 2 \, c_3 \ge 0$$

and

$$\omega_4 := \inf_{0 \le u \le (1+\kappa)\beta} \left(2 + 2 \, \varphi^{b'}(u) + K'(u) \, \varphi^\sharp(u) + (K(u) - 2) \, \varphi^{\sharp'}(u)\right)$$

$$\ge 2 - 2 \, c_1 + (2 - K_* \, \kappa \, \beta) \, c_2 - \left(|\varphi^\natural(\beta)| + c_3 \, \kappa \, \beta\right) K_* \ge 0.$$

Under our assumptions, ω_3 and ω_4 may well vanish however. (See an example in Remark 1.7.)

Case RC-1 : In the first subcase

$$\left[\tilde{V}(u^h(t_0))\right] = -2 \, |u_m - u_l|,$$

and in the second one

$$\left[\tilde{V}(u^h(t_0))\right] \le -\omega_2 \, |u_m - u_l|.$$

Case RC-2 :

$$\left[\tilde{V}(u^h(t_0))\right] \le -\omega_2 \, |u_m - u_l|.$$

Case RC-3 :

$$\left[\tilde{V}(u^h(t_0))\right] \le -\omega_2 \, |u_m - u_l|.$$

Case RN :

$$\left[\tilde{V}(u^h(t_0))\right] \le -\omega_4 \, |u_m - u_l| \le 0.$$

Case CR-1 :

$$\left[\tilde{V}(u^h(t_0))\right] = -K(u_l) \, |u_m - u_r|,$$

which vanishes for $u_l \le \beta$.

Case CR-2 :

$$\left[\tilde{V}(u^h(t_0))\right] = -2 \, |u_r - u_m|.$$

Case CR-3 :

$$\left[\tilde{V}(u^h(t_0))\right] = -K(u_l) \, |u_m - \varphi^\sharp(u_l)| - 2 \, |\varphi^\sharp(u_l) - \varphi^b(u_l)| \le 0.$$

Note in passing that the estimate becomes

$$\left[\tilde{V}(u^h(t_0))\right] \le -2 \, |u_m - u_r|$$

in the special case that $u_m = \varphi^\sharp(u_l)$ and $u_r = \varphi^b(u_l)$.

Case CR-4 :

$$\left[\tilde{V}(u^h(t_0))\right] = -K(u_l)\,|u_m - \varphi^\sharp(u_l)| - 2\,|\varphi^\sharp(u_l) - u_r| \le 0.$$

The estimate becomes

$$\left[\tilde{V}(u^h(t_0))\right] \le -2\,|u_m - u_r|$$

in the special case $u_m = \varphi^\sharp(u_l)$.

Case CC-1 : In the first subcase

$$\left[\tilde{V}(u^h(t_0))\right] = 0$$

and in the second one

$$\left[\tilde{V}(u^h(t_0))\right] \le -|u_r|\,|K(u_l) - K(u_m)|,$$

which vanishes as $u_r \to 0$.

Case CC-2 : In both subcases

$$\left[\tilde{V}(u^h(t_0))\right] \le -\omega_1\,|u_r - u_m|.$$

Case CN-1 : When $u_m \le \beta$ we have

$$\left[\tilde{V}(u^h(t_0))\right] \le 0.$$

When $u_m > \beta$ we find

$$\left[\tilde{V}(u^h(t_0))\right] \le -\omega_1\,|\varphi^\flat(u_m) - \varphi^\sharp(u_l)| = -\omega_1\,|u_r - \varphi^\sharp(u_l)|.$$

Case CN-2 :

$$\left[\tilde{V}(u^h(t_0))\right] \le -\omega_1\,|u_m - \varphi^\sharp(u_m)| \le -\frac{\omega_1}{2}\,|u_r - u_m|,$$

since $|u_m - \varphi^\flat(u_m)| \le 2\,|u_m - \varphi^\sharp(u_m)|$ by our assumptions (1.3) and (1.5).

Case CN-3 :

$$\left[\tilde{V}(u^h(t_0))\right] \le -\omega_3\,|u_l - u_m| \le 0.$$

Case NC : In both subcases

$$\left[\tilde{V}(u^h(t_0))\right] \le -\omega_1\,|u_r - \varphi^\sharp(u_l)|.$$

Case NN :

$$\left[\tilde{V}(u^h(t_0))\right] \le -\omega_1\,|\varphi^\sharp(u_l)| - \omega_1\,|\varphi^\sharp(u_m)|,$$

which is bounded away from zero. This completes the proof of Theorems 1.4 and 1.5.
□

2. A generalized weighted interaction potential

The (linear) functional described in Section 1 is non-increasing, but fails to be decreasing in some interaction cases. Our aim in the present section is to determine a (quadratic) functional which will be strictly decreasing at each interaction. Consider the **generalized weighted interaction potential**

$$\tilde{Q}(u) := \sum_{x<y} q(u_-(x), u_+(x))\, \tilde{\sigma}(u_-(x), u_+(x))\, \tilde{\sigma}(u_-(y), u_+(y)), \qquad (2.1)$$

where $\tilde{\sigma}$ is the generalized strength defined in (1.7) and q is a weight determined so that *a nonclassical shock located at x is regarded as being non-interacting with waves located on its right-hand side $y > x$.* This is motivated by the fact that nonclassical shocks for concave-convex equations are slow undercompressive. See also the classification in Section IV-3.

We generalize here the definition (IV-4.8) introduced first for classical solutions. Setting $u_\pm := u_\pm(x,t)$ we define the function $q(u_-, u_+)$ by

$$q(u_-, u_+) := \begin{cases} 1, & 0 \le u_- \le u_+, \\ \hat{\sigma}(\rho(u_-, u_+), u_+), & \varphi^\sharp(u_-) \le u_+ \le u_- \text{ and } u_- > 0, \\ 0, & u_+ = \varphi^\flat(u_-), \\ \hat{\sigma}(\rho(u_-, u_+), u_+), & u_- \le u_+ \le \varphi^\sharp(u_-) \text{ and } u_- < 0, \\ 1, & u_+ \le u_- \le 0. \end{cases} \qquad (2.2)$$

where $\rho(u_-, u_+) \ne u_-, u_+$ (when $u_- \ne u_+$) is defined by

$$\frac{f(\rho(u_-, u_+)) - f(u_-)}{\rho(u_-, u_+) - u_-} = \frac{f(u_+) - f(u_-)}{u_+ - u_-},$$

and $\hat{\sigma}(u, v)$ is a variant of the generalized strength defined earlier:

$$\hat{\sigma}(u, v) := \begin{cases} |v - u|, & uv \ge 0, \\ |u - (1 - K(u))\, v|, & uv \le 0. \end{cases} \qquad (2.3)$$

We suppose that the function K is defined by (1.13) to (1.15). The main result in this section is:

THEOREM 2.1. (Generalized interaction potential.) *Consider solutions with total variation less than some fixed constant V. Let f be a concave-convex flux-function f and $\delta, \lambda \in (0,1)$ be sufficiently small constants. Then, consider positive constants β, κ, c_1, c_1', and c_2, c_3 satisfying (1.2) and $c_1' < 1$ together with*

$$(1 + \kappa)\beta \le \delta,$$
$$\frac{2(1 - c_1 + c_2)\kappa\beta}{|\varphi^\sharp(\beta)| + (c_2 + c_3)\kappa\beta} \le \lambda, \qquad (2.4)$$

and let $\varphi^\flat$ be any kinetic function satisfying (1.3)–(1.5) and (1.10) on the interval $I_{\beta,\kappa}$. Then, for every constant C_* such that

$$C_* \delta V \le \lambda, \quad C_* \delta^2 \le \lambda\beta,$$

the piecewise constant approximations u^h of Section IV-3 (with range included in $I_{\beta,\kappa}$ and total variation less than V) satisfy at each interaction (for some uniform $c > 0$):

$$[\tilde{V}(u^h(t_0)) + C_* \tilde{Q}(u^h(t_0))] \le -c\,[\tilde{V}(u^h(t_0))]| - c\,q(u_l, u_m)\,|u_l - u_m|\,|u_m - u_r|. \qquad (2.5)$$

REMARK 2.2.

- In fact we have

$$[\tilde{V}(u^h(t_0)) + C_* \, \tilde{Q}(u^h(t_0))] \leq \begin{cases} -c \, |u_r - \varphi^\sharp(u_l)| & \text{in Case } NC, \\ -c \, q(u_l, u_m) \, |u_l - u_m| \, |u_m - u_r| & \text{in other cases.} \end{cases}$$

Observe that Case NC is the only interaction when a nonclassical shock meets some classical wave on its right-hand side: in that case, the interaction potential between the two incoming wave *vanishes* (according to the definition (1.19)) but the generalized total variation of the solution *decreases* strictly.

- Under the conditions (1.2) there always exist some β and δ (sufficiently small) such that the hypotheses (2.4) hold. This is clear since for $\kappa \to 0$ the second condition in (2.4) is trivial. On the other hand, κ cannot be arbitrary large since the same condition with $\kappa \to \infty$ becomes $2(1 - c_1 + c_2) \leq (c_2 + c_3)$ which would contradict (1.2). $\square$

PROOF. Note first that the second condition in (2.4) combined with (1.14)-(1.15) guarantees that

$$K_* \, \kappa \, \beta \leq \lambda \qquad (2.6)$$

and in particular that $K(u) \leq \lambda$. Note also that the function ρ satisfies

$$\begin{aligned} |\rho(u_l, u_r) - \rho(u_l, u_m)| &\leq (1 + O(\delta)) \, |u_m - u_r|, \\ |u_l - u_m - \rho(u_m, u_r) + \rho(u_l, u_r)| &\leq C \, \delta \, |u_l - u_m|. \end{aligned} \qquad (2.7)$$

When $f(u) = u^3$ these inequalities are obvious since $\rho(u, v) = -u - v$. For a general flux-function $\rho(u, v) \sim -u - v$ and these inequalities hold in a sufficiently small neighborhood of 0, at least.

We consider the same decomposition as in the proof of Theorem IV-4.3:

$$[\tilde{Q}(u^h(t_0))] = P_1 + P_2 + P_3, \qquad (2.8)$$

where P_1 contains products between the waves involved in the interaction, P_2 between waves which are not involved in the interaction, and P_3 products between these two sets of waves.

On one hand, we have

$$P_1 = -q(u_l, u_m) \, \tilde{\sigma}(u_l, u_m) \, \tilde{\sigma}(u_m, u_r) \leq -c \, q(u_l, u_m) \, |u_l - u_m| \, |u_m - u_r|,$$

since there is only one outgoing wave or else the two outgoing waves are regarded as non-interacting in view of the definition (2.2).

On the other hand, $P_2 = 0$ since the waves which are not involved in the interaction are not modified.

Call W_l and W_r the (*weighted*) total strength of waves located on the left- and right-hand side of the interaction point, respectively. Let us decompose P_3 accordingly, say

$$P_3 = P_{3l} + P_{3r}.$$

In view of the definition of the potential and the fact that the generalized strength diminishes at interaction (see (1.14)) the contribution to $\tilde{Q}$ involving waves located on the left-hand side of the interaction is non-increasing:

$$P_{3l} = W_l \, (\tilde{\sigma}(u_l, u_r) - \tilde{\sigma}(u_l, u_m) - \tilde{\sigma}(u_r, u_m)) \leq 0.$$

To deal with waves located on the right-hand side of the interaction point we set
$P_{3r} = \Omega_r\, W_r$ with

$$\Omega_r := q(u_l, u_m')\, \tilde{\sigma}(u_l, u_m') + q(u_m', u_r)\, \tilde{\sigma}(u_m', u_r) \\ - q(u_l, u_m)\, \tilde{\sigma}(u_l, u_m) - q(u_m, u_r)\, \tilde{\sigma}(u_m, u_r), \tag{2.9}$$

where we assume that the outgoing pattern contains a wave connecting u_l to some intermediate state u_m' plus a wave connecting to u_r. We will show that either $\Omega_r \le 0$ or else the term $\Omega_r\, W_r$ can be controlled by the cancellation determined in Theorem 1.4. We need only consider cases when one of the states under consideration at least is above the threshold β since, otherwise, the desired estimate was already established in Theorem IV-4.4.

The notation $O(\delta)$ refers to a term which can be made arbitrarily small since we are restricting attention to values sufficiently close to the origin. In view of the classification given in Section IV-3 we can distinguish between the following cases.

Case RC-1 : When $u_r \ge 0$, we obtain

$$\Omega_r = \left(\left(1 - K(\rho(u_l, u_r))\right) u_r - \rho(u_l, u_r)\right)(u_l - u_r) - (u_m - u_l) \\ - \left(\left(1 - K(\rho(u_m, u_r))\right) u_r - \rho(u_m, u_r)\right)(u_m - u_r) \\ = -\left(1 + (1 + K_* \delta)\, O(\delta)\right)|u_m - u_l| \le 0,$$

provided δ and $K_* \delta$ are sufficiently small. When $u_r < 0$, we obtain similarly

$$\Omega_r = \left(u_r - \rho(u_l, u_r)\right)\left(u_l - (1 - K(u_l))\, u_r\right) - (u_m - u_l) \\ - \left(u_r - \rho(u_m, u_r)\right)\left(u_m - (1 - K(u_m))\, u_r\right) \\ = -\left(1 + (1 + K_* \delta)\, O(\delta)\right)|u_m - u_l| \le 0.$$

Case RC-2 :

$$\Omega_r = \left(\varphi^\flat(u_l) - u_r\right) - (u_m - u_l) - \left(u_r - \rho(u_m, u_r)\right)\left(u_m - (1 - K(u_m))\, u_r\right) \\ \le \left(\varphi^\flat(u_l) - \varphi^\flat(u_m)\right) - (u_m - u_l) \\ \le -(1 - c_1)\,|u_m - u_l|.$$

Case RC-3 : Using $\varphi^\flat(u_m) \le \varphi^\flat(u_l)$ we obtain

$$\Omega_r = \left(\rho(\varphi^\flat(u_l), u_r) - (1 - K(\rho(\varphi^\flat(u_l), u_r)))\, u_r\right)\left(u_r - \varphi^\flat(u_l)\right) \\ - (u_m - u_l) - \left(u_r - \rho(u_m, u_r)\right)\left(u_m - (1 - K(u_m))\, u_r\right)$$

When $u_m \to u_l$ we have $u_r \to \varphi^\sharp(u_l)$ and $\rho(\varphi^\flat(u_l), u_r) \to u_l$, thus Ω_r converges toward

$$\left(u_l - (1 - K(u_l)\, \varphi^\sharp(u_l))\right)\left(\varphi^\sharp(u_l) - \varphi^\flat(u_l)\right) \\ - \left(\varphi^\sharp(u_l) - \varphi^\flat(u_l)\right)\left(u_l - (1 - K(u_l)\, \varphi^\sharp(u_l)\right),$$

which vanishes identically. Therefore, in the general case we have

$$\Omega_r = -\left(1 + (1 + K_* \delta)\, O(\delta)\right)|u_m - u_l| \le 0.$$

Case RN :

$$\Omega_r = \big(\varphi^\flat(u_l) - \varphi^\flat(u_m)\big) - (u_m - u_l)$$
$$\leq -(1 - c_1)\,|u_m - u_l|.$$

Case CR-1 :

$$\Omega_r = \big(u_r - \rho(u_l, u_r)\big)\big(u_l - \big(1 - K(u_l)\big)u_r\big)$$
$$- \big(u_m - \rho(u_l, u_m)\big)\big(u_l - \big(1 - K(u_l)\big)u_m\big) - (u_m - u_r)$$
$$= -\big(1 + O(\delta)\big)\,|u_m - u_r|.$$

Case CR-2 :

$$\Omega_r = \Big(\big(1 - K(\rho(u_l, u_r))\big)u_r - \rho(u_l, u_r)\Big)(u_l - u_r)$$
$$- \Big(\big(1 - K(\rho(u_l, u_m))\big)u_m - \rho(u_l, u_m)\Big)(u_l - u_m) - (u_r - u_m)$$
$$= -\Big(1 + (1 + K_*\delta)\,O(\delta)\Big)\,|u_m - u_r| \leq 0.$$

Case CR-3 :

$$\Omega_r = \big(\varphi^\flat(u_l) - u_r\big) - \big(u_m - \rho(u_l, u_m)\big)\big(u_l - \big(1 - K(u_l)\big)u_m\big) - (u_m - u_r)$$
$$= -\big|\varphi^\flat(u_l) - u_m\big| - \big|u_m - \rho(u_l, u_m)\big|\,\big|u_l - \big(1 - K(u_l)\big)u_m\big|.$$

Case CR-4 :

$$\Omega_r = \Big(\rho(\varphi^\flat(u_l), u_r) - \big(1 - K(\rho(\varphi^\flat(u_l), u_r))\big)u_r\Big)\big(u_r - \varphi^\flat(u_l)\big)$$
$$- \big(u_m - \rho(u_l, u_m)\big)\big(u_l - \big(1 - K(u_l)\big)u_m\big) - (u_m - u_r).$$

By construction, we have $\Omega_r = 0$ in the limiting case $u_r = u_m = \varphi^\sharp(u_l)$ since then $\rho(\varphi^\flat(u_l), u_r) = u_l$ and $\rho(u_l, u_m) = \varphi^\flat(u_l)$. Therefore, in the general case we find

$$\Omega_r = (1 + K_*\delta)\,O(\delta)\,\big|u_r - \varphi^\sharp(u_l)\big| + (1 + K_*\delta)\,O(\delta)\,\big|u_m - \varphi^\sharp(u_l)\big| - |u_m - u_r|$$
$$= -\Big(1 + (1 + K_*\delta)\,O(\delta)\Big)\,|u_m - u_r| \leq 0.$$

Case CC-1 : When $u_r \geq 0$ we have

$$\Omega_r = \Big(\big(1 - K(\rho(u_l, u_r))\big)u_r - \rho(u_l, u_r)\Big)(u_l - u_r)$$
$$- \Big(\big(1 - K(\rho(u_l, u_m))\big)u_m - \rho(u_l, u_m)\Big)(u_l - u_m)$$
$$- \Big(\big(1 - K(\rho(u_m, u_r))\big)u_r - \rho(u_m, u_r)\Big)(u_m - u_r)$$
$$= -\big(\rho(u_l, u_r) - \rho(u_l, u_m)\big)(u_l - u_m)$$
$$- \Big(u_l - u_m - \rho(u_m, u_r) + \rho(u_l, u_r)\Big)(u_m - u_r)$$
$$- \Big(-K(\rho(u_l, u_m))u_m + K(\rho(u_l, u_r))u_r\Big)(u_l - u_m)$$
$$+ \Big(-K(\rho(u_l, u_r)) + K(\rho(u_m, u_r))\Big)u_r(u_m - u_r),$$

therefore

$$\Omega_r \leq -\big(\rho(u_l, u_r) - \rho(u_l, u_m)\big)(u_l - u_m)$$
$$- \big(u_l - u_m - \rho(u_m, u_r) + \rho(u_l, u_r)\big)(u_m - u_r)$$
$$+ \big(K_*\delta + \max K\big)|u_l - u_m||u_m - u_r|$$
$$\leq -\big(1 - K_*\delta + \max K + O(\delta)\big)|u_l - u_m||u_m - u_r|$$

by (2.6)-(2.7).

When $u_r < 0$ we have along similar lines

$$\Omega_r = \big(u_r - \rho(u_l, u_r)\big)\big(u_l - (1 - K(u_l))u_r\big)$$
$$- \Big(\big(1 - K(\rho(u_l, u_m))\big)u_m - \rho(u_l, u_m)\Big)(u_l - u_m)$$
$$- \big(u_r - \rho(u_m, u_r)\big)\big(u_m - (1 - K(u_m))u_r\big)$$
$$= -\Big(u_m - u_r - \rho(u_l, u_m) + \rho(u_l, u_r)\Big)(u_l - u_m)$$
$$- \big(\rho(u_l, u_r) - \rho(u_m, u_r)\big)(u_m - u_r)$$
$$- \big(u_r - \rho(u_m, u_r)\big)K(u_m)u_r + \big(u_r - \rho(u_l, u_r)\big)K(u_l)u_r$$
$$+ K(\rho(u_l, u_m))u_m(u_l - u_m),$$

therefore

$$\Omega_r = -\Big(\rho(u_l, u_r) - \rho(u_l, u_m)\Big)(u_l - u_m)$$
$$- \big(u_l - u_m + \rho(u_l, u_r) - \rho(u_m, u_r)\big)(u_m - u_r)$$
$$+ \Big(K_*\delta + \big(2 + O(\delta)\big)\max K\Big)|u_l - u_m||u_m - u_r|$$
$$\leq -c\,|u_l - u_m||u_m - u_r|,$$

where we also used $|u_m|, |u_r| \leq |u_r - u_m|$.

Case CC-2 :   When $u_r \leq 0$ we have

$$\Omega_r = \big(u_r - \rho(u_l, u_r)\big)\big(u_l - (1 - K(u_l))u_r\big)$$
$$- \big(u_m - \rho(u_l, u_m)\big)\big(u_l - (1 - K(u_l))u_m\big)$$
$$- \Big(\rho(u_m, u_r) - \big(1 - K(\rho(u_m, u_r))\big)u_r\Big)(u_r - u_m)$$
$$= O(\delta)\,|u_r - u_m|,$$

which can be controlled by the cancelled strength $|u_r - u_m|$ in (1.16)–(1.17), provided $C_*\,\delta\,TV(u^h(t))$ is sufficiently small. When $u_r > 0$ we find

$$\Omega_r = \Big(\big(1 - K(\rho(u_l, u_r))\big)u_r - \rho(u_l, u_r)\Big)(u_l - u_r)$$
$$- \big(u_m - \rho(u_l, u_m)\big)\big(u_l - (1 - K(u_l))u_m\big)$$
$$- \big(\rho(u_m, u_r) - u_r\big)\big(u_m - (1 - K(u_m))u_r\big).$$

In the formal limit $u_r \to u_m$ (possible only if, simultaneously, $u_m \to 0$) we find $\Omega_r \to 0$. Therefore, using $|u_m|, |u_r| \leq |u_r - u_m|$ we obtain

$$\Omega_r = u_l\Big(\big(1 - K(\rho(u_l, u_r))\big)u_r - u_m + \rho(u_l, u_m) - \rho(u_l, u_r)\Big) + O(\delta)\,|u_r - u_m|$$
$$= O(\delta)\,|u_r - u_m|,$$

which can be controlled by the cancelled strength.

Case CN-1 : The case $u_m \leq \beta$ is the same as the case CC-1 with $u_r < 0$. When $u_m > \beta$ we have

$$\Omega_r = \left(\varphi^\flat(u_m) - \rho(\varphi^\flat(u_m), u_l)\right)\left(u_l - (1 - K(u_l))\,\varphi^\flat(u_m)\right)$$
$$- \left(\left(1 - K(\rho(u_l, u_m))\right)u_m - \rho(u_l, u_m)\right)(u_l - u_m)$$

But, when u_l decreases toward u_m, the value $\varphi^\sharp(u_l)$ increases and one reaches first equality in the inequality $\varphi^\sharp(u_l) \leq u_r = \varphi^\flat(u_m)$, and one can check that Ω_r is non-positive when $\varphi^\sharp(u_l) = u_r$: this is actually a special case of Case CN-3 treated below. Hence, by continuity we find

$$\Omega_r \leq O(\delta)\left|\varphi^\flat(u_m) - \varphi^\sharp(u_l)\right| = O(\delta)\left|u_r - \varphi^\sharp(u_l)\right|,$$

which can be controlled by the cancelled strength.

Case CN-2 :

$$\Omega_r = \left(\left(1 - K(\rho(u_l, \varphi^\flat(u_m)))\right)\varphi^\flat(u_m) - \rho(u_l, \varphi^\flat(u_m))\right)\left(u_l - \varphi^\flat(u_m)\right)$$
$$- \left(u_m - \rho(u_l, u_m)\right)\left(u_l - (1 - K(u_l))\,u_m\right)$$
$$\leq O(\delta)\left|u_r - u_m\right|,$$

which is controlled by the cancelled strength.

Case CN-3 : Since $|\varphi^\flat(u_l) - \varphi^\flat(u_m)| \leq |u_l - u_m|$ we have

$$\Omega_r = \left(\rho(\varphi^\flat(u_l), \varphi^\flat(u_m)) - \left(1 - K(\rho(\varphi^\flat(u_l), \varphi^\flat(u_m)))\right)\varphi^\flat(u_m)\right)\left(\varphi^\flat(u_m) - \varphi^\flat(u_l)\right)$$
$$- \left(\left(1 - K(\rho(u_l, u_m))\right)u_m - \rho(u_l, u_m)\right)(u_l - u_m)$$
$$\leq -|u_l - u_m|\left(-\rho(\varphi^\flat(u_l), \varphi^\flat(u_m)) - \rho(u_l, u_m)\right.$$
$$+ \left(1 - K(\rho(u_l, u_m))\right)u_m - \left(1 - K(\rho(\varphi^\flat(u_l), \varphi^\flat(u_m)))\right)\varphi^\flat(u_m)\Big).$$

But we have

$$\left(1 - K(\rho(u_l, u_m))\right)u_m - \left(1 - K(\rho(\varphi^\flat(u_l), \varphi^\flat(u_m)))\right)\varphi^\flat(u_m).$$
$$\geq u_m\left(1 - \max K\right) \geq 0.$$

and for δ sufficiently small

$$-\rho(\varphi^\flat(u_l), \varphi^\flat(u_m)) - \rho(u_l, u_m) \geq \left(1 + O(\delta)\right)\left(u_l + \varphi^\flat(u_l) + u_m + \varphi^\flat(u_m)\right) \geq 0.$$

Therefore $\Omega_r \leq 0$.

Case NC : When $u_r \leq 0$ we find

$$\Omega_r = \left(u_r - \rho(u_l, u_r)\right)\left(u_l - (1 - K(u_l))\,u_r\right)$$
$$- \left(\rho(\varphi^\flat(u_l), u_r) - \left(1 - K(\rho(\varphi^\flat(u_l), u_r))\right)u_r\right)\left(u_r - \varphi^\flat(u_l)\right).$$

In the limiting case $u_r = \varphi^\sharp(u_l)$ we have $\rho(u_l, u_r) = \varphi^\flat(u_l)$, $\rho(\varphi^\flat(u_l), u_r) = u_l$, and thus $\Omega_r = 0$. Therefore, in the general case we find

$$\Omega_r = O(\delta)\left|u_r - \varphi^\sharp(u_l)\right|,$$

which can be controlled by the cancelled strength.

When $u_r \geq 0$ we find

$$\Omega_r = \Big(\big(1 - K(\rho(u_l, u_r)) \big) u_r - \rho(u_l, u_r) \Big) (u_l - u_r)$$

$$- \big(\rho(\varphi^\flat(u_l), u_r) - u_r \big) \Big(\big(1 - K(\varphi^\flat(u_l)) \big) u_r - \varphi^\flat(u_l) \Big)$$

Formally, if $u_r \to \varphi^\sharp(u_l)$ we would find $\rho(u_l, u_r) = \varphi^\flat(u_l)$, $\rho(\varphi^\flat(u_l), u_r) = u_l$, and $\Omega_r = 0$. Therefore, in the general case we have

$$\Omega_r = O(\delta) \, |u_r - \varphi^\sharp(u_l)|,$$

which can be controlled by the cancellation.

Case NN : We have $u_m = \varphi^\flat(u_l)$ and $u_r = \varphi^\flat(u_m)$, thus

$$\Omega_r = \Big(\big(1 - K(\rho(\varphi^\flat(u_m), u_l)) \big) \varphi^\flat(u_m) - \rho(\varphi^\flat(u_m), u_l) \Big) (u_l - \varphi^\flat(u_m))$$

$$= O(\delta) \, |u_r - u_l| = O(\delta^2),$$

which, thanks to our condition $C_* \, \delta^2 \leq \lambda \beta$, is controlled by the cancelled strength which is

$$|u_r - \varphi^\sharp(u_l)| \leq |\varphi^\natural(\beta)| = O(\beta)$$

This completes the proof of Theorem 2.1. $\square$

3. Existence theory

We now turn to the Cauchy problem for the system of conservation laws

$$\partial_t u + \partial_x f(u) = 0, \quad u = u(x, t) \in \mathcal{U} \tag{3.1}$$

under the usual assumptions of strict hyperbolicity in $\mathcal{U} = \mathcal{B}(\delta_0)$. Following Chapters VI and VII, we assume that (1.1) admits genuinely nonlinear or concave-convex characteristic fields. (The analysis can be extended to linearly degenerate or convex-concave fields.) Throughout this section we strongly rely on the notations and assumptions introduced in Sections VI-2 to VI-4. In particular, a parameter μ_i is provided for each i-wave family and, for each concave-convex characteristic field, it is normalized so that

$$\mu_i(u) = 0 \text{ if and only if } \nabla \lambda_i(u) \cdot r_i(u) = 0.$$

Based on the parameter μ_i we defined the critical value $\mu_i^\natural$ in Lemma VI-2.3. On the other hand in Section VI-4, to determine the admissible nonclassical shock waves in each concave-convex i-family we prescribed *a kinetic function* $\mu_i^\flat$. Finally, from the parameter $\mu_i^\flat$ we defined the companion value $\mu_i^\sharp$ (see the formula (VI-4.3)).

Beyond the assumptions made in Section VI-4 we also postulate the existence of a **threshold value** for the mapping $\mu_i^\flat$, that is: there exists $\beta_i : \mathcal{U} \setminus \mathcal{M}_i \mapsto \mathbb{R}$ which is defined and smooth away from the *critical manifold*

$$\mathcal{M}_i := \Big\{ u \in \mathcal{U} \, / \, \nabla \lambda_i(u) \cdot r_i(u) = 0 \Big\}$$

and satisfies the following four properties for some constant $\beta_i^* > 0$:
- $\nabla \beta_i \cdot r_i < 0$ on $\mathcal{M}_i^- \cup \mathcal{M}_i^+$, where

$$\mathcal{M}_i^\pm := \Big\{ u \in \mathcal{U} \, / \, \nabla \lambda_i \cdot r_i(u) \lesseqgtr 0 \Big\},$$

- $\beta_i(u) \in \left(-2\beta_i^*, -\beta_i^*/2\right)$ for all $u \in \mathcal{M}_i^+$, while $\beta_i(u) \in \left(\beta_i^*/2, 2\beta_i^*\right)$ for all $u \in \mathcal{M}_i^-$,
- and

$$\mu_i^\flat(u) = \mu_i^\flat(u) \quad \text{when } |\mu_i(u)| \le |\beta_i(u)|. \tag{3.2}$$

- Additionally, the kinetic function $\mu_i^\flat$ is smooth everywhere but possibly only *Lipschitz continuous* along the **threshold manifolds**

$$\mathcal{N}_i^\pm := \left\{u \in \mathcal{M}_i^\pm \ / \ \mu_i(u) = \beta_i(u)\right\}.$$

The condition (3.2) means that, in a neighborhood of the manifold $\mathcal{M}_i$ the nonclassical Riemann solution described in Section VI-4 reduces to the classical one (Section VI-2). Our assumptions cover the examples of interest arising in continuum physics; see Remark III-5.4 for instance.

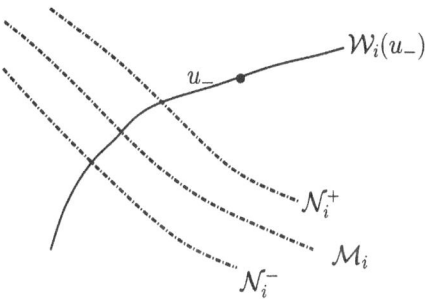

Figure VIII-2 : Critical and threshold manifolds.

To solve the Cauchy problem associated with the system (3.1) we follow the strategy developed in Chapter VII, the novelty being that we now rely on the *nonclassical Riemann solver*. We will only stress here the main differences with the classical case. The approximate solution may contain classical and nonclassical shock fronts, rarefaction fronts, and artificial fronts. By definition, an approximate **nonclassical wave front** is a propagating discontinuity connecting two states satisfying the Hugoniot relations and the kinetic relation (up to possible errors of order $O(h)$).

Based on the study in Section 1 above we introduce a **generalized total variation functional** for systems, defined for piecewise constant functions $u : \mathbb{R} \to \mathcal{U}$ made of single wave fronts. Given such a function $u = u(x)$ we set

$$\tilde{V}(u) := \sum_x \tilde{\sigma}\left(u_-(x), u_+(x)\right),$$

where the summation is over the points of discontinuity of u. The **generalized strength** $\tilde{\sigma}(u_-, u_+)$ will be defined shortly so that

$$\frac{1}{C}\,|u_+ - u_-| \le \tilde{\sigma}(u_-, u_+) \le C\,|u_+ - u_-|$$

for some uniform constant $C > 1$, implying that the functional $\tilde{V}$ is equivalent to the standard total variation:

$$\frac{1}{C} TV(u) \leq \tilde{V}(u) \leq C\, TV(u). \tag{3.3}$$

Depending upon the kind of i-wave connecting u_- to u_+, we define the strength $\tilde{\sigma}(u_-, u_+) = \tilde{\sigma}_i(u_-, u_+)$ as follows:

$$\tilde{\sigma}(u_-, u_+) := \begin{cases} |\mu_i(u_+) - \mu_i(u_-)| & \text{genuinely nonlinear } i\text{-field,} \\ |u_+ - u_-| & \text{artificial front,} \end{cases} \tag{3.4}$$

and, for every concave-convex i-characteristic field,

$$\tilde{\sigma}(u_-, u_+)$$
$$:= \begin{cases} |\mu_i(u_+) - \mu_i(u_-)|, & \mu_i(u_+)\,\mu_i(u_-) \geq 0, \\ |\mu_i(u_-) - \left(1 - K_i(u_-)\right)\mu_i(u_+)|, & \mu_i(u_+)\,\mu_i(u_-) \leq 0 \\ & \text{and } |\mu_i(u_+)| \leq |\mu_i^{\sharp}(u_-)|, \\ |\mu_i(u_-) + \mu_i^{\flat}(u_-) - \left(2 - K_i(u_-)\right)\mu_i^{\sharp}(u_-)|, & \mu_i(u_+) = \mu_i^{\flat}(u_-), \end{cases} \tag{3.5}$$

which is the natural generalization of the definition (1.7) introduced for scalar equations. Here, the mapping $K_i : \mathcal{U} \to I\!\!R$ is Lipschitz continuous and is given (by analogy with (1.15)) by

$$K_i(u) := \begin{cases} -K_i^* \left(\mu_i(u) + \beta_i(u)\right), & \mu_i(u) \leq \beta_i(u) \leq 0, \\ 0, & |\mu_i(u)| \leq |\beta_i(u)|, \\ K_i^* \left(\mu_i(u) - \beta_i(u)\right), & \mu_i(u) \geq \beta_i(u) \geq 0, \end{cases} \tag{3.6}$$

where $K_i^* \in [0, 1)$ are sufficiently small constants. Observe that when all waves are classical and remain within the region $|\mu_i(u)| \leq |\beta_i(u)|$ the generalized strength coincides with the strength defined earlier in Section VII-1.

Next, following the discussion in Section 2 above let us introduce the **generalized interaction potential**

$$\tilde{Q}_M(u) := \tilde{Q}_s(u) + M\,\tilde{Q}_d(u) \tag{3.7}$$

with

$$\tilde{Q}_d(u) := \sum_{x<y} \tilde{\sigma}\big(u_-(x), u_+(x)\big)\, \tilde{\sigma}\big(u_-(y), u_+(y)\big) \tag{3.8}$$

and

$$\tilde{Q}_s(u) := \sum_{x<y} q\big(u_-(x), u_+(x)\big)\, \tilde{\sigma}\big(u_-(x), u_+(x)\big)\, \tilde{\sigma}\big(u_-(y), u_+(y)\big), \tag{3.9}$$

in which the summation is done over all pairs of jumps in the function $u = u(x)$. In $\tilde{Q}_d(u)$ we count all products between waves of *different* families provided the left-hand wave is faster than the right-hand one. In $\tilde{Q}_s(u)$, we include products between waves of the *same* characteristic family, say i-waves. We define the weight by

$$q(u_-, u_+) := 1, \quad \text{genuinely nonlinear fields,}$$

and for concave-convex i-field by

$$
q(u_-, u_+) := \begin{cases}
1, & 0 \leq \mu_i(u_-) \leq \mu_i(u_+), \\
\hat{\sigma}_i\big(\rho_i(u_-, u_+), u_+\big), & \mu_i^\sharp(u_-) \leq \mu_i(u_+) \leq \mu_i(u_-) \text{ and } \mu_i(u_-) > 0, \\
0, & \mu_i(u_+) = \mu_i^\flat(u_-), \\
\hat{\sigma}_i\big(\rho_i(u_-, u_+), u_+\big), & \mu_i(u_-) \leq \mu_i(u_+) \leq \mu_i^\sharp(u_-) \text{ and } \mu_i(u_-) < 0, \\
1, & \mu_i(u_+) \leq \mu_i(u_-) \leq 0,
\end{cases}
\tag{3.10}
$$

where

$$
\hat{\sigma}_i(u, v) := \begin{cases}
|\mu_i(v) - \mu_i(u)|, & \mu_i(v)\,\mu_i(u) \geq 0, \\
|\mu_i(u) - \big(1 - K_i(u)\big)\,\mu_i(v)|, & \mu_i(v)\,\mu_i(u) \leq 0,
\end{cases}
\tag{3.11}
$$

which is the natural generalization of (2.2)-(2.3). Given any two distinct vectors u_- and u_+ satisfying the Hugoniot conditions, we denoted by $\rho_i(u_-, u_+) \neq u_{\pm}$ the solution of

$$
-\overline{\lambda}_i(u_-, u_+) \big(\rho_i(u_-, u_+) - u_-\big) + f\big(\rho_i(u_-, u_+)\big) - f(u_-) = 0.
$$

(See the discussion in Lemma VI-2.5).

We follow the general strategy in Chapter VII. To initial data $u_0 : \mathbb{R} \to \mathcal{U}$ we associate a sequence of piecewise constant approximations $u_0^h : \mathbb{R} \to \mathcal{U}$ containing at most C/h jumps and such that

$$
\begin{aligned}
u_0^h &\to u_0 \quad \text{in the } L^1 \text{ norm}, \\
TV(u_0^h) &\to TV(u_0) \quad \text{as } h \text{ tends to zero}.
\end{aligned}
\tag{3.12}
$$

At each discontinuity on the line $t = 0$ we solve a Riemann problem using the non-classical solver constructed in Section VI-4 and based on prescribed kinetic functions $\mu_i^\flat$. Locally in time, the corresponding approximate solution $u^h : \mathbb{R} \times \mathbb{R}_+ \to \mathcal{U}$ is made of admissible (classical or nonclassical) shock fronts and rarefaction fronts (with strength $\leq h$), only. To extend the solution further in time, approximate interaction solvers are considered, an accurate one and a rough one, which we use depending on the size of the incoming waves and of some threshold $\mathcal{E}(h) \to 0$, as was explained in Section VII-2.

THEOREM 3.1. (Existence theory for nonclassical entropy solutions.) *Consider the strictly hyperbolic system of conservation laws (3.1) defined in $\mathcal{U} = \mathcal{B}(\delta_0)$ together with some sufficiently small $\delta_1 < \delta_0$. Suppose that each characteristic field of (3.1) is either genuinely nonlinear or concave-convex. Then, there exist a constant $c, C > 0$ such that the following result holds for $0 < C' < C''$.*

Let $\delta_2 \leq \delta_1$. Let $c_1, c_2, \ldots$ be constants that are sufficiently close to $1/2$ and satisfy

$$
c_1 \geq c_4, \quad c_2 \geq c_3.
\tag{3.13}
$$

For each concave-convex i-family let $\mu_i^\flat(u)$ be a kinetic function satisfying the threshold condition (3.2) for some $\beta_i(u)$ satisfying

$$
C' \delta_2^2 \leq \beta_i(u) \leq C'' \delta_2,
$$

together with the inequalities

$$
-c_1 \leq \frac{\nabla \mu_i^\flat(u) \cdot r_i(u)}{\nabla \mu_i^\natural(u) \cdot r_i(u)} \leq -c_4, \quad -c_2 \leq \frac{\nabla \mu_i^\sharp(u) \cdot r_i(u)}{\nabla \mu_i^\natural(u) \cdot r_i(u)} \leq -c_3.
\tag{3.14}
$$

Given some initial data $u_0 : I\!R \to \mathcal{B}(\delta_2)$ with small total variation:

$$TV(u_0) < c, \tag{3.13}$$

the approximation scheme based on the corresponding nonclassical Riemann solver generates a globally defined sequence $u^h : I\!R \times I\!R_+ \to \mathcal{B}(\delta_0)$ such that for suitable constants $C_*, M, K_i^* > 0$ the function

$$\tilde{V}(u^h(t)) + C_* \, \tilde{Q}_M(u^h(t)) \text{ is decreasing} \tag{3.16}$$

at each interaction time. The sequence u^h converges almost everywhere to a weak solution $u : I\!R \times I\!R_+ \to \mathcal{B}(\delta_0)$ of (3.1) with

$$TV(u(t)) \le C \, TV(u_0), \qquad t \ge 0.$$

Observe that the result in Theorem 3.1 applies to solutions with total variation less than a fixed constant c. The threshold $\beta_i(u)$ can be taken to be sufficiently small so that nonclassical shocks can exist in the ball $\mathcal{B}(\delta_2)$. However, for δ_2 fixed, the inequality $C'' \delta_2^2 \le \beta_i(u)$ prevents the threshold $\beta_i(u)$ to become arbitrary small; in fact, as $\beta_i \to 0$ the size δ_2 of the neighborhood $\mathcal{B}(\delta_2)$ shrinks to the origin. Classical entropy solutions are covered by Theorem 3.1 by taking $\mu_i^\flat(u) = \mu_i^\sharp(u) = \mu_i^\natural(u)$ and recalling Lemma VI-2.4.

The key to the proof of Theorem 3.1 is deriving a uniform bound on the total variation of u^h, which is based on the following generalization of Theorem VII-1.1.

THEOREM 3.2. (Wave interaction estimates for nonclassical solutions.) *For all u_l, u_m, and $u_r \in \mathcal{B}(\delta_1)$ we have the following property. Suppose that u_l is connected to u_m by a i-wave front and that u_m is connected to u_r by a j-wave front ($1 \le i, j \le N$). Then the outgoing wave strengths $\tilde{\sigma}_k(u_l, u_r)$ satisfy ($1 \le k \le N$)*

$$\tilde{\sigma}_k(u_l, u_r) \le \tilde{\sigma}_k(u_l, u_m) + \tilde{\sigma}_k(u_m, u_r) + O(1) \, \tilde{Q}_-(u_l, u_m, u_r),$$

$$= \begin{cases} \tilde{\sigma}_i(u_l, u_m) + O(1) \, \tilde{Q}_-(u_l, u_m, u_r), & k = i \ne j, \\ \tilde{\sigma}_j(u_m, u_r) + O(1) \, \tilde{Q}_-(u_l, u_m, u_r), & k = j \ne i, \\ \tilde{\sigma}_i(u_l, u_m) + \tilde{\sigma}_j(u_m, u_r) + O(1) \, \tilde{Q}_-(u_l, u_m, u_r), & k = j = i, \\ O(1) \, \tilde{Q}_-(u_l, u_m, u_r), & \text{otherwise,} \end{cases} \tag{3.17}$$

*where the **generalized interaction potential** between the two incoming waves is defined as*

$$\tilde{Q}_-(u_l, u_m, u_r) = \begin{cases} \tilde{\sigma}_i(u_l, u_m) \, \tilde{\sigma}_j(u_m, u_r), & i > j, \\ q(u_l, u_m) \, \tilde{\sigma}_i(u_l, u_m) \, \tilde{\sigma}_i(u_m, u_r), & i = j. \end{cases} \tag{3.18}$$

$\square$

4. Pointwise regularity properties

In this section we state without proof some regularity properties of the solution obtained in Theorem 3.1.

THEOREM 4.1. (Structure of shock curves). *Let* $u = u(x, t)$ *be a solution of* (3.1) *given by Theorem 3.1. For each (sufficiently small)* $\varepsilon > 0$ *there exists finitely many Lipschitz continuous curves,* $x = z_k(t)$ *for* $t \in (\underline{t}_k^\varepsilon, \overline{t}_k^\varepsilon)$, $k = 1, \ldots, K_\varepsilon$, *such that the following holds. For each* k *and all (but countably many) times* $t_0 \in (\underline{t}_k^\varepsilon, \overline{t}_k^\varepsilon)$ *the derivative* $\dot{z}_k(t_0)$ *and the left- and right-hand limits*

$$u_-(z_k(t_0), t_0) := \lim_{\substack{(x,t) \to (z_k(t_0), t_0) \\ x < z_k(t_0)}} u(x, t), \qquad u_+(z_k(t_0), t_0) := \lim_{\substack{(x,t) \to (z_k(t_0), t_0) \\ x > z_k(t_0)}} u(x, t)$$

(4.1)

exist and determine a shock wave with strength $|\sigma_k(t)| \geq \varepsilon/2$, *satisfying the Rankine-Hugoniot relations: it is either a classical shock satisfying Lax shock inequalities or a nonclassical shock satisfying the kinetic relation. Moreover, the mappings* $t \mapsto \sigma_k(t)$ *and of* $t \mapsto \dot{z}_k(t)$ *are of uniformly bounded (with respect to* ε*) total variation. At each point* (x_0, t_0) *outside the set*

$$\mathcal{J}_\varepsilon(u) := \left\{ (z_k(t), t) \, / \, t \in (\underline{t}_k^\varepsilon, \overline{t}_k^\varepsilon), \, k = 1, \ldots, K_\varepsilon \right\}$$

and outside a finite set $\mathcal{I}_\varepsilon(u)$, *the function* u *has small oscillation:*

$$\limsup_{(x,t) \to (x_0, t_0)} |u(x, t) - u(x_0, t_0)| \leq 2\varepsilon.$$

(4.2)

□

We have also:

THEOREM 4.2. (Regularity of solutions). *Let* u *be a solution of* (1.1) *given by Theorem 3.1. Then there exists an (at most) countable set* $\mathcal{I}(u)$ *of* **interaction points** *and an (at most) countable family of Lipschitz continuous* **shock curves**

$$\mathcal{J}(u) := \left\{ (z_k(t), t) \, / \, t \in (\underline{T}_k, \overline{T}_k), \, k = 1, 2, \ldots \right\}$$

such that the following holds. For each k *and each* $t \in (\underline{T}_k, \overline{T}_k)$ *such that* $(z_k(t), t) \notin \mathcal{I}(u)$, *the left- and right-hand limits in* (4.1) *exist at* $(z_k(t), t)$; *the shock speed* $\dot{z}_k(t)$ *also exists and satisfies the Rankine-Hugoniot relations. The corresponding propagating discontinuity is either a classical shock satisfying Lax shock inequalities or a nonclassical shock satisfying the kinetic relation. Moreover,* u *is continuous at each point outside the set* $\mathcal{J}(u) \cup \mathcal{I}(u)$. □

CHAPTER IX

CONTINUOUS DEPENDENCE OF SOLUTIONS

In this chapter, we investigate the L^1 *continuous dependence* of solutions for systems of conservation laws. We restrict attention to solutions generated in the limit of piecewise approximate solutions and we refer to Chapter X for a discussion of the uniqueness of *general* solutions with bounded variation. In Section 1 we outline a general strategy based on a L^1 *stability result* for a class of linear hyperbolic systems with discontinuous coefficients. The main result in Theorem 1.5 shows that the sole source of instability would be the presence of *rarefaction-shocks*. In Section 2 we apply the setting to systems with genuinely nonlinear characteristic fields; see Theorem 2.3. One key observation here is that rarefaction-shocks never arise from comparing two classical entropy solutions to systems of conservation laws. In Section 3 we provide a *sharp version* of the continuous dependence estimate which shows that the L^1 distance between two solutions is "strictly decreasing"; see Theorem 3.2. Finally, in Section 4 we state the generalization to nonclassical entropy solutions.

1. A class of linear hyperbolic systems

For solutions $u = u(x,t)$ and $v = v(x,t)$ of the system of conservation laws

$$\partial_t u + \partial_x f(u) = 0, \quad u = u(x,t) \in \mathcal{U}, \quad x \in I\!R, \, t \geq 0, \tag{1.1}$$

we want to establish the L^1 continuous dependence estimate

$$\|u(t) - v(t)\|_{L^1(I\!R)} \leq C \, \|u(0) - v(0)\|_{L^1(I\!R)}, \quad t \geq 0, \tag{1.2}$$

for some uniform constant $C > 0$. In (1.1), $\mathcal{U} := \mathcal{B}(\delta) \subset I\!R^N$ is a ball with sufficiently small radius δ, and the flux $f : \mathcal{U} \to I\!R^N$ is a given smooth mapping. We assume that $Df(u)$ is strictly hyperbolic for all $u \in \mathcal{U}$, with eigenvalues

$$\lambda_1(u) < \ldots < \lambda_N(u)$$

and left- and right-eigenvectors $l_j(u)$ and $r_j(u)$ $(1 \leq j \leq N)$, respectively, normalized so that

$$l_i(u) \, r_j(u) = 0 \text{ if } i \neq j, \text{ and } l_i(u) \, r_i(u) = 1.$$

To motivate the results in this section we outline our general strategy of proof to derive the estimate (1.2). Consider any **averaging matrix** $\overline{A} = \overline{A}(u,v)$ satisfying, by definition,

$$\begin{aligned} \overline{A}(u,v) \, (v-u) &= f(v) - f(u), \\ \overline{A}(u,v) &= \overline{A}(v,u), \quad u,v \in \mathcal{U}. \end{aligned} \tag{1.3}$$

For instance, one could choose

$$\overline{A}(u,v) = \int_0^1 Df\big((1-\theta)\,u + \theta\,v\big) \, d\theta.$$

We denote by $\overline{\lambda}_j(u,v)$ the (real and distinct) eigenvalues of the matrix $\overline{A}(u,v)$ and by $\overline{l}_j(u,v)$ and $\overline{r}_j(u,v)$ corresponding left- and right-eigenvectors, normalized in the standard way.

Let $u = u(x,t)$ and $v = v(x,t)$ be two entropy solutions of (1.1). Clearly, the function

$$\psi := v - u \tag{1.4}$$

is a solution of

$$\partial_t \psi + \partial_x \big(\overline{A}(u,v)\,\psi\big) = 0. \tag{1.5}$$

Therefore, to establish (1.2) it is sufficient to derive the L^1 *stability property*

$$\|\psi(t)\|_{L^1(\mathbb{R})} \leq C\,\|\psi(0)\|_{L^1(\mathbb{R})}, \quad t \geq 0 \tag{1.6}$$

for a class of matrices $A = A(x,t)$ and a class of solutions $\psi = \psi(x,t)$ of the **linear hyperbolic system**

$$\partial_t \psi + \partial_x \big(A(x,t)\,\psi\big) = 0, \quad \psi = \psi(x,t) \in \mathbb{R}^N, \tag{1.7}$$

covering the situation of interest (1.4) and (1.5).

The present section is devoted, precisely, to deriving (1.6) for solutions of (1.7) (and, more generally, of (1.12) below.) As we will see, the *characteristic curves* (in the (x,t)-plane) associated with the matrix-valued function A will play a major role here. We begin with some assumption and notation. Throughout we restrict attention to piecewise constant functions $A = A(x,t)$ and piecewise constant solutions $\psi = \psi(x,t)$. By definition, A admits finitely many polygonal lines of discontinuity and finitely many interaction times $t \in \mathcal{I}(A)$ at which (or simplicity in the presentation) we assume that exactly two discontinuity lines meet. The set of points of discontinuity of A is denoted by $\mathcal{J}(A)$. At each $(x,t) \in \mathcal{J}(A)$ we can define the left- and right-hand limits $A_\pm(x,t)$ and the corresponding discontinuity speed $\lambda^A(x,t)$. On the other hand, the matrix $A(x,t)$ is assumed to be strictly hyperbolic at each point (x,t), with eigenvalues denoted by $\lambda_j^A(x,t)$ and left- and right-eigenvectors denoted by $l_j^A(x,t)$ and $r_j^A(x,t)$, respectively, $(1 \leq j \leq N)$. We also use the notation $l_{j\pm}^A(x,t)$ and $r_{j\pm}^A(x,t)$ for the limits at a point of discontinuity.

We suppose that the eigenvalues are totally separated in the following sense: There exist disjoint intervals $\big[\lambda_j^{\min}, \lambda_j^{\max}\big]$ $(j = 1,\ldots,N)$ having sufficiently small length (that is, $\big|\lambda_j^{\max} - \lambda_j^{\min}\big| << 1$), such that

$$\lambda_j^A(x,t) \in \big[\lambda_j^{\min}, \lambda_j^{\max}\big]. \tag{1.8}$$

Similarly, we assume at each $(x,t) \in \mathcal{J}(A)$ there exists some index i such that

$$\lambda^A(x,t) \in \big[\lambda_i^{\min}, \lambda_i^{\max}\big], \tag{1.9}$$

and we refer to the propagating discontinuity located at this point (x,t) as a i-**wave front**. More precisely, we assume that the matrix A may also contain **artificial wave fronts** which do not fulfill (1.9) but, by definition, propagate at a *fixed and constant speed* λ_{N+1} satisfying

$$\lambda_N^{\max} < \lambda_{N+1}. \tag{1.10}$$

Such a propagating discontinuity is also called a $(N+1)$-**wave front**.

DEFINITION 1.1. Depending on the respective values of the left- and right-hand limits $\lambda_{i\pm}^A = \lambda_{i\pm}^A(x,t)$ and the propagation speed $\lambda^A = \lambda^A(x,t)$, an i-wave front $(1 \leq i \leq N)$ located at some point $(x,t) \in \mathcal{J}(A)$ is called

- a **Lax front** if
$$\lambda_{i-}^A \geq \lambda^A \geq \lambda_{i+}^A,$$

- a **slow undercompressive front** if
$$\min(\lambda_{i-}^A, \lambda_{i+}^A) > \lambda^A,$$

- a **fast undercompressive front** if
$$\max(\lambda_{i-}^A, \lambda_{i+}^A) < \lambda^A,$$

- or a **rarefaction-shock front** if
$$\lambda_{i-}^A \leq \lambda^A \leq \lambda_{i+}^A.$$

Note that Definition 1.1 is slightly ambiguous: So, for definiteness, an i-wave front having $\lambda_{i-}^A = \lambda = \lambda_{i+}^A$ (but $A_-(x,t) \neq A_+(x,t)$ if $(x,t) \in \mathcal{J}(A)$) will be called a Lax front (rather than a rarefaction-shock front). We will use the notation $\mathcal{L}(A)$, $\mathcal{S}(A)$, $\mathcal{F}(A)$, $\mathcal{R}(A)$, and $\mathcal{A}(A)$ for the sets of all Lax, slow undercompressive, fast undercompressive, rarefaction-shock, and artificial fronts, respectively. When it will be necessary to specify which family the wave front belongs to, we will use the corresponding notation $\mathcal{L}_i(A)$, $\mathcal{S}_i(A)$, $\mathcal{F}_i(A)$, and $\mathcal{R}_i(A)$, respectively $(1 \leq i \leq N)$. Finally, we denote by $\mathcal{J}_i(A)$ the set of all i-wave fronts $(1 \leq i \leq N+1)$, so that we have

$$\mathcal{J}(A) = \mathcal{L}(A) \cup \mathcal{S}(A) \cup \mathcal{F}(A) \cup \mathcal{R}(A) \cup \mathcal{A}(A),$$
$$\mathcal{J}_i(A) = \mathcal{L}_i(A) \cup \mathcal{S}_i(A) \cup \mathcal{F}_i(A) \cup \mathcal{R}_i(A), \quad 1 \leq i \leq N,$$
$$\mathcal{J}_{N+1}(A) = \mathcal{A}(A).$$

Suppose that we are given a vector-valued function $g = g(x,t)$ consisting (for each time t) of finitely many Dirac masses located on the discontinuity lines of A, that is,

$$\mathcal{J}(g) \subset \mathcal{J}(A), \tag{1.11}$$

where, by extension, $\mathcal{J}(g)$ denotes the set of locations of Dirac masses in g. Let $\mathcal{M}(\mathbb{R})$ be the space of all bounded measures on $\mathbb{R}$ and, for each time t, denote by $\|g(t)\|_{\mathcal{M}(\mathbb{R})}$ the sum of all Dirac masses in $g(t)$. Then, consider piecewise constant solutions $\psi = \psi(x,t)$ of the **linear hyperbolic system with measure right-hand side**

$$\partial_t \psi + \partial_x (A\psi) = g. \tag{1.12}$$

The source term g will be necessary later (Section 2) to handle *approximate* solutions of the systems of conservation laws (1.1).

To derive the estimate (1.6) we introduce a *weighted* L^1 norm which will be non-increasing in time for the solutions of (1.12). So, given a piecewise constant function $\psi = \psi(x,t)$, define its **characteristic components** $\alpha = (\alpha_1, \dots, \alpha_N)$ by

$$\psi(x,t) = \sum_{j=1}^{N} \alpha_j(x,t)\, r_j^A(x,t). \tag{1.13}$$

With any piecewise constant "weight"

$$w = (w_1, \dots, w_N) \quad \text{such that } w_j > 0\, (1 \leq j \leq N),$$

we associate the **weighted L^1 norm** of ψ

$$\||\psi(t)\||_{w(t)} := \int_{I\!R} \sum_{j=1}^{N} |\alpha_j(x,t)| \, w_j(x,t) \, dx. \qquad (1.14)$$

Note that the weighted norm depends upon both w and A. As long as there exists uniform constants $w^{\min}$ and $w^{\max}$ (independent of the number of discontinuity lines in A and ψ but possibly dependent upon the L^∞ norms or total variation of A and ψ) such that

$$0 < w^{\min} \le w_j(x,t) \le w^{\max} \quad \text{for all } j \text{ and } (x,t), \qquad (1.15)$$

the weighted norm (1.14) is clearly *equivalent* to the standard L^1 norm.

It is convenient to a priori assume that the lines of discontinuity in w are either lines of discontinuity in A or else characteristic lines associated with the matrix A. In other words, the weight w satisfies the following property at all (*but finitely many*) points (x,t):

> If $(x,t) \in \mathcal{J}(w) \setminus \mathcal{J}(A)$, then the discontinuity speed $\lambda^w(x,t)$
> coincides with one of the characteristic speeds $\lambda_i^A(x,t)$ (1.16)
> and $w_{j+}(x,t) = w_{j-}(x,t)$ for all j but $j = i$.

To begin with, we derive now a *closed formula* for the time-derivative of the weighted norm.

LEMMA 1.2. *For each solution ψ of (1.12) and at all but finitely many times t we have*

$$\frac{d}{dt} \||\psi(t)\||_{w(t)} = \sum_{(x,t) \in \mathcal{J}(A)} \sum_{j=1}^{N} \beta_{j-}(x,t)\, w_{j-}(x,t) + \beta_{j+}(x,t)\, w_{j+}(x,t), \qquad (1.17)$$

where

$$\begin{aligned}
\beta_{j-}(x,t) &:= \left(\lambda^A(x,t) - \lambda_{j-}^A(x,t) \right) |\alpha_{j-}(x,t)|, \\
\beta_{j+}(x,t) &:= \left(\lambda_{j+}^A(x,t) - \lambda^A(x,t) \right) |\alpha_{j+}(x,t)|.
\end{aligned} \qquad (1.18)$$

PROOF. Consider the family of polygonal lines of discontinuity $t \mapsto y_k(t)$ (k describing a finite set of integers) in any of the functions A, ψ, and w. In the forthcoming calculation, we exclude all of the interaction times $\tilde{\mathcal{I}}$ of the vector-valued function (A, ψ, w). For instance, we exclude times when a discontinuity line in A, for instance, crosses a discontinuity line in ψ while the speeds λ^A and λ^w are distinct. The discontinuity lines are straight lines in any interval disjoint from $\tilde{\mathcal{I}}$ and the following calculation makes sense.

In each interval $\big(y_k(t), y_{k+1}(t)\big)$ all of the functions are constant and we can write with obvious notation

$$\int_{y_k(t)}^{y_{k+1}(t)} \sum_{j=1}^{N} |\alpha_j(x,t)| \, w_j(x,t) \, dx = \big(y_{k+1}(t) - y_k(t)\big) \sum_{j=1}^{N} |\alpha_j| \, w_j,$$

hence

$$\int_{y_k(t)}^{y_{k+1}(t)} \sum_{j=1}^{N} |\alpha_j(x,t)| \, w_j(x,t) \, dx = \sum_{j=1}^{N} \left(y_{k+1}(t) - \lambda^A_{j,k+1-}(t) \, t \right) |\alpha_{j,k+1-}(t)| \, w_{j,k+1-}(t)$$

$$+ \sum_{j=1}^{N} \left(\lambda^A_{j,k+}(t) \, t - y_k(t) \right) |\alpha_{j,k+}(t)| \, w_{j,k+}(t),$$

where

$$\lambda^A_{j,k+}(t) := \lambda^A_{j+}(y_k(t),t) = \lambda^A_{j-}\left(y_{k+1}(t),t \right) =: \lambda^A_{j,k+1-}(t).$$

After summing over j we arrive at

$$\|\|\psi(t)\|\|_{w(t)} = \sum_{k} \sum_{j=1}^{N} \left(y_k(t) - \lambda^A_{j,k-}(t) \, t \right) |\alpha_{j,k-}(t)| \, w_{j,k-}(t)$$

$$+ \left(\lambda^A_{j,k+}(t) \, t - y_k(t) \right) |\alpha_{j,k+}(t)| \, w_{j,k+}(t).$$

At all t but interaction times we can differentiate this identity with respect to t:

$$\frac{d}{dt} \|\|\psi(t)\|\|_{w(t)} = \sum_{k} \sum_{j=1}^{N} \left(y_k'(t) - \lambda^A_{j,k-}(t) \right) |\alpha_{j,k-}(t)| \, w_{j,k-}(t) \tag{1.19}$$

$$+ \left(\lambda^A_{j,k+}(t) - y_k'(t) \right) |\alpha_{j,k+}(t)| \, w_{j,k+}(t).$$

In view of the conditions (1.11) and (1.16) on g and w, respectively, we need to distinguish between three cases only, as follows:

• If A has a jump discontinuity at $(y_k(t),t)$ propagating at the speed λ^A and associated with some i-family, then we have

$$y_k'(t) = \lambda^A(y_k(t),t),$$

which leads to the desired terms in (1.17) and (1.18).

• If both A and ψ are continuous but w is discontinuous at $(y_k(t),t)$, we deduce from (1.16) that the speed $y_k'(t)$ coincides with some characteristic speed of the matrix A: All the components w_j but the i-component w_i are continuous. In (1.19) the latter is multiplied by

$$\lambda^A_{i\pm}(y_k(t),t) - y_k'(t) = \lambda^A_i(y_k(t),t) - \lambda^A_i(y_k(t),t) = 0.$$

Again the corresponding term in (1.19) vanishes identically.

• Finally, if A is continuous near $(y_k(t),t)$ while ψ contains a discontinuity propagating at some speed Λ, we find from (1.11) and (1.12):

$$\left(-\Lambda + A(y_k(t),t) \right) \left(\psi_+(y_k(t),t) - \psi_-(y_k(t),t) \right) = 0,$$

which implies that Λ is an i-eigenvalue of the matrix $A(x,t)$ and that the jump of ψ is an i-eigenvector. Hence, all of the components α_j, but possibly the i-component α_i, are continuous. The coefficient in front of α_i is

$$\lambda^A_{i+}(y_k(t),t) - y_k'(t) = \lambda^A_{i+}(y_k(t),t) - \Lambda = 0.$$

Hence, the corresponding term in (1.19) vanishes identically.

We conclude that it is sufficient in (1.17) and (1.18) to sum up over jumps in $\mathcal{J}(A)$, only. This completes the proof of Lemma 1.2. $\qquad\square$

Next, in view of (1.18) we observe that for each i-wave front located some point of discontinuity (x,t) we have (dropping the variable (x,t) for simplicity)

$$\begin{aligned} \pm\beta_{j\pm} \leq 0, & \quad j < i, \\ \pm\beta_{j\pm} \geq 0, & \quad j > i, \end{aligned} \tag{1.20i}$$

and

$$\begin{aligned} \beta_{i\pm} \leq 0, & \quad \text{Lax front,} \\ \pm\beta_{i\pm} \geq 0, & \quad \text{slow undercompressive,} \\ \pm\beta_{i\pm} \leq 0, & \quad \text{fast undercompressive,} \\ \beta_{i\pm} \geq 0, & \quad \text{rarefaction-shock.} \end{aligned} \tag{1.20ii}$$

Let us introduce some more notation. We assume that the matrix $A = A(x,t)$ is associated with a scalar-valued function $\varepsilon^A = \varepsilon^A(x,t)$, called the **strength** of the propagating discontinuity located at $(x,t) \in \mathcal{J}(A)$, such that

$$\frac{1}{C}\left|A_+(x,t) - A_-(x,t)\right| \leq \varepsilon^A(x,t) \leq C\left|A_+(x,t) - A_-(x,t)\right|. \tag{1.21}$$

for some uniform constant $C \geq 1$.

LEMMA 1.3. *The coefficients introduced in* (1.18) *satisfy*

$$|\beta_{j+}| = |\beta_{j-}| + O(\varepsilon^A)\sum_{k=1}^{N}|\beta_{k-}| + O(g), \quad 1 \leq j \leq N, \tag{1.22}$$

where g denotes simply the (constant) mass of the measure source-term in (1.12) *along the line of discontinuity under consideration.*

PROOF. The Rankine-Hugoniot relation for the system (1.12) reads

$$-\lambda^A\left(\psi_+ - \psi_-\right) + A_+\psi_+ - A_-\psi_- = g.$$

Thus, we have

$$\sum_{j=1}^{N}\left(\lambda^A - \lambda_{j+}^A\right)\alpha_{j+}\, r_{j+}^A = \sum_{j=1}^{N}\left(\lambda^A - \lambda_{j-}^A\right)\alpha_{j-}\, r_{j-}^A - g.$$

Multiplying by l_{j+}^A and using the normalization $l_{i+}^A\, r_{j+}^A = 0$ if $i \neq j$ and $l_{i+}^A\, r_{i+}^A = 1$, we arrive at

$$\left(\lambda^A - \lambda_{j+}^A\right)\alpha_{j+} = \left(\lambda^A - \lambda_{j-}^A\right)\alpha_{j-} + \sum_{k=1}^{N}\left(\lambda^A - \lambda_{k-}^A\right)\alpha_{k-}\, l_{j+}^A\left(r_{k-}^A - r_{k+}^A\right) - l_{j+}^A\, g. \tag{1.23}$$

From (1.23) we deduce that $(1 \leq j \leq N)$

$$\left(\lambda^A - \lambda_{j+}^A\right)\alpha_{j+} = \left(\lambda^A - \lambda_{j-}^A\right)\alpha_{j-} + O(|A_+ - A_-|)\sum_{k=1}^{N}\left|\left(\lambda^A - \lambda_{k-}^A\right)\alpha_{k-}\right| + O(g), \tag{1.24}$$

which yields (1.22) in view of (1.18) and (1.21). $\qquad\square$

Lemma 1.3 shows that the components $|\beta_{j-}|$ and $|\beta_{j+}|$ coincide "up to first-order" in ε^A. Therefore, in view of the signs determined in (1.20) it becomes clear that the natural constraints to place on the weight in order for the right-hand side of (1.17) to be ("essentially") non-positive are the following ones: For each i-wave front and each $1 \le j \le N$

$$
w_{j+} - w_{j-} \quad
\begin{cases}
\ge 0, & j < i, \\
\le 0, & j = i \text{ and slow undercompressive}, \\
\ge 0, & j = i \text{ and fast undercompressive}, \\
\le 0, & j > i.
\end{cases}
\tag{1.25}
$$

Indeed, if such a weight exists, then from (1.17), (1.20), (1.22), and (1.25) we can immediately derive the following estimate away from interaction times:

$$
\begin{aligned}
\frac{d}{dt} \| |\psi(t)| \|_{w(t)} &+ \sum_{\substack{(x,t) \in \mathcal{J}_i(A) \\ 1 \le i \ne j \le N}} \big| w_{j-}(x,t) - w_{j+}(x,t) \big| \, |\beta_{j-}(x,t)| \\
&+ \sum_{\substack{(x,t) \in \mathcal{L}_i(A) \\ 1 \le i \le N}} \big(w_{i-}(x,t) + w_{i+}(x,t) \big) \, |\beta_{i-}(x,t)| \\
&+ \sum_{\substack{(x,t) \in \mathcal{S}_i(A) \cup \mathcal{F}_i(A) \\ 1 \le i \le N}} \big| w_{i-}(x,t) - w_{i+}(x,t) \big| \, |\beta_{i-}(x,t)| \\
&= \sum_{\substack{(x,t) \in \mathcal{R}_i(A) \cup \mathcal{A}(A) \\ 1 \le i \le N}} \big(w_{i-}(x,t) + w_{i+}(x,t) \big) \, |\beta_{i-}(x,t)| \\
&+ \sum_{\substack{(x,t) \in \mathcal{J}(A) \\ 1 \le j \le N}} O(\varepsilon^A(x,t)) \, |\beta_{j-}(x,t)| + O(1) \, \|g(t)\|_{\mathcal{M}(\mathbb{R})}.
\end{aligned}
\tag{1.26}
$$

To control the remainder arising in the right-hand side of (1.26) our strategy will be to choose now $\big| w_{j-}(x,t) - w_{j+}(x,t) \big| = K \varepsilon^A$ with a sufficiently large $K > 0$, so that the favorable term in the left-hand side of (1.26) becomes greater than the last term in the right-hand side. Recall that a weight satisfying (1.25) was indeed determined for scalar conservation laws, in Chapter V. An additional difficulty arises here to treat systems of conservation laws: The conditions (1.25) are somewhat too restrictive and must be relaxed. Strictly speaking, it would be possible to exhibit a weight w satisfying (1.25) for systems for every choice of matrices A and solutions ψ. But, such a weight would strongly depend on the number of lines of discontinuity in A and ψ and the corresponding estimate (1.6) would not be valid for a general class of piecewise constant solutions. Alternatively, it would not be difficult to construct a weight w independent of the number of lines of discontinuity in A and ψ but exhibiting (uncontrolled) jump discontinuities in time at each interaction.

To weaken (1.25), our key observation is the following one:

The conditions (1.25) on the jump $w_{j+} - w_{j-}$ are not necessary
for those components $\beta_{j\pm}$ which are "small" (in the sense (1.27), below)
compared to other components.

DEFINITION 1.4. Consider a solution $\psi = \psi(x,t)$ of (1.12) together with its characteristic components $\alpha_{j\pm}$ and $\beta_{j\pm}$ defined by (1.13) and (1.18), respectively $(1 \leq j \leq N)$. For each $j = 1, \ldots, N$ we shall say that the component β_{j-} is **dominant** if and only if for some uniform constant $C > 0$

$$|\beta_{j-}| \geq C \, \varepsilon^A \sum_{k=1}^{N} |\beta_{k-}| + C \, |g|. \tag{1.27}$$

For each i-wave front the i-characteristic components $\alpha_{i\pm}$ are said to be **dominant** if and only if for some uniform constant $c > 0$

$$c \, |\alpha_{i\pm}| \geq \sum_{k \neq i} |\alpha_{k\pm}|. \tag{1.28}$$

Later, we will need that the constant c in (1.28) is sufficiently small.
We now prove the main result in this section.

THEOREM 1.5. (L^1 stability for linear hyperbolic systems.) *Consider a matrix-valued function $A = A(x,t)$ satisfying the assumptions given above and whose strength ε^A is sufficiently small. Let $\psi = \psi(x,t)$ be a piecewise constant solution of (1.12). Suppose that there exists a weight satisfying the following strengthened version of the conditions (1.25) but for dominant components β_{j-} only:*

$$w_{j+} - w_{j-} = \begin{cases} K \, \varepsilon^A, & j < i, \\ K \, \varepsilon^A \text{ or } -K \, \varepsilon^A, & j = i \text{ and undercompressive}, \\ -K \, \varepsilon^A, & j = i, \text{ slow undercompressive}, \, \alpha_{i\pm} \text{ dominant}, \\ K \, \varepsilon^A, & j = i, \text{ fast undercompressive}, \, \alpha_{i\pm} \text{ dominant}, \\ -K \, \varepsilon^A, & j > i. \end{cases} \tag{1.29}$$

Then, for some sufficiently large K and uniform constants $C_1, C_2 > 0$ the weighted norm of ψ satisfies the inequality (for all $t \geq 0$)

$$\|\psi(t)\|_{w(t)} + C_1 \int_0^t \Big(\mathbf{D}_2(s) + \mathbf{D}_3(s) \Big) \, ds \leq \|\psi(0)\|_{w(0)} + C_2 \int_0^t \mathbf{R}(s) \, ds,$$

where the dissipation terms and the remainder are defined by

$$\mathbf{D}_2(s) := \sum_{\substack{(x,s) \in \mathcal{L}_i(A) \\ 1 \leq i \leq N}} \big| \beta_{i-}(x,s) \big|,$$

$$\mathbf{D}_3(s) := \sum_{\substack{(x,s) \in \mathcal{J}(A) \\ 1 \leq j \leq N}} \varepsilon^A(x,s) \big| \beta_{j-}(x,s) \big| \, ds$$

and

$$\mathbf{R}(s) := \|g(s)\|_{\mathcal{M}(\mathbb{R})} + TV(\psi(s)) \sup_{\substack{(x,\tau) \in \mathcal{R}(A) \\ \tau \in (0,t)}} \varepsilon^A(x,\tau)$$

$$+ \|\psi(s)\|_{L^\infty(\mathbb{R})} \sup_{\tau \in [0,t]} \sum_{(x,\tau) \in \mathcal{A}(A)} \varepsilon^A(x,\tau).$$

Since the weighted norm is equivalent to the standard L^1 norm, the estimate in Theorem 1.5 is also equivalent to

$$\|\psi(t)\|_{L^1(\mathbb{R})} + \int_0^t \left(\widehat{\mathbf{D}}_2(s) + \widehat{\mathbf{D}}_3(s)\right) ds \leq C_3 \|\psi(0)\|_{L^1(\mathbb{R})} + C_3 \int_0^t \widehat{\mathbf{R}}(s)\, ds, \quad (1.30)$$

for some uniform constant $C_3 > 0$, where now the corresponding dissipation terms and remainder are

$$\widehat{\mathbf{D}}_2(s) := \sum_{\substack{(x,s)\in\mathcal{L}_i(A) \\ 1\leq i \leq N}} \left| l_{i-}(x,s)\left(A_-(x,s) - \lambda(x,s)\right)\psi_-(x,s)\right|,$$

$$\widehat{\mathbf{D}}_3(s) := \sum_{(x,s)\in\mathcal{J}(A)} |A_+(x,s) - A_-(x,s)| \left|\left(A_-(x,s) - \lambda(x,s)\right)\psi_-(x,s)\right|,$$

and

$$\widehat{\mathbf{R}}(s) := \|g(s)\|_{\mathcal{M}(\mathbb{R})} + TV(\psi(s)) \sup_{\substack{(x,\tau)\in\mathcal{R}(A) \\ \tau\in(0,t)}} |A_+(x,\tau) - A_-(x,\tau)|$$

$$+ \|\psi(s)\|_{L^\infty(\mathbb{R})} \sup_{\tau\in(0,t)} \sum_{(x,\tau)\in\mathcal{A}(A)} |A_+(x,\tau) - A_-(x,\tau)|.$$

The following important remarks concerning (1.30) are in order:
- Only rarefaction-shocks, artificial fronts, and the source-term g may amplify the L^1 norm. In particular, in the special case that

$$g = 0, \quad \mathcal{R}(A) = \mathcal{A}(A) = \emptyset,$$

(1.30) implies the solutions ψ of the Cauchy problem associated with (1.12) are *unique and stable*. In the following sections, (1.30) will be applied with a sequence of approximate solutions (of (1.1)) for which the last three terms in the right-hand side of (1.30) precisely vanish in the limit.
- Theorem 1.5 provides a *sharp* bound on the decay of the L^1 norm. Note that the left-hand side of (1.30) contains *cubic terms* associated with undercompressive wave fronts of A, and *quadratic terms* associated with Lax fronts.

PROOF. We now consider each kind of i-wave front successively, and we derive a corresponding estimate for the boundary term

$$B := \sum_{j=1}^N \beta_{j-}\, w_{j-} + \beta_{j+}\, w_{j+}. \tag{1.31}$$

Throughout we often use that by (1.27)

$$\sum_j |\beta_{j-}| \leq C \sum_{\beta_{j-}\text{dominant}} |\beta_{j-}| + O(g), \tag{1.32}$$

provided ε^A is sufficiently small. In view of (1.22) we have

$$B = \sum_{j=1}^N \left(\beta_{j-}\, w_{j-} + \text{sgn}(\beta_{j+})\,|\beta_{j-}|\, w_{j+}\right) + O(\varepsilon^A) \sum_{k=1}^N |\beta_{k-}| + O(g),$$

that is, since by (1.20i) the signs of the $\beta_{j\pm}$ is determined for all $j \neq i$,

$$B = \left(w_{i-}\,\mathrm{sgn}(\beta_{i-}) + w_{i+}\,\mathrm{sgn}(\beta_{i+})\right)|\beta_{i-}|$$

$$- \sum_{j<i}(w_{j+} - w_{j-})\,|\beta_{j-}| - \sum_{j>i}(w_{j-} - w_{j+})\,|\beta_{j-}| + O(\varepsilon^A)\sum_{k=1}^{N}|\beta_{k-}| + O(g).$$

In view of (1.29), for $j \neq i$ the *dominant* components β_{j-} are associated with a *favorable* sign of the jump $w_{j+} - w_{j-}$. On the other hand, non-dominant components β_{j-} for $j \neq i$ can be collected in the first-order remainder. So, we obtain

$$B = \left(w_{i-}\,\mathrm{sgn}(\beta_{i-}) + w_{i+}\,\mathrm{sgn}(\beta_{i+})\right)|\beta_{i-}| - K\,\varepsilon^A \sum_{\substack{j\neq i \\ \text{dominant}}} |\beta_{j-}|$$

$$+ O(\varepsilon^A)\sum_{k=1}^{N}|\beta_{k-}| + O(g),$$

where the first sum above contains dominant components β_{j-} only and, relying on (1.32), can absorb the first-order remainder, provided we choose K sufficiently large so that $K\,\varepsilon^A$ dominates $O(\varepsilon^A)$. We obtain

$$B \leq \left(w_{i-}\,\mathrm{sgn}(\beta_{i-}) + w_{i+}\,\mathrm{sgn}(\beta_{i+})\right)|\beta_{i-}| + O(\varepsilon^A)|\beta_{i-}| - \frac{K}{2}\,\varepsilon^A \sum_{j\neq i}|\beta_{j-}| + O(g). \quad (1.33)$$

It remains to deal with the term $|\beta_{i-}|$, which can be assumed to be *dominant* otherwise the argument above would also apply to the i-component and we would arrive to the desired estimate for B. We distinguish now between four main cases: Lax, undercompressive, rarefaction-shock, and artificial fronts.

Case 1 : If the i-wave is an i-*Lax front*, we have $\mathrm{sgn}(\beta_{j-}) = \mathrm{sgn}(\beta_{j+}) = -1$ and therefore, by (1.33) and (1.15),

$$B \leq -2\,w^{\min}\,|\beta_{i-}| + O(\varepsilon^A)\,|\beta_{i-}| - \frac{K}{2}\,\varepsilon^A \sum_{j\neq i}|\beta_{j-}| + O(g).$$

So we obtain

$$B \leq -w^{\min}\,|\beta_{i-}| - \frac{K}{2}\,\varepsilon^A \sum_{j\neq i}|\beta_{j-}| + O(g),$$

and, therefore, for ε^A sufficiently small

$$B \leq -\frac{K}{2}\,\varepsilon^A \sum_{j=1}^{N}|\beta_{j-}| - \frac{w^{\min}}{2}\,|\beta_{i-}| + O(g) \quad \text{for Lax fronts.} \quad (1.34)$$

Case 2 : Next, we consider an *undercompressive front* and we prove that

$$B \leq -\frac{K}{3}\,\varepsilon^A \sum_{j=1}^{N}|\beta_{j-}| + O(g) \quad \text{for undercompressive fronts.} \quad (1.35)$$

Suppose, for instance, that the front is slow undercompressive.

First of all, the case that both $\alpha_{i\pm}$ and β_{i-} are all dominant is simple, since $\operatorname{sgn}(\beta_{i-}) = -1$ and $\operatorname{sgn}(\beta_{i+}) = 1$ while $w_{i+} - w_{i-} = -K\,\varepsilon^A$. Therefore, from (1.33) we have

$$B \le -K\,\varepsilon^A\,|\beta_{i-}| + O(\varepsilon^A)\,|\beta_{i-}| - \frac{K}{2}\,\varepsilon^A \sum_{j\ne i} |\beta_{j-}| + O(g),$$

which yields (1.35) by choosing K sufficiently large.

Second, we already pointed out that the case that β_{i-} is non-dominant is obvious in view of (1.32).

Thus, it remains to consider the case where one of the characteristic components $\alpha_{i\pm}$ is non-dominant. Suppose, for instance, that α_{i-} is non-dominant and that β_{i-} is dominant. Using that (see (1.28))

$$|\alpha_{i-}| \le c \sum_{k\ne i} |\alpha_{k-}|$$

and the condition $|w_{i+} - w_{i-}| = K\,\varepsilon^A$ from (1.29) (since β_{i-} is dominant), we obtain

$$
\begin{aligned}
|\beta_{i-}|\,\big|w_{i-} - w_{i+}\big| = K\,\varepsilon^A\,|\beta_{i-}| &\le O(\varepsilon^A)\,K\,(\lambda_i^{\max} - \lambda_i^{\min})\,|\alpha_{i-}| \\
&\le O(\varepsilon^A)\,K\,(\lambda_i^{\max} - \lambda_i^{\min}) \sum_{j\ne i} |\alpha_{j-}|, \\
&\le O(\varepsilon^A)\,K\,(\lambda_i^{\max} - \lambda_i^{\min}) \sum_{j\ne i} |\beta_{j-}|.
\end{aligned}
$$

On the other hand, since $\lambda_i^{\max} - \lambda_i^{\min} \ll 1$ by assumption, we observe that

$$O(\varepsilon^A)\,K\,(\lambda_i^{\max} - \lambda_i^{\min}) < \frac{K}{4}\,\varepsilon^A.$$

Therefore, we conclude that the term $|\beta_{i-}|\,\big|w_{i-} - w_{i+}\big|$ can be controlled by the term

$$-\frac{K}{2}\,\varepsilon^A \sum_{j\ne i} |\beta_{j-}|$$

in (1.33), and we arrive again at the inequality (1.35).

Case 3 : Consider next the case of an *i-rarefaction-shock*, for which no constraint has been imposed on the component w_i. Here, we will show that

$$B \le C\,\varepsilon^A\,|\psi_+ - \psi_-| - \frac{K}{3}\,\varepsilon^A \sum_{j=1}^{N} |\beta_{j-}| + O(g) \quad \text{for rarefaction-shocks.} \tag{1.36}$$

From (1.33) we get

$$
\begin{aligned}
B &\le 2\,w^{\max}\,|\beta_{i-}| + O(\varepsilon^A)\,|\beta_{i-}| - \frac{K}{2}\,\varepsilon^A \sum_{j\ne i} |\beta_{j-}| + O(g) \\
&\le 3\,w^{\max}\,|\beta_{i-}| - \frac{K}{2}\,\varepsilon^A \sum_{j\ne i} |\beta_{j-}| + O(g)
\end{aligned}
\tag{1.37}
$$

and we distinguish between two subcases:

– If $\alpha_{i-}\,\alpha_{i+} \geq 0$, then by (1.24) in the proof of Lemma 1.3 we have

$$(\lambda^A - \lambda^A_{i+})\,\alpha_{i+} + (\lambda^A_{i-} - \lambda^A)\,\alpha_{i-} = O(\varepsilon^A) \sum_{j=1}^{N} |\beta_{j-}| + O(g).$$

The two terms on the left-hand side above have the *same* sign, therefore

$$|\beta_{i+}| + |\beta_{i-}| = \left|(\lambda^A - \lambda^A_{i+})\,\alpha_{i+}\right| + \left|(\lambda^A_{i-} - \lambda^A)\,\alpha_{i-}\right| = O(\varepsilon^A) \sum_{j=1}^{N} |\beta_{j-}| + O(g).$$

From (1.37) and by absorbing the term $O(\varepsilon^A)$ above by taking K sufficiently large, we deduce that

$$B \leq -\frac{K}{3}\,\varepsilon^A \sum_{j \neq i} |\beta_{j-}| + O(g),$$

which –using once more the previous inequality– implies (1.36).

– If $\alpha_{i-}\,\alpha_{i+} < 0$, then we observe that

$$
\begin{aligned}
|\beta_{i-}| = |\lambda^A_{i-} - \lambda^A|\,|\alpha_{i-}| &\leq |\lambda^A_{i+} - \lambda^A_{i-}|\,|\alpha_{i+} - \alpha_{i-}| \\
&\leq O(\varepsilon^A)\,|\psi_+ - \psi_-|.
\end{aligned}
\tag{1.38}
$$

We conclude that

$$
\begin{aligned}
B &\leq O(\varepsilon^A)\,|\psi_+ - \psi_-| - \frac{K}{2}\,\varepsilon^A \sum_{j \neq i} |\beta_{j-}| + O(g) \\
&\leq O(\varepsilon^A)\,|\psi_+ - \psi_-| - \frac{K}{2}\,\varepsilon^A \sum_{j=1}^{N} |\beta_{j-}| + O(g),
\end{aligned}
$$

where, in the latter, the estimate (1.38) on $|\beta_{i-}|$ was used once more. This proves (1.36).

Case 4 : Finally, it is easy to derive

$$B \leq C \max\big(|\psi_-|, |\psi_+|\big)\,\varepsilon^A + O(g) \quad \text{for artificial fronts.} \tag{1.39}$$

The estimate (1.30) follows from (1.34)–(1.36) and (1.39), and the proof of Theorem 1.5 is completed. $\qquad\square$

2. L^1 Continuous dependence estimate

We show here that the framework developed in Section 1 applies to (classical) entropy solutions of conservation laws. For simplicity, we assume that the system under consideration is genuinely nonlinear. Theorem 1.5 provides us with the desired L^1 continuous dependence estimate (1.2), provided we can exhibit a weight-function satisfying the requirements of Section 1, especially the conditions (1.29). Our first key observation is:

THEOREM 2.1. (First fundamental property.) *Consider the strictly hyperbolic system* (1.1) *defined in the ball* $\mathcal{U} = \mathcal{B}(\delta)$ *with* $\delta > 0$ *sufficiently small. Suppose that all of the characteristic fields of* (1.1) *are genuinely nonlinear. Let* $\overline{A}$ *be an averaging matrix satisfying the two conditions* (1.3). *Then, if* $u = u(x,t)$ *and* $v = v(x,t)$ *are any two piecewise constant functions made of classical shock fronts associated with* (1.1) *(and defined in some region of the* (x,t)-*plane, say), the averaging matrix*

$$A(x,t) := \overline{A}(u(x,t), v(x,t))$$

cannot contain rarefaction-shocks.

PROOF. Since $\overline{\lambda}_j(u,u) = \lambda_j(u)$ and $\overline{A}$ is symmetric (see (1.3)) we have

$$\nabla_1 \overline{\lambda}_j(u,u) = \nabla_2 \overline{\lambda}_j(u,u) = \frac{1}{2} \nabla \lambda_j(u), \quad 1 \leq j \leq N. \tag{2.1}$$

The function u and v play completely symmetric roles. Consider, for instance, a shock wave in the solution u connecting a left-hand state u_- to a right-hand state u_+, and suppose that the solution v is constant in a neighborhood of this shock. It may also happen that two shocks in u and v have the same speed and are superimposed in the (x,t)-plane, locally. However, this case is not generic and can be removed by an arbitrary small perturbation of the data. (Alternatively, this case can also be treated by the same arguments given now for a single shock, provided we regard the wave pattern as the superposition of two shocks, one in u while v remains constant, and another in v while u remains constant.)

According to the results in Chapter VI, u_+ is a function of the left-hand state u_- and of some parameter along the Hugoniot curve, denoted here by ε. For some index i we have

$$u_+ =: u_+(\varepsilon) = u_- + \varepsilon\, r_i(u_-) + O(\varepsilon^2), \tag{2.2}$$

By convention, $\varepsilon < 0$ for classical shocks satisfying Lax shock inequality

$$\lambda_i(u_-) > \lambda_i(u_+),$$

since we imposed the normalization $\nabla \lambda_i \cdot r_i \equiv 1$. We claim that the *averaging wave speed* $\overline{\lambda}_i$ is *decreasing* across the shock, that is,

$$\overline{\lambda}_i(u_-, v) > \overline{\lambda}_i(u_+, v), \tag{2.3}$$

uniformly in v. In particular, this implies that the inequalities characterizing a rarefaction-shock (that is, $\overline{\lambda}_i(u_-, v) \leq \overline{\lambda}_i(u_-, u_+) \leq \overline{\lambda}_i(u_+, v)$ in Definition 1.1) cannot hold simultaneously, which is precisely the desired property on $\overline{A}$.

Indeed, by using the expansion (2.2) the inequality (2.3) is equivalent to saying

$$\overline{\lambda}_i(u_-, v) > \overline{\lambda}_i(u_-, v) + \varepsilon\, \nabla_1 \overline{\lambda}_i(u_-, v) \cdot r_i(u_-) + O(\varepsilon^2),$$

or

$$\nabla_1 \overline{\lambda}_i(u_-, v) \cdot r_i(u_-) + O(\varepsilon) > 0.$$

Expanding in term of $|v - u_-|$ and using (2.1), we arrive at the equivalent condition

$$\nabla \lambda_i(u_-) \cdot r_i(u_-) + O\big(|\varepsilon| + |v - u_-|\big) > 0.$$

However, since $\nabla \lambda_i \cdot r_i \equiv 1$ and $|\varepsilon| + |v - u_-| \leq O(\delta)$, the above inequality, and thus (2.3), holds for δ sufficiently small. This completes the proof of Theorem 2.1. $\qquad \square$

We will also need the following observation which, in fact, motivated us in formulating Definition 1.4. Note that the entropy condition does not play a role here.

THEOREM 2.2. (Second fundamental property.) *Under the same assumptions as in Theorem 2.1, consider now two piecewise constant functions u and v made of classical shock fronts and rarefaction fronts (see Section VII-1 for the definition), and set (see (1.13))*

$$\psi := v - u = \sum_{j=1}^{N} \alpha_j \, r_j^A. \tag{2.4}$$

Recall the notation

$$\lambda_j^A(x,t) = \overline{\lambda}_j(u(x,t), v(x,t)), \quad r_j^A(x,t) = \overline{r}_j(u(x,t), v(x,t)).$$

Then, for each wave front of A propagating at the speed λ^A and associated with a wave front in the solution u we have

$$\operatorname{sgn}(\alpha_{i\pm}) = \operatorname{sgn}(\lambda_{i\mp}^A - \lambda^A) \quad \text{if } \alpha_{i\pm} \text{ are dominant.} \tag{2.5}$$

The opposite sign is found for wave fronts of A associated with wave fronts in the solution v.

PROOF. We use the same notation as in the proof of Theorem 2.1. Consider an i-wave front connecting u_- to u_+ while the other solution v is locally constant. Using the decomposition

$$v - u_\pm = \sum_{j=1}^{N} \alpha_{j\pm} \, \overline{r}_j(u_\pm, v),$$

we can write

$$\begin{aligned}
\lambda_{i+}^A - \lambda^A &= \overline{\lambda}_i(u_+, v) - \overline{\lambda}_i(u_-, u_+) \\
&= \nabla_2 \overline{\lambda}_i(u_-, u_+) \cdot (v - u_-) + O(|v - u_-|^2) \\
&= \alpha_{i-} \left(1/2 + O(|u_+ - u_-| + |v - u_-|) \right) + O(1) \sum_{j \neq i} |\alpha_{j-}| \\
&= \left(1/2 + O(\delta) \right) \alpha_{i-} + O(1) \sum_{j \neq i} |\alpha_{j-}|.
\end{aligned}$$

This proves that, when $c \, |\alpha_{i-}| \geq \sum_{j \neq i} |\alpha_{j-}|$ where c is sufficiently small, the terms $\lambda_{i+}^A - \lambda^A$ and α_{i-} have the same sign for wave fronts associated with wave fronts in the solution u. The opposite sign is found for waves associated with wave fronts in the solution v. The calculation for λ_{i-}^A is completely similar. $\qquad\square$

Finally, we arrive at the main result in this section.

THEOREM 2.3. (Continuous dependence of classical entropy solutions.) *Consider the strictly hyperbolic system (1.1) in the ball $\mathcal{U} = \mathcal{B}(\delta)$ with small radius $\delta > 0$. Suppose that all of its characteristic fields are genuinely nonlinear. Let u^h and v^h be two sequences of piecewise constant, wave front tracking approximations (see Chapter VII) and denote by u and v the corresponding classical entropy solutions obtained in the limit $h \to 0$. Then, for some uniform constant $C > 0$ we have the inequality*

$$\|v^h(t) - u^h(t)\|_{L^1(\mathbb{R})} \leq C \, \|v^h(0) - u^h(0)\|_{L^1(\mathbb{R})} + O(h), \quad t \geq 0, \tag{2.6}$$

which, in the limit, yields the L^1 continuous dependence estimate

$$\|u(t) - v(t)\|_{L^1(I\!R)} \leq C \|u(0) - v(0)\|_{L^1(I\!R)}, \quad t \geq 0. \tag{2.7}$$

PROOF. **Step 1 : Convergence analysis.** Recall from Section VII-2 that u^h and v^h have uniformly bounded total variation

$$TV(u^h(t)) + TV(v^h(t)) \leq C, \quad t \geq 0, \tag{2.8}$$

and converge almost everywhere toward u and v, respectively. Each wave front in u^h (and similarly for v^h) is one of the following:

- A *Lax shock front* satisfying the Rankine-Hugoniot jump conditions and Lax shock inequalities and propagating at the speed $\lambda = \overline{\lambda}_i(u^h_-(x,t), u^h_+(x,t)) + O(h)$.
- A *rarefaction front* violating both of Lax shock inequalities and having *small strength*, i.e.,

$$|u^h_+(x,t) - u^h_-(x,t)| \leq C h \quad \text{for rarefaction fronts}. \tag{2.9}$$

For some $i = 1, \ldots, N$, $u^h_+(x,t)$ lies on the i-rarefaction curve issuing from $u^h_-(x,t)$. The jump propagates at the speed $\lambda = \overline{\lambda}_i(u^h_-(x,t), u^h_+(x,t)) + O(h)$.

- An *artificial front* propagating at a large fixed speed λ_{N+1}. Denote by $\mathcal{A}$ the set of all artificial fronts. The total strength of waves in $\mathcal{A}$ vanishes with h, precisely

$$\sum_{x/\,(x,t)\in\mathcal{A}} |u^h_+(x,t) - u^h_-(x,t)| \leq C h, \quad t \geq 0. \tag{2.10}$$

Moreover, the Glimm interaction estimates (Section VII-1) hold at each interaction point. The interaction of two waves of different families $i \neq i'$ produces two principal waves of the families i and i', plus small waves in other families $j \neq i, i'$ whose total strength is quadratic with respect to the strengths of the incoming waves. The interaction of two i-waves generates one principal i-wave, plus small waves in other families $j \neq i$.

Furthermore, by an arbitrary small change of the propagation speeds we can always assume that:

- At each interaction time there is exactly one interaction between either two fronts in u^h or else two fronts in v^h.
- The polygonal lines of discontinuity in u^h and v^h cross at finitely many points and do not coincide on some non-trivial time interval.

We now apply the general strategy described at the beginning of Section 1. Observe that the approximate solutions u^h and v^h satisfy systems of equations of the form

$$\partial_t u^h + \partial_x f(u^h) = g^{1,h}, \quad \partial_t v^h + \partial_x f(v^h) = g^{2,h}, \tag{2.11}$$

where $g^{1,h}$ and $g^{2,h}$ are measures on $I\!R \times I\!R_+$ induced by the facts that rarefaction fronts do not satisfy the Rankine-Hugoniot relations and that fronts do not propagate with their exact wave speed. Define

$$A^h := \overline{A}(u^h, v^h), \quad \psi^h = v^h - u^h, \quad g^h := g^{2,h} - g^{1,h}.$$

The matrix A^h satisfies our assumptions in Section 1 (uniformly in h) and the function ψ^h is a solution of

$$\partial_t \psi^h + \partial_x \left(A^h \, \psi^h \right) = g^h. \tag{2.12}$$

Observe here that any wave interaction point (x, t) for the averaging matrix A^h

- is either a point of wave interaction for u^h while v^h remains constant, or vice-versa,
- or else a point where both u^h and v^h contain single wave fronts crossing at (x, t).

We define the wave strength ε^A for the matrix A^h (see (1.21)) to be the (modulus of) usual strength of the wave fronts in u^h or in v^h, whichever carries the jump. Relying on the assumption of genuine nonlinearity, we now prove that the strength ε^A is equivalent to the standard jump of A^h, that is, uniformly in v

$$\frac{1}{C} |u_+ - u_-| \le |\overline{A}(u_+, v) - \overline{A}(u_-, v)| \le C |u_+ - u_-| \tag{2.13}$$

for any shock connecting a left-hand state u_- to a right-hand state u_+ and for some constant $C \ge 1$.

To derive (2.13), we simply regard $\overline{A}(u_+, v) - \overline{A}(u_-, v)$ as a function $\psi(\varepsilon)$ of the wave strength parameter ε along the j-Hugoniot curve. Using the expansion (2.2) we find

$$\psi'(0) = D_{u_1} \overline{A}(u_-, v) \, r_j(u_-).$$

Since u_-, u_+, and v remain in a small neighborhood of a given point in $\mathbb{R}^N$, it is enough to check that $B := D_{u_1} \overline{A}(u, u) \, r_j(u) \ne 0$ for every u and j. But, since $\overline{A}$ is symmetric (see (1.3)) we have $D_{u_1} \overline{A}(u, u) = D_{u_2} \overline{A}(u, u) = DA(u)/2$. Multiplying the identity

$$\left(D^2 f(u) \, r_j(u) - \nabla \lambda_j(u) \cdot r_j(u) \right) r_j(u) = \left(\lambda_j(u) - Df(u) \right) D r_j(u) \, r_j(u)$$

by the left-eigenvector $l_j(u)$ yields

$$l_j(u) \, B \, r_j(u) = l_j(u) \left(D_{u_1} \overline{A}(u, u) \, r_j(u) \right) r_j(u) = \frac{1}{2} \nabla \lambda_j(u) \cdot r_j(u) = \frac{1}{2}.$$

This shows that the gradient of $\psi(\varepsilon)$ does not vanish, for small ε at least, and establishes (2.13) and thus (1.21).

Next, we claim that $g^h \to 0$ strongly as locally bounded measures, precisely

$$\int_0^T \|g^{1,h}(t)\|_{\mathcal{M}(\mathbb{R})} \, dt \to 0 \quad \text{for every } T > 0. \tag{2.14}$$

We can decompose the contributions to the measure source-term $g^{1,h}$ in three sets. First, shock waves satisfy (2.12) almost exactly, with

$$g^{1,h} = O(1) \, O(h) \, \varepsilon^A \text{ locally.}$$

Second, for an i-rarefaction front connecting two states u_- and u_+ and propagating at some speed $\lambda = \overline{\lambda}_i(u_-, u_+)$ we find

$$g^{1,h} = \gamma^h \, \delta_{x = \lambda \, t},$$
$$\gamma^h := -\left(\overline{\lambda}_i(u_-, u_+) + O(h) \right) (u_+ - u_-) + f(u_+) - f(u_-).$$

But, in view of (2.9) we have

$$|\gamma^h| \le C\,|u_+ - u_-|^2 + C\,h\,|u_+ - u_-| \le O(h)\,|u_+ - u_-|.$$

Summing over all rarefaction waves in $I\!R \times [0, T]$, for any fixed T we find

$$\int_0^T \|g^{1,h}(t)\|_{\mathcal{M}(I\!R)}\,dt \le O(h), \tag{2.15}$$

which gives (2.14). Finally, using the same notation and for artificial waves we obtain

$$|\gamma^h| := |-\lambda_{N+1}\,(u_+ - u_-) + f(u_+) - f(u_-)| \le C\,|u_+ - u_-|,$$

and so again (2.15) thanks to the estimate (2.10). This establishes (2.14).

We conclude that the assumptions in Theorem 1.5 are satisfied and, provided we can construct a suitable weight function, we obtain the stability estimate (1.30). Thanks to Theorem 2.1 the set of rarefaction-shocks in A^h is included in the union of the sets of rarefaction fronts and artificial fronts in u^h and v^h. From the estimates (2.14) it then follows that the last three terms in the right-hand side of (1.30) converge to zero with $h \to 0$ and, therefore, that (2.6) holds. Since

$$\psi^h \to \psi := v - u \qquad \text{almost everywhere}$$

the desired L^1 stability estimate (2.7) is obtained in the limit.

Step 2 : Reduction step. To simplify the notation we drop the exponent h in what follows. In Step 3 below, we will construct a weight satisfying (1.15) and (1.16) together with the following constraint: for each i-discontinuity associated with the solution u and for each $j = 1, \dots, N$, provided $\alpha_{j-}\,\alpha_{j+} > 0$, we will impose

$$w_{j+} - w_{j-} = \begin{cases} K\,\varepsilon^A, & j < i, \\ -K\,\varepsilon^A, & j = i, \text{ undercompressive, and } \alpha_{i\pm} > 0, \\ K\,\varepsilon^A, & j = i, \text{ undercompressive, and } \alpha_{i\pm} < 0, \\ -K\,\varepsilon^A, & j > i. \end{cases} \tag{2.16}$$

For discontinuities associated with the solution v similar conditions should hold, simply exchanging the conditions $\alpha_{i\pm} > 0$ and $\alpha_{i\pm} < 0$ in the two cases $j = i$. Note that no constraint is imposed when $\alpha_{j-}\,\alpha_{j+} < 0$.

Let us here check that the conditions (2.16) do imply the conditions (1.29) needed in Step 1 in order to apply Theorem 1.5. Indeed, consider an i-discontinuity for which some component β_{j-} is *dominant*. Then, it follows from (1.24) that

$$\text{sgn}\big((\lambda^A - \lambda_{j+}^A)\,\alpha_{j+}\big) = \text{sgn}\big((\lambda^A - \lambda_{j-}^A)\,\alpha_{j-}\big). \tag{2.17}$$

When $j \ne i$, the term $\lambda^A - \lambda_{j+}^A$ has the same sign as $\lambda^A - \lambda_{j-}^A$, so from (2.17) we deduce that

$$\text{sgn}(\alpha_{j+}) = \text{sgn}(\alpha_{j-}),$$

so that (2.16) can be applied, which implies the conditions in (1.29) for $j \ne i$.

On the other hand, for the i-component of an *undercompressive* i-wave, again the terms $\lambda^A - \lambda_{i+}^A$ and $\lambda^A - \lambda_{i-}^A$ have the same sign, therefore by (2.17) and provided β_{i-} is *dominant* we have again

$$\text{sgn}(\alpha_{i+}) = \text{sgn}(\alpha_{i-}).$$

Relying on Theorem 2.2 and restricting also attention to *dominant* components $\alpha_{i\pm}$ we see that

$$\begin{aligned} \alpha_{i-}, \alpha_{i+} > 0, &\quad \text{slow undercompressive,} \\ \alpha_{i-}, \alpha_{i+} < 0, &\quad \text{fast undercompressive.} \end{aligned} \tag{2.18}$$

Therefore, in view of (2.18), the conditions in (2.16) for $j = i$ imply the corresponding ones in (1.29).

Step 3 : Constructing the weight-function. It remains to determine a weight $w = w(x, t)$ satisfying (1.15), (1.16), and (2.16). The construction of each component of w will be analogous to what was done in the proof of Theorem V-2.3 with scalar conservation laws. Given some index j, the (piecewise constant) component w_j will be uniquely defined –up to some (sufficiently large) additive constant– if we prescribe its jumps $(w_{j+}(x, t) - w_{j-}(x, t))$. Additionally, the weight can be made positive and uniformly bounded away from zero, provided we guarantee that the sums of all jumps contained in any arbitrary interval remain uniformly bounded, i.e.,

$$\Big| \sum_{x \in (a,b)} (w_{j+}(x, t) - w_{j-}(x, t)) \Big| \leq C, \tag{2.19}$$

where the constant $C > 0$ is independent of h, t and the interval (a, b).

The function w_j will be made of *a superposition of elementary jumps* propagating along discontinuity lines or characteristic lines of the matrix A. It will be convenient to refer to some of these jumps as **particles** and **anti-particles**, generalizing here a terminology introduced in Section V-2.

Decompose the (x, t)-plane in regions where the characteristic component α_j keeps a constant sign. For simplicity in the presentation we may assume that α_j never vanishes. Call Ω_+ a region in which $\alpha_j > 0$. (The arguments are completely similar in a region where $\alpha_j < 0$.) Observe that no constraint is imposed by (2.16) along the boundary of Ω_+. (As a matter of fact, the boundary is made of fronts which either are Lax or rarefaction-shock fronts of the j-family or else have a non-dominant component β_{j-}.) The weight w_j is made of finitely many particles and anti-particles, with the possibility that several of them occupy the same location. However, within Ω_+ a single particle will travel together with each i-discontinuity for $i \neq j$ and with each undercompressive i-discontinuity. That is, in Ω_+ we require that for each i-discontinuity with $i \neq j$

$$w_{j+}(x, t) - w_{j-}(x, t) = \begin{cases} K\,\varepsilon(x, t) & \text{if } j < i, \\ -K\,\varepsilon(x, t) & \text{if } j > i, \end{cases} \tag{2.20}$$

and for each *undercompressive* i-discontinuity:

$$w_{j+}(x, t) - w_{j-}(x, t) = \begin{cases} -K\,\varepsilon(x, t) & \text{for a jump in the solution } u, \\ K\,\varepsilon(x, t) & \text{for a jump in the solution } v. \end{cases} \tag{2.21}$$

To construct the weight we proceed in the following way. First of all, we note that the weight w_j can be defined locally near the initial time $t = 0+$ before the first interaction time: we can guarantee that (2.20)-(2.21) hold at each discontinuity satisfying $\alpha_{j-}\,\alpha_{j+} > 0$: each propagating discontinuity carries a particle with mass $\pm K\,\varepsilon$ determined by (2.20)-(2.21) while, in order to compensate for it, an *anti-particle* with opposite strength $\mp K\,\varepsilon$ is introduced and propagate together with discontinuities satisfying $\alpha_{j-}\,\alpha_{j+} > 0$. (Alternatively, w_j could be taken to be continuous at the latter.) We now describe the generation, dynamics, and cancellation of particles and anti-particles within Ω_+.

When a wave with strength (in modulus) ε enters the region Ω_+ it generates a *particle* with strength $\pm K \varepsilon$ determined by (2.20)-(2.21) and propagating together with the entering wave and, in order to compensate for the change, it also generates an *anti-particle* with opposite strength $\mp K \varepsilon$ propagating along the boundary of Ω_+.

When a wave exits the region Ω_+ its associated *particle* remains stuck along with the boundary of Ω_+.

At this stage we have associated one particle and one anti-particle to each wave in Ω_+. Then, as a given wave passes through regions where α_j remains constant, it creates an *oscillating train of particles and anti-particles*. Clearly, the property (2.19) holds with a constant C of the order of the sum of the total variations of the solutions u and v.

But, particles and anti-particles can also be generated by *cancellation* and *interaction* effects. When two waves with strength ε and ε' (associated with the solution u, say) meet within a region Ω_+ their strengths are modified in agreement with Glimm's interaction estimates. Basically, the change in strength is controlled by the amount of cancellation and interaction θ^u. For the interaction of a shock and rarefaction wave of the same family we have $\theta^u = |\varepsilon - \varepsilon'| + \varepsilon \varepsilon'$, while, in all other cases, $\theta^u = \varepsilon \varepsilon'$. New waves with strengths $\varepsilon \varepsilon'$ may also arise from the interaction.

To carry away the extra mass of order θ^u and for each of the outgoing waves we introduce particles leaving from the point of interaction and propagating with the local j-characteristic speed. The oscillating train of particles and anti-particles associated with each of the incoming fronts is also decomposed so that, after the interaction time, we still have waves with attached particles and associated anti-particles. Additionally, new particles are attached to the new waves and new anti-particles propagating with the local j-characteristic speed are introduced.

More precisely, in the above construction the local j-characteristic speed is used whenever their is no j-front or else the front is *undercompressive*; otherwise the particle or anti-particle under consideration propagates with the j-front. We use here the fact that the constraint in (2.21) concerns undercompressive fronts only.

Finally, we impose that a particle and its associated anti-particle cancel out whenever they come to occupy the same location: Their strengths add up to 0 and they are no longer accounted for.

In conclusion, to each wave with strength ε within a region Ω_+ we have associated an *oscillating train of particles and anti-particles* with mass $\pm K \varepsilon$, propagating along lines of discontinuity or characteristic lines of the matrix A. Additionally, we have oscillating trains of particles and anti-particles associated with interaction and cancellation measures θ^u and θ^v.

In turn, the estimate (2.19) holds: for every interval (a, b) we have

$$\left| \sum_{x \in (a,b)} (w_{j+}(x,t) - w_{j-}(x,t)) \right| \leq C_1 \left(TV(u^h(t)) + TV(v^h(t)) + \sum_{t' \leq t} \theta^u(x',t') + \theta^v(x',t') \right),$$

where the sums are over interaction points (x', t'). Since the total amount of cancellation and interaction is controlled by the initial total variation, we arrive at

$$\left| \sum_{x \in (a,b)} (w_{j+}(x,t) - w_{j-}(x,t)) \right| \leq C_2 \left(TV(u^h(0)) + TV(v^h(0)) \right),$$

which yields the uniform bounds (1.15), provided the total variation of the solutions is sufficiently small. This completes the proof of Theorem 2.3.   □

REMARK 2.4. For certain systems the L^1 continuous dependence estimate (2.7) remains valid even for solutions with "large" amplitude. Consider for instance the system of nonlinear elasticity

$$\begin{aligned}
\partial_t v - \partial_x \sigma(w) &= 0, \\
\partial_t w - \partial_x v &= 0,
\end{aligned} \qquad (2.22)$$

where the stress-strain function $\sigma(w)$ is assumed to be increasing and convex so that the system (2.22) is hyperbolic and genuinely nonlinear. It is not difficult to check that the fundamental properties discovered in Theorems 2.1 and 2.2 are valid for solutions of (2.22) having *arbitrary large* amplitude.   □

3. Sharp version of the continuous dependence estimate

In this section, we derive a sharp version of the L^1 continuous dependence property of entropy solutions established in Theorem 2.3. This version keeps track of the dissipation terms which account for the "strict decrease" of the L^1 distance between two solutions. Throughout this section, the flux-function is defined on $\mathcal{U} := \mathcal{B}(\delta)$ and admits genuinely nonlinear fields.

To state this result we introduce some concept of "wave measures" associated with a function of bounded variation $u : I\!R \to \mathcal{U}$. Denote by $\mathcal{J}(u)$ the set of jump points of u. The vector-valued measure $\mu(u) := \dfrac{du}{dx}$ can be decomposed as

$$\mu(u) = \mu^a(u) + \mu^c(u),$$

where $\mu^a(u)$ and $\mu^c(u)$ are the corresponding atomic and continuous parts, respectively. For $i = 1, \ldots, N$ the i-*wave measures* associated with u are, by definition, the signed measures $\mu_i(u)$ satisfying

$$\mu_i(u) = \mu_i^a(u) + \mu_i^c(u), \qquad (3.1a)$$

where on one hand the continuous part $\mu_i^a(u)$ is characterized by

$$\int_{I\!R} \varphi \, d\mu_i^c(u) = \int_{I\!R} \varphi \, l_i(u) \cdot d\mu^c(u) \qquad (3.1b)$$

for every continuous function φ with compact support, and on the other hand the atomic part $\mu_i^a(u)$ is concentrated on $\mathcal{J}(u)$ and is characterized by

$$\mu_i^a(u)(\{x_0\}) = \gamma_i(u_-, u_+), \quad x_0 \in \mathcal{J}(u), \qquad (3.1c)$$

$\gamma_i(u_-, u_+)$ being the strength of the i-wave within the Riemann solution associated with the left and right-data $u_\pm = u(x_0\pm)$. It is easily checked that the functional

$$V(u; x) := \sum_{i=1}^{N} |\mu_i(u)| ((-\infty, x)), \quad x \in I\!R,$$

is "equivalent" to the total variation of u

$$TV_{-\infty}^x(u) = \int_{(-\infty, x)} \left| \frac{du}{dx} \right|, \quad x \in I\!R,$$

in the sense that

$$C_- \, TV_{-\infty}^x(u) \leq V(u; x) \leq C_+ \, TV_{-\infty}^x(u), \quad x \in \mathbb{R},$$

where the positive constants $C_\pm$ depend on the flux-function f and the parameter δ, only.

We also consider the *measure of potential interaction* associated with the function $u = u(x)$ and defined by

$$q(u)(I) := \sum_{1 \leq i < j \leq N} \left(|\mu_j(u)| \otimes |\mu_i(u)| \right) \left(\{(x, y) \in I \times I \,/\, x < y\} \right)$$

$$+ \sum_{i=1}^{N} \left(\mu_i^-(u) \otimes |\mu_i(u)| \right) \left(\{(x, y) \in I \times I \,/\, x \neq y\} \right) \tag{3.2}$$

for every interval $I \subset \mathbb{R}$, where we have called $\mu_i^\pm(u)$ the positive and the negative parts of the wave measure, respectively:

$$\mu_i(u) =: \mu_i^+(u) - \mu_i^-(u), \quad \mu_i^\pm(u) \geq 0,$$
$$|\mu_i(u)| := \mu_i^+(u) + \mu_i^-(u),$$

Finally, the *modified wave measures* are by definition

$$\tilde{\mu}_i(u) := |\mu_i(u)| + c\, q(u),$$

where $c > 0$ is a (sufficiently) small constant. Observe that $\tilde{\mu}_i$ are indeed bounded measures and satisfy

$$|\mu_i(u)| + (1 - c\,\delta) \sum_{j=1}^{N} |\mu_j(u)| \leq \tilde{\mu}_i(u) \leq |\mu_i(u)| + (1 + c\,\delta) \sum_{j=1}^{N} |\mu_j(u)|,$$

where $\delta := V(u; +\infty)$. Like for the μ_i's, they determine a functional which is completely "equivalent" to the total variation of u.

The advantage of the measures $\tilde{\mu}_i(u)$ is their lower semi-continuity, as is the total variation functional. (See the bibliographical notes for a reference.)

LEMMA 3.1. (Lower semi-continuity of the modified wave measures.) *The functionals* $u \mapsto \tilde{\mu}_i(u)$ $(i = 1, \dots, N)$ *are lower semi-continuous with respect to the L^1 convergence, that is, if $u^h : \mathbb{R} \to \mathcal{U}$ is a sequence of functions with uniformly bounded variation converging almost everywhere to a function (of bounded variation) $u : \mathbb{R} \to \mathcal{U}$, then for every interval $I \subset \mathbb{R}$*

$$\tilde{\mu}_i(u)(I) \leq \liminf_{h \to 0} \tilde{\mu}_i(u^h)(I).$$

□

We now define some nonconservative products associated with two given functions of bounded variation, $u, v : \mathbb{R} \to \mathcal{U}$. Recall that $\overline{A}(u, v)$ denotes the averaging matrix defined in (1.3). We set

$$\omega_{ij}(u, v) := \left| l_j(u) \cdot \left(\overline{A}(u, v) - \lambda_i(u) \right) (v - u) \right|.$$

For $i, j = 1, \dots, N$ the *dissipation measures* $\nu_{ij}(u, v)$ are defined as follows:

$$\nu_{ij}(u, v) = \nu_{ij}^a(u, v) + \nu_{ij}^c(u, v), \tag{3.3a}$$

where the atomic and continuous parts are uniquely characterized by the following
two conditions (with an obvious notation):

$$\nu_{ij}^c(u,v) := \omega_{ij}(u,v)\, d\tilde{\mu}_i^c(u) \tag{3.3b}$$

and

$$\nu_{ij}^a(u,v)\big(\{x\}\big) = \omega_{ij}(u_-,v_-)\, \tilde{\mu}_i^a(u)\big(\{x\}\big) \tag{3.3c}$$

for $x \in \mathcal{J}(u)$, where $u_- := u(x-)$, etc., and $\overline{\lambda}_i(u_-,u_+)$ denotes the (smallest for
definiteness) wave speed of the i-wave (fan) in the Riemann solution corresponding
to the left and right-data $u_\pm$. Finally, the (a priori formal) nonconservative product
$\omega_{ij}(u,v)\, d\tilde{\mu}_i(u)$ is now well-defined as

$$\omega_{ij}(u,v)\, d\tilde{\mu}_i(u) := \nu_{ij}(u,v).$$

In the following we use the above definition for functions u and v depending on
time also, and so we use the obvious notation $\omega_{ij}(u,v,t)$, etc.

THEOREM 3.2. (Sharp L^1 continuous dependence.) *Consider the strictly hyperbolic
system of conservation laws (1.1) where the flux-function $f : \mathbb{R} \to \mathcal{U} = \mathcal{B}(\delta)$ has
genuinely nonlinear fields and δ is sufficiently small. Then, there exist constants
$c > 0$ and $C \geq 1$ such that for any two entropy solutions $u,v : \mathbb{R} \times \mathbb{R}_+ \to \mathcal{U}$ of sufficiently small total variation, that is, $TV(u), TV(v) < c$, the* **sharp L^1 continuous
dependence estimate**

$$\|v(t) - u(t)\|_{L^1(\mathbb{R})} + \int_0^t \big(\mathbf{D}_2(s) + \mathbf{D}_3(s)\big)\, ds \leq C\, \|v(0) - u(0)\|_{L^1(\mathbb{R})} \tag{3.4}$$

holds for all $t \geq 0$, where

$$\mathbf{D}_2(s) := \sum_{i=1}^N \sum_{(x,s)\in\mathcal{L}_i(\overline{A})} \Big(\omega_{ii}(u_-(x),v_-(x),s) + \omega_{ii}(v_-(x),u_-(x),s)\Big),$$

$$\mathbf{D}_3(s) := \int_{\mathbb{R}} \sum_{i,j=1}^N \Big(\omega_{ij}(u,v,s)\, d\tilde{\mu}_i(u,s) + \omega_{ij}(v,u,s)\, d\tilde{\mu}_i(v,s)\Big).$$

*Here, $\mathcal{L}_i(\overline{A})$ denotes the set of all Lax i-discontinuities associated with the matrix
$\overline{A}(u,v)$, in other words, points (x,t) where the shock speed λ satisfies*

$$\overline{\lambda}_i(u_-,v_-) \geq \lambda \geq \overline{\lambda}_i(u_+,v_+).$$

Let us point out the following important features of the sharp estimate (3.4):
- Each jump in u or in v contribute to the strict decrease of the L^1 distance.
- The contribution of each jump is *cubic* in nature.
- Furthermore, Lax discontinuities provides a stronger, *quadratic* decay.

The rest of this section is devoted to proving Theorem 3.2. Denote by u^h wave-
front tracking approximations with uniformly bounded total variation and converging
to some entropy solution $u : \mathbb{R} \times \mathbb{R}_+ \to \mathcal{U}$. The local uniform convergence of the
sequence u^h was discussed in Section VII-4. Recall that for all but countably many
times t, the functions $x \mapsto u^h(x) := u^h(x,t)$ satisfy:
- If x_0 is a point of continuity of u, then for every $\varepsilon > 0$ there exists $\eta > 0$ such
  that for all sufficiently small h

$$|u^h(x) - u(x_0)| < \varepsilon \quad \text{for each } x \in (x_0 - \eta, x_0 + \eta). \tag{3.5}$$

- If x_0 is a point of jump discontinuity of u, then there exists a sequence $x_0^h \in \mathbb{R}$ converging to x_0 such that: for every $\varepsilon > 0$ there exists $\eta > 0$ such that for all sufficiently small h

$$
\begin{aligned}
|u^h(x) - u_-(x_0)| &< \varepsilon \quad \text{for each } x \in (x_0^h - \eta, x_0^h), \\
|u^h(x) - u_+(x_0)| &< \varepsilon \quad \text{for each } x \in (x_0^h, x_0^h + \eta).
\end{aligned}
\tag{3.6}
$$

To each approximate solution u^h we then associate the *(approximate) i-wave measures* $\mu_i^{u,h}(t)$. By a standard compactness theorem, there exist some limiting measures $\mu_i^{u,\infty}(t)$ and $\tilde{\mu}_i^{u,\infty}(t)$ such that, for all but countably many times t,

$$
\mu_i^{u,h}(t) \rightharpoonup \mu_i^{u,\infty}(t)
\tag{3.7a}
$$

and

$$
\tilde{\mu}_i^{u,h}(t) \rightharpoonup \tilde{\mu}_i^{u,\infty}(t)
\tag{3.7b}
$$

in the weak sense of bounded measures. Actually the convergence in (3.7a) holds in a stronger sense. We state with proof (see the bibliographical notes) the following important property of wave measures:

THEOREM 3.3. (Convergence of the *i*-wave measures.) *For $i = 1, \dots, N$ and for all but countably many times t, the atomic parts of the measures $\mu_i^{u,\infty}(t)$ and $\mu_i(u,t)$ coincide. In other words, we have*

$$
\mu_i^{u,\infty}(t)(\{x_0\}) = \mu_i(u,t)(\{x_0\}) = \gamma_i(u_-, u_+)
\tag{3.8}
$$

at each jump point x_0 of the function $x \mapsto u(x,t)$, with $u_\pm = u(x_0\pm, t)$.

Two main observations needed in the proof of Theorem 3.2 are summarized in the following preparatory lemmas:

LEMMA 3.4. (Convergence of dissipation measures.) *Let u^h and v^h be wave front tracking approximations associated with two entropy solutions u and v respectively. Then for each $i, j = 1, \dots, N$ and for all but countably many times t, we have*

$$
C_0 \int_{\mathbb{R}} \omega_{ij}(u, v, t) \, d\tilde{\mu}_i^{u,\infty}(t) \leq \lim_{h \to 0} \int_{\mathbb{R}} \omega_{ij}(u^h, v^h, t) \, d\tilde{\mu}_i^{u,h}(t)
\tag{3.9}
$$

for some constants $C_0 > 0$.

LEMMA 3.5. (Convergence of dissipation measures on Lax shocks.) *Consider the sets $\mathcal{L}_i(\overline{A}^h)$ and $\mathcal{L}_i(\overline{A})$ of all Lax shock discontinuities in the averaging matrices $\overline{A}^h$ and $\overline{A}$ associated with the approximate and exact solutions, respectively. Then, for each $i = 1, \dots, N$ and for all but countably many times t, we have*

$$
\lim_{h \to 0} \sum_{(x,t) \in \mathcal{L}_i(\overline{A}^h)} \omega_{ii}(u_-^h(x), v_-^h(x), t) = \sum_{(x,t) \in \mathcal{L}_i(\overline{A})} \omega_{ii}(u_-(x), v_-(x), t).
$$

Furthermore, the following estimate is a direct consequence of the definition (3.3):

LEMMA 3.6. *For all functions of bounded variation $u, \hat{u}, v, \hat{v}, w$ defined on $\mathbb{R}$ we have*

$$
\left| \int_\alpha^\beta \omega_{ij}(u, v) \, d\mu_i(w) - \int_\alpha^\beta \omega_{ij}(\hat{u}, \hat{v}) \, d\mu_i(w) \right|
$$
$$
\leq C \left(\|\hat{u} - u\|_{L^\infty(\alpha,\beta)} + \|\hat{v} - v\|_{L^\infty(\alpha,\beta)} \right) TV_{[\alpha,\beta]}(w),
$$

where the constant $C > 0$ only depends on the range of the functions under consideration.

PROOF OF LEMMA 3.4. Throughout the proof, a time $t > 0$ is fixed at which the results (3.5) and (3.6) hold. Given $\varepsilon > 0$ we select finitely many jumps in u and v, located at $y_1, y_2, \ldots y_n$, so that

$$\sum_{\substack{x \neq y_k \\ k=1,2,\ldots,n}} |u_+(x) - u_-(x)| + |v_+(x) - v_-(x)| < \varepsilon. \tag{3.10}$$

To each point y_k we associate the corresponding point y_k^h in u^h or in v^h. To simplify the notation we restrict our attention to the case

$$y_k < y_k^h < y_{k+1} < y_{k+1}^h \quad \text{for all } k,$$

the other cases being entirely similar. By the local uniform convergence property, for h sufficiently small we have

$$(a) \quad \sum_{k=1}^{n} \sum_{\pm} |u_\pm^h(y_k^h) - u_\pm(y_k)| + |v_\pm^h(y_k^h) - v_\pm(y_k)| \leq \varepsilon,$$

$$(b) \quad \sum_{k=1}^{n} |u^h(x) - u(y_k+)| + |v^h(x) - v(y_k+)| \leq \varepsilon, \quad x \in \left(y_k, y_k^h\right), \tag{3.11}$$

for the "large" jumps and, in regions of "approximate continuity",

$$|u^h(x) - u(x)| + |v^h(x) - v(x)| \leq 2\varepsilon, \quad x \in \left(y_k^h, y_{k+1}\right) \subseteq \left(y_k, y_{k+1}\right). \tag{3.12}$$

Based on (3.10) it is not difficult to construct some functions u_ε and v_ε that are continuous everywhere except possibly at the points y_k and such that the following conditions hold:

$$TV\left(u_\varepsilon; I\!\!R \setminus \{y_1, \ldots, y_n\}\right) \leq C\,TV\left(u; I\!\!R \setminus \{y_1, \ldots, y_n\}\right),$$

$$\|u - u_\varepsilon\|_\infty \leq \frac{C}{j}\varepsilon, \quad TV\left(u - u_\varepsilon; I\!\!R \setminus \{y_1, \ldots, y_n\}\right) \leq C\varepsilon, \tag{3.13}$$

as well a as completely similar statement with u replaced with v, where C is independent of ε.

For a constant C_0 to be determined, consider the decomposition

$$\Omega(h) = \int_{I\!\!R} \omega_{ij}(u^h, v^h)\, d\tilde{\mu}_i^{u,h} - C_0 \int_{I\!\!R} \omega_{ij}(u, v)\, d\tilde{\mu}_i^{u,\infty}$$

$$= \left(\sum_{k=0}^{n} \int_{(y_k^h, y_{k+1}^h)} + \sum_{k=1}^{n} \int_{\{y_k^h\}}\right) \omega_{ij}(u^h, v^h)\, d\tilde{\mu}_i^{u,h}$$

$$- C_0 \left(\sum_{k=0}^{n} \int_{(y_k, y_{k+1})} + \sum_{k=1}^{n} \int_{\{y_k\}}\right) \omega_{ij}(u, v)\, d\tilde{\mu}_i^{u,\infty},$$

that is,

$$\Omega(h) = \sum_{k=1}^{n} \left(\int_{\{y_k^h\}} \omega_{ij}(u^h, v^h) \, d\tilde{\mu}_i^{u,h} - C_0 \int_{\{y_k\}} \omega_{ij}(u, v) \, d\tilde{\mu}_i^{u,\infty} \right)$$

$$+ \sum_{k=0}^{n} \left(\int_{(y_k^h, y_{k+1}^h)} \omega_{ij}(u^h, v^h) \, d\tilde{\mu}_i^{u,h} - C_0 \int_{(y_k, y_{k+1})} \omega_{ij}(u, v) \, d\tilde{\mu}_i^{u,\infty} \right) \qquad (3.14)$$

$$= \Omega_1(h) + \Omega_2(h).$$

Here $y_0^h = y_0 = -\infty$ and $y_{n+1}^h = y_{n+1} = +\infty$.

In view of Theorem 3.3 we have

$$\left| \mu_i^{u,\infty}(\{y_k\}) \right| = \left| \gamma_i(u_{k-}, u_{k+}) \right|.$$

Passing to the limit in the inequality $\mu_i^{u,h} \leq \tilde{\mu}_i^{u,h}$ we find $\mu_i^{u,\infty} \leq \tilde{\mu}_i^{u,\infty}$ and therefore

$$\mu_i^{u,\infty}(\{y_k\}) \leq \tilde{\mu}_i^{u,\infty}(\{y_k\}),$$

with $u_{k\pm} = u_\pm(y_k)$. On the other hand, by definition,

$$\left| \mu_i^{u,h}(\{y_k^h\}) \right| = \left| \gamma_i(u_{k-}^h, u_{k+}^h) \right|.$$

Using that u^h is a piecewise constant function, we have

$$(1 - c\delta) \, \mu_i^{u,h}(\{y_k^h\}) \leq \tilde{\mu}_i^{u,h}(\{y_k^h\}) \leq (1 + c\delta) \, \mu_i^{u,h}(\{y_k^h\})$$

with $u_{k\pm}^h = u_\pm^h(y_k^h)$. Therefore we arrive at the following key inequality

$$\Omega_1(h) \leq (1 + c\delta) \sum_{k=1}^{n} \omega_{ij}(u_{k-}^h, v_{k-}^h) \left| \gamma_i(u_{k-}^h, u_{k+}^h) \right| - C_0 \, \omega_{ij}(u_{k-}, v_{k-}) \left| \gamma_i(u_{k-}, u_{k+}) \right|.$$

Thus, choosing C_0 large enough so that $C_0 \geq (1 + c\delta)$ we find

$$\Omega_1(h) \leq (1 + c\delta) \, \hat{\Omega}_1(h)$$

with

$$\hat{\Omega}_1(h) := \sum_{k=1}^{n} \omega_{ij}(u_{k-}^h, v_{k-}^h) \left| \gamma_i(u_{k-}^h, u_{k+}^h) \right| - \omega_{ij}(u_{k-}, v_{k-}) \left| \gamma_i(u_{k-}, u_{k+}) \right|.$$

But, since the functions γ_i are locally Lipschitz continuous

$$\left| \hat{\Omega}_1(h) \right| \leq C \sum_{k=1}^{n} \sum_{\pm} |u_{k\pm}^h - u_{k\pm}| + |v_{k\pm}^h - v_{k\pm}| \leq C\varepsilon \qquad (3.15)$$

by the property (3.11a), provided h is sufficiently small. So

$$\limsup_{h \to 0} \Omega_1(h) \leq 0.$$

Using that $C_0 \geq 1$ and relying on the inequalities $y_k < y_k^h < y_{k+1} < y_{k+1}^h$ for all k, we consider the following decomposition:

$$\Omega_2(h) \leq \hat{\Omega}_2(h) := \sum_{k=0}^{n} \int_{(y_k^h, y_{k+1})} \omega_{ij}(u^h, v^h) \, d\tilde{\mu}_i^{u,h} - \omega_{ij}(u,v) \, d\tilde{\mu}_i^{u,\infty}$$
$$- \sum_{k=0}^{n} \int_{(y_k, y_k^h]} \omega_{ij}(u,v) \, d\tilde{\mu}_i^{u,\infty} + \sum_{k=0}^{n} \int_{[y_{k+1}, y_{k+1}^h)} \omega_{ij}(u^h, v^h) \, d\mu_i^{u,h}$$
$$=: \Omega_{2,1}(h) + \Omega_{2,2}(h) + \Omega_{2,3}(h). \tag{3.16}$$

We will show that $\hat{\Omega}_2(h) \to 0$.

Consider first $\Omega_{2,2}(h)$ and $\Omega_{2,3}(h)$ which are somewhat simpler to handle:

$$\Omega_{2,2}(h) = \sum_{k=0}^{n} \int_{(y_k, y_k^h]} \left(-\omega_{ij}(u(y), v(y)) + \omega_{ij}(u(y_k+), v(y_k+)) \right) d\tilde{\mu}_i^{u,\infty}(y)$$
$$- \sum_{k=0}^{n} \int_{(y_k, y_k^h]} \omega_{ij}(u(y_k+), v(y_k+)) \, d\tilde{\mu}_i^{u,\infty}(y).$$

Therefore, with Lemma 3.6 we obtain

$$\left| \Omega_{2,2}(h) \right| \leq C \int_{I\!R} d\mu_i^{u,\infty} \left(\sup_{y \in (y_k, y_k^h]} |u(y) - u(y_k+)| + \sup_{x \in (y_k, y_k^h]} |v(y) - v(y_k+)| \right)$$
$$+ C \sum_{k=0}^{n} \int_{(y_k, y_k^h]} d\tilde{\mu}_i^{u,\infty}.$$

Since $y_k^h \to y_k$, we have for h sufficiently small

$$\left| \Omega_{2,2}(h) \right| \leq C\varepsilon. \tag{3.17}$$

A similar argument for $\Omega_{2,3}(h)$, but introducing now the left-hand values $u_-(y_k)$ and $v_-(y_k)$ and relying on (3.11b), shows that

$$\left| \Omega_{2,3}(h) \right| \leq C\varepsilon. \tag{3.18}$$

Next consider the decomposition

$$\omega_{ij}(u^h, v^h) \, d\tilde{\mu}_i^{u,h} - \omega_{ij}(u,v) \, d\tilde{\mu}_i^{u,\infty}$$
$$= \left(\omega_{ij}(u^h, v^h) \, d\tilde{\mu}_i^{u,h} - \omega_{ij}(u,v) \, d\tilde{\mu}_i^{u,h} \right) + \left(\omega_{ij}(u,v) \, d\tilde{\mu}_i(u^h) - \omega_{ij}(u_\varepsilon, v_\varepsilon) \, d\tilde{\mu}_i^{u,h} \right)$$
$$+ \left(\omega_{ij}(u_\varepsilon, v_\varepsilon) \, d\tilde{\mu}_i^{u,h} - \omega_{ij}(u_\varepsilon, v_\varepsilon) \, d\tilde{\mu}_i^{u,\infty} \right) + \left(\omega_{ij}(u_\varepsilon, v_\varepsilon) \, d\tilde{\mu}_i^{u,\infty} - \omega_{ij}(u,v) \, d\tilde{\mu}_i^{u,\infty} \right).$$

With obvious notation, this yields a decomposition of the form

$$\Omega_{2,1}(h) = M_1(h) + M_2(h) + M_3(h) + M_4(h). \tag{3.19}$$

Using Lemma 3.6 and the consequence (3.11b) of the local convergence property, we obtain

$$|M_1(h)| \leq C \sum_{k=0}^{n} \int_{(y_k^h, y_{k+1}^h)} d\tilde{\mu}_i^{u,h} \left(\sup_{(y_k^h, y_{k+1}^h)} |u^h - u| + \sup_{(y_k^h, y_{k+1}^h)} |v^h - v| \right)$$

$$\leq C' \sum_{k=0}^{n} \left(\sup_{(y_k^h, y_{k+1}^h)} |u^h - u| + \sup_{(y_k^h, y_{k+1}^h)} |v^h - v| \right) \leq C' \, \varepsilon. \tag{3.20}$$

Next using Lemma 3.6 and (3.13) we have

$$|M_2(h)| \leq C \sum_{k=0}^{n} \int_{(y_k^h, y_{k+1}^h)} d\tilde{\mu}_i^{u,h} \left(\sup_{(y_k^h, y_{k+1}^h)} |u - u_\varepsilon| + \sup_{(y_k^h, y_{k+1}^h)} |v - v_\varepsilon| \right)$$

$$\leq C' \, \varepsilon. \tag{3.21}$$

Dealing with $M_4(h)$ is similar:

$$|M_4(h)| \leq C \sum_{k=0}^{n} \int_{(y_k^h, y_{k+1}^h)} d\tilde{\mu}_i^{u,\infty} \left(\sup_{(y_k^h, y_{k+1}^h)} |u - u_\varepsilon| + \sup_{(y_k^h, y_{k+1}^h)} |v - v_\varepsilon| \right)$$

$$\leq C \varepsilon. \tag{3.22}$$

Finally to treat $M_3(h)$ we observe that, since u_ε and v_ε are continuous functions on each interval (y_k^h, y_{k+1}^h) and since $\tilde{\mu}_i^{u,h}$ is a sequence of bounded measures converging weakly toward $\tilde{\mu}_i^{u,\infty}$, we have for all h sufficiently small

$$|M_3(h)| \leq \varepsilon. \tag{3.23}$$

Combining (3.19)–(3.23) we get

$$|\Omega_{2,1}(h)| \leq C \varepsilon. \tag{3.24}$$

Combining (3.17), (3.18), and (3.24) we obtain

$$|\hat{\Omega}_2(h)| \leq C \varepsilon$$

and thus, with (3.14)-(3.15),

$$\Omega(h) \leq C \varepsilon \quad \text{for all } h \text{ sufficiently small.}$$

Since ε is arbitrary, this completes the proof of Lemma 3.4. $\qquad\square$

PROOF OF LEMMA 3.5. Fixing some $i = 1, \ldots, N$ and excluding countably many times t only, we want to show that

$$\lim_{h \to 0} \sum_{x \in \mathcal{L}_i(\overline{A}^h)} \omega_{ii}(u_-^h(x), v_-^h(x)) = \sum_{x \in \mathcal{L}_i(\overline{A})} \omega_{ii}(u_-(x), v_-(x)). \tag{3.25}$$

Let y_k for $k = 1, 2, \ldots$ be the jump points in u or v. Denote by y_k^h the corresponding jump points in u^h or v^h. Extracting a subsequence if necessary we can always assume that for each k either $y_j^h \in \mathcal{L}_i(\overline{A}^h)$ for all h, or else $y_j^h \notin \mathcal{L}_i(\overline{A}^h)$ for all h. Then we consider the following three sets: Denote by J_1 the set of indices k such that $y_k^h \in \mathcal{L}_i(\overline{A}^h)$ and $y_k \in \mathcal{L}_i(\overline{A})$. Let J_2 the set of indices k such that $y_k^h \notin \mathcal{L}_i(\overline{A}^h)$ and $y_k \in \mathcal{L}_i(\overline{A})$. Finally J_3 is the set of indices k such that $y_k^h \in \mathcal{L}_i(\overline{A}^h)$ and $y_k \notin \mathcal{L}_i(\overline{A})$.

First of all the local convergence property (3.6) implies

$$\sum_{k \in J_1} \omega_{ii}\big(u_-^h(y_k^h), v_-^h(y_k^h)\big) \longrightarrow \sum_{k \in J_1} \omega_{ii}\big(u_-(y_k), v_-(y_k)\big). \tag{3.26}$$

(Indeed, given $\varepsilon > 0$, choose finitely many jump points as in (3.10) and impose (3.11a) with ε replaced with $\varepsilon\,|u_+(y_k) - u_-(y_k)|$.)

On the other hand for indices in $J_2 \cup J_3$ we claim that

$$\sum_{k \in J_2 \cup J_3} \omega_{ii}\big(u_-^h(y_k^h), v_-^h(y_k^h)\big) \longrightarrow 0, \tag{3.27}$$

while

$$\sum_{k \in J_2 \cup J_3} \omega_{ii}\big(u_-(y_k), v_-(y_k)\big) = 0. \tag{3.28}$$

Indeed, for each $k \in J_2$, y_k is a Lax discontinuity but y_k^h is not. Extracting a subsequence if necessary, the Lax inequalities are violated on the left or on the right side of y_k^h for all h. Assuming that it is the case on the left side, we have

$$\overline{\lambda}_i\big(u_-(y_k), v_-(y_k)\big) - \overline{\lambda}_i\big(u_-(y_k), u_+(y_k)\big) \geq 0$$

while

$$\overline{\lambda}_i\big(u_-(y_k^h), v_-(y_k^h)\big) - \overline{\lambda}_i\big(u_-(y_k^h), u_+(y_k^h)\big) \leq 0$$

for all h. We have denoted here by $\overline{\lambda}_i(u,v)$ the i-eigenvalue of the matrix $\overline{A}(u,v)$. But the latter converges toward the former by the local uniform convergence, which proves that

$$\overline{\lambda}_i\big(u_-(y_k), v_-(y_k)\big) - \overline{\lambda}_i\big(u_-(y_k), u_+(y_k)\big) = 0$$

and, by the genuine nonlinearity condition, $v_-(y_k) = u_+(y_k)$. In this case, we finally get

$$\omega_{ii}(u_-(y_k), v_-(y_k)) = 0.$$

$$\square$$

PROOF OF THEOREM 3.2. From the analysis in Sections 1 and 2 (Theorems 1.5 and 2.3) it follows immediately that for all $t \geq 0$

$$\|v(t) - u(t)\|_{L^1(\mathbb{R})} + \int_0^t \limsup_{h \to 0} \sum_{(x,s) \in \mathcal{J}(\overline{A}^h)} |\overline{A}_+^h(x,s) - \overline{A}_-^h(x,s)|$$

$$\big|\big(\overline{A}_-^h(x,s) - \overline{\lambda}^h(x,s)\big)\big(v_-^h(x,s) - u_-^h(x,s)\big)\big|\, ds$$

$$+ \int_0^t \limsup_{h \to 0} \sum_{\substack{(x,s) \in \mathcal{L}_i(\overline{A}^h) \\ 0 \leq s \leq t,\ 1 \leq i \leq N}} |\overline{l}_{i-}^h(x,s) \cdot \big(\overline{A}_-^h(x,s) - \overline{\lambda}^h(x,s)\big)\big(v_-^h(x,s) - u_-^h(x,s)\big)|\, ds$$

$$\leq C\,\|v(0) - u(0)\|_{L^1(\mathbb{R})},$$

since the contributions from the rarefaction fronts and artificial waves vanish as $h \to 0$. Here $\overline{l}_i^h$ denote the left-eigenvectors of the matrix $\overline{A}^h$.

Using that

$$|\overline{A}_+^h(x,s) - \overline{A}_-^h(x,s)| \geq c\,|u_+^h(x,s) - u_-^h(x,s)|$$

for discontinuities in u^h and

$$\left| \left(\overline{A}_-^h(x,s) - \overline{\lambda}^h(x,s) \right) \left(v_-^h(x,s) - u_-^h(x,s) \right) \right|$$

$$\geq c \sum_{j=1}^N \left| \overline{l}_{i-}^h(x,s) \cdot \left(\overline{A}_-^h(x,s) - \overline{\lambda}^h(x,s) \right) \left(v_-^h(x,s) - u_-^h(x,s) \right) \right|,$$

we arrive at

$$\|v(t) - u(t)\|_{L^1(\mathbb{R})}$$

$$+ \int_0^t \int_{\mathbb{R}} \sum_{ij=1}^N \limsup_{h \to 0} \left(\omega_{ij}(u^h, v^h, s) \, d\tilde{\mu}_i^{u,h}(s) + \omega_{ij}(v^h, u^h, s) \, d\tilde{\mu}_i^{v,h}(s) \right) ds$$

$$+ \int_0^t \sum_{i=1}^N \limsup_{h \to 0} \sum_{(x,s) \in \mathcal{L}_i(\overline{A}^h)} \left(\omega_{ii}(u_-^h(x), v_-^h(x), s) + \omega_{ii}(v_-^h(x), u_-^h(x), s) \right) ds$$

$$\leq C \, \|v(0) - u(0)\|_{L^1(\mathbb{R})}.$$

$$(3.29)$$

Applying Lemmas 3.4 and 3.5 we deduce from (3.29) that

$$\|v(t) - u(t)\|_{L^1(\mathbb{R})}$$

$$+ \int_0^t \int_{\mathbb{R}} \sum_{ij=1}^N \left(\omega_{ij}(u, v, s) \, d\tilde{\mu}_i^{u,\infty}(s) + \omega_{ij}(v, u, s) \, d\tilde{\mu}_i^{v,\infty}(s) \right) ds$$

$$+ \int_0^t \sum_{i=1}^N \sum_{(x,s) \in \mathcal{L}_i(\overline{A})} \left(\omega_{ii}(u_-(x), v_-(x), s) + \omega_{ii}(v_-(x), u_-(x), s) \right) ds$$

$$\leq C' \, \|v(0) - u(0)\|_{L^1(\mathbb{R})}.$$

$$(3.30)$$

To conclude we use Lemma 3.1 for instance for u: by lower semi-continuity we have

$$\tilde{\mu}_i(u,s) \leq \tilde{\mu}_i^{u,\infty}(s) \qquad (3.31)$$

in the sense of measures, for all but countably many times s and for some constant $c > 0$. Finally, it is clear from the definition of nonconservative products that (3.31) implies the same inequality on the nonconservative products. This completes the proof of Theorem 3.2.   □

4. Generalizations

Nonclassical entropy solutions determined in Chapter VIII should satisfy an analogue of the stability results derived in Sections 2 and 3. In particular, we conjecture that:

THEOREM 4.1. (L^1 continuous dependence of nonclassical solutions.) *Under the notations and assumptions in Theorem VIII-3.1, any two nonclassical entropy solutions $u = u(x,t)$ and $v = v(x,t)$, generated by wave front tracking and based on a prescribed kinetic relation, satisfy the L^1 continuous dependence property*

$$\|u(t) - v(t)\|_{L^1(\mathbb{R})} \leq C_* \, \|u(0) - v(0)\|_{L^1(\mathbb{R})}, \quad t \geq 0. \qquad (4.1)$$

where the constant $C_* > 0$ depends on the kinetic function and the L^∞ norm and total variation of the solutions under consideration.

CHAPTER X

UNIQUENESS OF ENTROPY SOLUTIONS

In this chapter, we establish a general uniqueness theorem for nonlinear hyperbolic systems. Solutions are sought in the space of functions with bounded variation, slightly restricted by the so-called *tame variation condition* (Definition 1.1). The results of existence and continuous dependence established in previous chapters covered solutions obtained as limits of piecewise constant approximate solutions with uniformly bounded total variation (in Chapters IV and V for scalar conservation laws and in Chapters VII to IX for systems). Our purpose now is to cover general functions with bounded variation and to establish a *general uniqueness theory* for hyperbolic systems of conservation laws.

It is convenient to introduce a very general notion, the (Φ, ψ)–*admissible entropy solutions*, based on prescribed sets of *admissible discontinuities* Φ and *admissible speeds* ψ. Roughly speaking, we supplement the hyperbolic system with the "dynamics" of elementary propagating discontinuities. The definition encompasses not only classical and nonclassical solutions of conservative systems but also solutions of hyperbolic systems that need not be in conservative form. Under certain natural assumptions on the prescribed sets Φ and ψ we prove in Theorem 3.1 that the associated Cauchy problem admits one solution depending L^1 continuously upon its initial data, at most. In turn, our framework yields the *uniqueness* for the Cauchy problem in each situation when the *existence* of one solution depending L^1 continuously upon its initial data is also known; see Theorems 4.1 and 4.3.

1. Admissible entropy solutions

Consider **a nonlinear hyperbolic system** of partial differential equations **in non-conservative form**

$$\partial_t u + A(u)\, \partial_x u = 0, \quad u = u(x,t) \in \mathcal{U}, \ x \in \mathbb{R}, \ t > 0, \tag{1.1}$$

where the $N \times N$ matrix $A(u)$ depends smoothly upon u and need not be the Jacobian matrix of some vector-valued mapping. All values u under consideration belong to an open and bounded subset $\mathcal{U} \subset \mathbb{R}^N$; interestingly enough, this set need not be small nor connected. For each u in $\mathcal{U}$, the matrix $A(u)$ is assumed to admit N real (but not necessarily distinct) eigenvalues

$$\lambda_1(u) \leq \ldots \leq \lambda_N(u)$$

and basis of left- and right-eigenvectors $l_j(u)$, $r_j(u)$, $1 \leq j \leq N$, normalized such that $l_i(u)\, r_j(u) = \delta_{ij}$. It is also assumed that there exists a bound λ^∞ for the wave speeds:

$$\sup_{\substack{1 \leq j \leq N \\ u \in \mathcal{U}}} \left| \lambda_j(u) \right| < \lambda^\infty.$$

We are interested in the Cauchy problem associated with (1.1), in a class of
functions with bounded variation. By definition, a **space-like segment** is a set of
the form

$$\Gamma := \Big\{(x,t)\,/\,x \in [x_1,x_2],\quad t = \gamma(x) := \alpha\,x + \beta\Big\}$$

for some $x_1 < x_2$ and $\alpha, \beta \in \mathbb{R}$ with $|\alpha| < 1/\lambda^\infty$. In particular, an **horizontal
segment** is a set of the form

$$\tilde\Gamma := \Big\{(x,t_0)\,/\,x \in [y_1,y_2]\Big\}$$

for some $y_1 < y_2$ and some fixed time t_0.

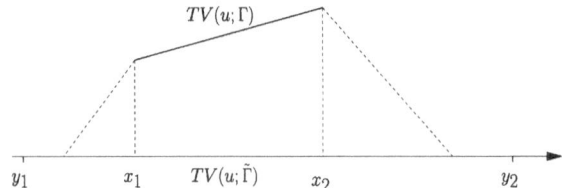

$$TV(u;\Gamma)$$

$$y_1 \qquad\qquad x_1 \qquad TV(u;\tilde\Gamma) \qquad x_2 \qquad\qquad\qquad y_2$$

Figure X-1 : The tame variation condition.

By definition, the segment Γ lies **inside the domain of determinacy** of $\tilde\Gamma$ if
we have

$$[x_1,x_2] \subseteq [y_1,y_2],\quad 0 \le \gamma(x_1) - t_0 \le \frac{x_1 - y_1}{\lambda^\infty},\quad 0 \le \gamma(x_2) - t_0 \le \frac{y_2 - x_2}{\lambda^\infty}.$$

By definition, the **total variation along the segment** Γ of some function $u = u(x,t)$
is the total variation of its restriction to the segment Γ and is denoted by $TV(u;\Gamma)$.

The class of functions under consideration in this chapter is defined as follows.
Our condition (1.2) below requires that the total variation does not grow too wildly
as time increases. From now on, some positive constant κ is fixed. (See Figure X-1.)

DEFINITION 1.1. (Notion of tame variation.) A map $u : \mathbb{R} \times [0,\infty) \to \mathcal{U}$ is said to be
a **function with tame variation** if $u = u(x,t)$ is a bounded, Lebesgue measurable
function and for every space-like segment Γ the restriction of u to the segment $\tilde\Gamma$ is a
function with bounded variation satisfying

$$TV(u;\Gamma) \le \kappa\,TV(u;\tilde\Gamma),\tag{1.2}$$

provided the segment Γ is inside the domain of determinacy of the horizontal segment
$\tilde\Gamma$. $\square$

In particular, any function with tame variation satisfies

$$TV(u(t)) \le \kappa\,TV(u(0)) < \infty,\quad t \in \mathbb{R}_+.\tag{1.3}$$

For instance (Theorem 4.1 below), solutions of (1.1) obtained as limits of wave front
tracking approximations are functions with tame variation.

We recall here some standard properties of functions with bounded variation.
(Additional results can be found in the appendix.) If a function $u = u(x,t)$ has

tame variation, it has also **bounded variation in both variables** (x, t) on each set $\mathbb{R} \times (0, T)$, that is, the distributional derivatives $\partial_t u$ and $\partial_x u$ are Radon measures in $\mathbb{R} \times (0, T)$. For such a function there exists a decomposition of the form

$$\mathbb{R} \times \mathbb{R}_+ = \mathcal{C}(u) \cup \mathcal{J}(u) \cup \mathcal{I}(u),$$

where $\mathcal{C}(u)$ is the set of **points of L^1 approximate continuity**, $\mathcal{J}(u)$ is the set of **points of approximate jump**, and $\mathcal{I}(u)$ is the set of **interaction points** of the function u. The latter is negligible in the sense that $\mathcal{H}_1(\mathcal{I}(u)) = 0$, where $\mathcal{H}_1$ denotes the one-dimensional Hausdorff measure in the plane. At each point $(x_0, t_0) \in \mathcal{J}(u)$ there exist **left- and right-approximate limits** $u_\pm(x_0, t_0)$ and a **propagation speed** $\lambda^u(x_0, t_0)$ such that, setting

$$\overline{u}(x, t) := \begin{cases} u_-(x_0, t_0), & x < x_0 + \lambda^u(x_0, t_0)\,(t - t_0), \\ u_+(x_0, t_0), & x > x_0 + \lambda^u(x_0, t_0)\,(t - t_0), \end{cases} \tag{1.4}$$

we have

$$\lim_{h \to 0} \frac{1}{h^2} \int_{t_0-h}^{t_0+h} \int_{x_0-h}^{x_0+h} \left| u(x, t) - \overline{u}(x, t) \right| dx\, dt = 0. \tag{1.5}$$

The **right-continuous representative** u_+ of u is defined $\mathcal{H}_1$–almost everywhere as

$$u_+(x, t) := \begin{cases} u(x, t), & (x, t) \in \mathcal{C}(u), \\ u_+(x, t), & (x, t) \in \mathcal{J}(u), \end{cases}$$

and the nonconservative product $A(u_+)\,\partial_x u$ is the radon measure such that, for every Borel set $B \subset \mathbb{R} \times (0, T)$,

$$\iint_B A(u_+)\,\partial_x u = \iint_{B \cap \mathcal{C}(u)} A(u)\,\partial_x u + \int_{B \cap \mathcal{J}(u)} A(u_+)\,(u_+ - u_-)\, d\mathcal{H}_1. \tag{1.6}$$

To define the notion of entropy solutions for (1.1) we prescribe a family of **admissible discontinuities**

$$\Phi \subset \mathcal{U} \times \mathcal{U}$$

and a family of **admissible speeds**

$$\psi : \Phi \to \left(-\lambda^\infty, \lambda^\infty \right)$$

satisfying the following **consistency property** for all pairs $(u_-, u_+) \in \Phi$:

$$\left| \left(A(u_+) - \psi(u_-, u_+) \right)(u_+ - u_-) \right| \leq \eta(|u_+ - u_-|)\,|u_+ - u_-|, \tag{1.7}$$

where the function $\eta = \eta(\varepsilon) \geq 0$ is increasing and satisfies $\eta(\varepsilon) \to 0$ as $\varepsilon \to 0$.

DEFINITION 1.2. (General concept of entropy solution.) Let $\Phi \subset \mathcal{U} \times \mathcal{U}$ be a set of admissible jumps and $\psi : \Phi \to \left(-\lambda^\infty, \lambda^\infty \right)$ be a family of admissible speeds satisfying (1.7). A function u with tame variation is called a (Φ, ψ)–**admissible entropy solution** of (1.1) or, in short, an **entropy solution** if the following two conditions hold:

- The restriction of the measure $\partial_t u + A(u_+)\,\partial_x u$ to the set of points of approximate continuity of u vanishes identically, that is,

$$\iint_B \partial_t u + A(u_+)\,\partial_x u = 0 \quad \text{for every Borel set } B \subset \mathcal{C}(u). \tag{1.8}$$

- At each point of approximate jump $(x, t) \in \mathcal{J}(u)$ the limits $u_\pm(x, t)$ and the speed $\lambda^u(x, t)$ satisfy

$$(u_-(x, t), u_+(x, t)) \in \Phi, \quad \lambda^u(x, t) = \psi(u_-(x, t), u_+(x, t)). \qquad (1.9)$$

□

From (1.6), (1.8), and (1.9) we deduce that if u is an entropy solution then, for every Borel set B,

$$\iint_B \left(\partial_t u + A(u_+) \, \partial_x u \right)$$

$$= \iint_{B \cap \mathcal{C}(u)} \left(\partial_t u + A(u) \, \partial_x u \right) + \int_{B \cap \mathcal{J}(u)} \left(-\lambda^u + A(u_+) \right) (u_+ - u_-) \, d\mathcal{H}_1 \qquad (1.10)$$

$$= \int_{B \cap \mathcal{J}(u)} \left(A(u_+) - \psi(u_-, u_+) \right) (u_+ - u_-) \, d\mathcal{H}_1.$$

REMARK 1.3.
- Roughly speaking, (1.7) guarantees that, as the wave strength $|u_+ - u_-|$ vanishes, the propagating discontinuity connecting u_- to u_+ approaches a trivial solution of the linear hyperbolic system $\partial_t u + A(u_+) \, \partial_x u = 0$ (specifically, $\psi(u_-, u_+) = \lambda_i(u_+)$ and $u_- = u_+ + \alpha \, r_i(u_+)$ for some integer i and real α). Examples of admissible jumps and speeds are discussed below (Section 4).
- The right-continuous representative is chosen for definiteness only. Choosing u_- in (1.8) leads to a completely equivalent definition of solution.

□

Some important consequences of the tame variation condition are now derived.

LEMMA 1.4. *Let $u : \mathbb{R} \times \mathbb{R}_+ \to \mathcal{U}$ be a function with tame variation.*
- *Then, at every point $(x_0, t_0) \in \mathcal{J}(u)$, the L^1 approximate traces $u_-(x_0, t_0)$ and $u_+(x_0, t_0)$ of u considered as a function with bounded variation in two variables coincide with the traces of the one-variable function $x \mapsto u(x, t_0)$ at the point x_0.*
- *Moreover, u is L^1 Lipschitz continuous in time, i.e.,*

$$\|u(t_2) - u(t_1)\|_{L^1(\mathbb{R})} \le M \, |t_2 - t_1|, \quad t_1, t_2 \in \mathbb{R}_+, \qquad (1.11)$$

where the Lipschitz constant is $M := 2\lambda^\infty \, \kappa(\kappa + 1) TV(u(0))$.

PROOF. Given some point $(x_0, t_0) \in \mathcal{J}(u)$ and $\varepsilon > 0$, in view of (1.5) we have

$$\frac{1}{h^2} \int_{t_0-h}^{t_0+h} \int_{x_0-h}^{x_0+h} \left| u(x, t) - \overline{u}(x, t) \right| dx dt \le \varepsilon$$

for all sufficiently small h and, in particular,

$$\frac{1}{h^2} \int_{t_0}^{t_0+h \min(1, 1/(2\lambda^\infty))} \int_{x_0 - \lambda^\infty (t - t_0)}^{x_0 - h + \lambda^\infty (t - t_0)} \left| u(x, t) - u_-(x_0, t_0) \right| dx dt$$

$$+ \frac{1}{h^2} \int_{t_0}^{t_0+h \min(1, 1/(2\lambda^\infty))} \int_{x_0 + \lambda^\infty (t - t_0)}^{x_0 + h - \lambda^\infty (t - t_0)} \left| u(x, t) - u_+(x_0, t_0) \right| dx dt \le \varepsilon. \qquad (1.12)$$

On the other hand, denoting by $\tilde{u}_-$ and $\tilde{u}_+$ the traces of the function $x \mapsto u(x, t_0)$ at the point x_0, we can always choose h sufficiently small so that

$$TV(u(t_0); (x_0 - h, h)) + TV(u(t_0); (x_0, x_0 + h)) \leq \frac{\varepsilon}{\kappa}.$$

(See Figure X-2.)

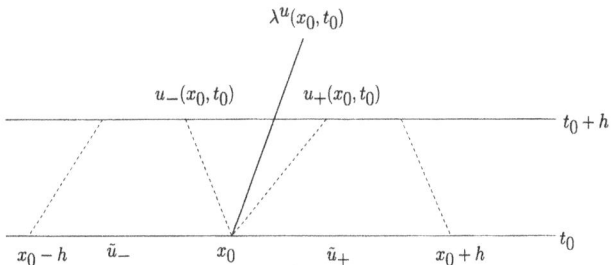

Figure X-2 : The two notions of traces coincide.

With the tame variation condition (1.2) we deduce that

$$\left| u(x, t) - \tilde{u}_- \right| \leq \varepsilon, \quad t \in (t_0, t_0 + h),$$
$$x \in (x_0 - h + \lambda^\infty(t - t_0), x_0 - \lambda^\infty(t - t_0)),$$
$$\left| u(x, t) - \tilde{u}_+ \right| \leq \varepsilon, \quad t \in (t_0, t_0 + h),$$
$$x \in (x_0 + \lambda^\infty(t - t_0), x_0 + h - \lambda^\infty(t - t_0)).$$

Comparing with this pointwise estimate with the integral estimate (1.12), since ε is arbitrary we conclude that

$$\tilde{u}_\pm = u_\pm(x_0, t_0).$$

We now check that the map $t \mapsto u(t)$ is Lipschitz continuous. Consider any interval $[t_1, t_2]$ and set $\tau := (t_2 - t_1) > 0$. At every point $x \in I\!\!R$ we can apply (1.2) by taking $\tilde{\Gamma}$ to be the segment with endpoints $(t_1, x - \tau \lambda^\infty)$, $(t_1, x + \tau \lambda^\infty)$, and Γ to be the segment with endpoints (t_2, x), $(t_1, x + \tau \lambda^\infty)$. This yields

$$\left| u(t_2, x) - u(t_1, x) \right| \leq \left| u(t_2, x) - u(t_1, x + \tau \lambda^\infty) \right| + \left| u(t_1, x + \tau \lambda^\infty) - u(t_1, x) \right|$$
$$\leq (\kappa + 1) \, TV\big(u(t_1); [x - \tau \lambda^\infty, x + \tau \lambda^\infty]\big)$$
$$= G(x + \tau \lambda^\infty) - G(x - \tau \lambda^\infty),$$

where $G(x) := TV\big(u(t_1); (-\infty, x]\big)$. After integration one finds

$$\int_{-\infty}^{\infty} \left| u(t_2, x) - u(t_1, x) \right| dx \leq 2\tau \lambda^\infty (\kappa + 1) \, TV\big(u(t_1)\big)$$

$$\leq 2(t_2 - t_1) \lambda^\infty \kappa(\kappa + 1) \, TV\big(u(0)\big),$$

where we also used (1.3). This completes the proof of Lemma 1.4.   □

2. Tangency property

Our aim is proving that the Cauchy problem associated with (1.1) has at most one solution depending continuously on its initial data. In the present section we derive the following key estimate:

THEOREM 2.1. (Tangency property.) *Consider the hyperbolic system* (1.1) *together with prescribed admissible jumps* Φ *and speeds* ψ *satisfying the property* (1.7). *Let* u *and* v *be two entropy solutions with tame variation. Denote by* $\tilde{\mathcal{I}} \subset I\!\!R_+$ *the projection on the t-axis of the set* $\mathcal{I}(u) \cup \mathcal{I}(v)$ *of all interaction points of* u *or* v. *Then, at each time* $t_0 \notin \tilde{\mathcal{I}}$ *such that*

$$u(t_0) = v(t_0)$$

we have the **tangency property**

$$\lim_{\substack{t \to t_0 \\ t > t_0}} \frac{1}{t - t_0} \big\| u(t) - v(t) \big\|_{L^1(I\!\!R)} = 0. \tag{2.1}$$

REMARK 2.2.

- Since $\mathcal{H}_1(\mathcal{I}(u)) = \mathcal{H}_1(\mathcal{I}(v)) = 0$ the set $\tilde{\mathcal{I}}$ is of Lebesgue measure zero. Since these points of wave interaction in u or in v are excluded in Theorem 2.1, the existence and the uniqueness of the solution of the associated Riemann problem is completely irrelevant to the derivation of (2.1).
- In view of (1.11), the weaker estimate

$$\big\| u(t) - v(t) \big\|_{L^1(I\!\!R)}$$
$$\leq \big\| u(t) - u(t_0) \big\|_{L^1(I\!\!R)} + \big\| u(t_0) - v(t_0) \big\|_{L^1(I\!\!R)} + \big\| v(t) - v(t_0) \big\|_{L^1(I\!\!R)}$$
$$\leq C \left(t - t_0 \right)$$

  is valid for *every* time $t \geq t_0$ at which $u(t_0) = v(t_0)$.

$\square$

The proof of Theorem 2.1 will rely on two technical observations. Lemma 2.3 below provides us with a control of the space averages of a function by its space and time averages. Lemma 2.4 provides us with a control of the L^1 norm of a function from its integrals on arbitrary intervals. The first observation will be used near large discontinuities of the solutions (Step 1 below) while the second one will be useful in regions where the solutions have small oscillations (Step 3).

LEMMA 2.3. *Let* $w = w(\xi, \tau)$ *be a bounded and measurable function satisfying the* L^1 *Lipschitz continuity property*

$$\big\| w(\tau_2) - w(\tau_1) \big\|_{L^1(I\!\!R)} \leq K \left| \tau_2 - \tau_1 \right|, \quad \tau_1, \tau_2 \in I\!\!R_+$$

for some constant $K > 0$. *Then, for each* $h > 0$ *we have*

$$\frac{1}{h} \int_{-h}^{h} \big| w(\xi, h) \big| d\xi \leq \sqrt{2K} \left(\frac{1}{h^2} \int_0^h \int_{-h}^{h} |w| \, d\xi d\tau \right)^{1/2},$$

whenever the right-hand side is less than K.

PROOF. Given $h, h' > 0$ we can write

$$\int_{-h}^{h} |w(\xi, h) - w(\xi, h')| \, d\xi \leq K |h - h'|,$$

thus

$$\frac{1}{h} \int_{-h}^{h} |w(\xi, h)| \, d\xi \leq \frac{1}{h} \int_{-h}^{h} |w(\xi, h')| \, d\xi + \frac{K}{h} |h - h'|.$$

For each $\varepsilon \in (0, 1)$ integrate the above inequality on the interval $h' \in (h - \varepsilon h, h)$:

$$\frac{1}{h} \int_{-h}^{h} |w(h)| \, d\xi \leq \frac{1}{\varepsilon h^2} \int_{h-\varepsilon h}^{h} \int_{-h}^{h} |w| \, d\xi d\tau + \frac{K}{h} \frac{\varepsilon h}{2} \tag{2.2}$$

$$\leq \frac{1}{\varepsilon h^2} \int_{0}^{h} \int_{-h}^{h} |w| \, d\xi d\tau + \frac{K\varepsilon}{2}.$$

The optimal value for ε is the one minimizing the right-hand side of (2.2), that is,

$$\varepsilon_* = \sqrt{\frac{2}{K}} \left(\frac{1}{h^2} \int_{0}^{h} \int_{-h}^{h} |w| \, d\xi d\tau \right)^{1/2}.$$

The condition $\varepsilon < 1$ is equivalent to saying that the right-hand side of the desired inequality is less than K. The conclusion then follows from (2.2). $\qquad \square$

LEMMA 2.4. *For each function w in $L^1((a, b); \mathbb{R}^N)$ we have*

$$\|w\|_{L^1(a,b)} = \sup_{a < z_1 < z_2 < \ldots < b} \sum_{k=1,2,\ldots} \left| \int_{z_k}^{z_{k+1}} w(x) \, dx \right|, \tag{2.3}$$

where the supremum is taken over all finite sub-divisions of the interval (a, b). $\qquad \square$

The formula (2.3) is obvious if w is piecewise constant. The general case follows by approximation (in the L^1 norm) by piecewise constant functions.

PROOF OF THEOREM 2.1. Let u and v be two (Φ, ψ)–admissible entropy solutions of (1.1), having tame variation and satisfying, for some time $t_0 \notin \tilde{\mathcal{I}}$,

$$u(t_0) = v(t_0).$$

Given $\varepsilon > 0$ arbitrary, we want to estimate the integral $\|u(t_0 + h) - v(t_0 + h)\|_{L^1(\mathbb{R})}$ by $O(h \varepsilon)$, which will establish (2.1). We decompose the proof in several steps.

Step 1 : *Estimate near large jumps.* Let $x_1, x_2, \ldots, x_p$ be the finite set of all large jumps in $u(t)$ such that

$$|u_+(x_k, t_0) - u_-(x_k, t_0)| \geq \varepsilon, \quad 1 \leq k \leq p. \tag{2.4}$$

Since $(x_k, t_0) \notin \mathcal{I}(u)$ by assumption, we have $(x_k, t_0) \in \mathcal{J}(u)$. Since u is an entropy solution (see (1.9) in Definition 1.2) the pair $(u_{k-}, u_{k+}) := (u_-(x_k, t_0), u_+(x_k, t_0))$ belongs to the set Φ of admissible jumps and, therefore, the corresponding speed $\psi(u_{k-}, u_{k+})$ is well-defined and coincides with the shock speed in the solution u at that point. Precisely, for all $t \geq t_0$ and all x, we define

$$\bar{u}_k(x, t) = \begin{cases} u_{k-}, & x - x_k < \psi(u_{k-}, u_{k+}) \, (t - t_0), \\ u_{k+}, & x - x_k > \psi(u_{k-}, u_{k+}) \, (t - t_0). \end{cases}$$

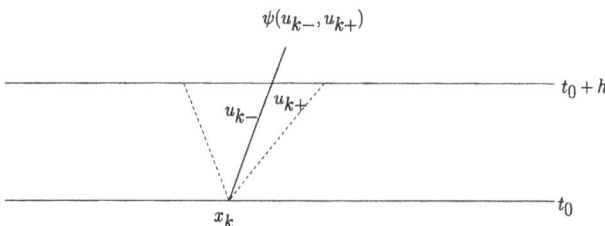

Figure X-3 : Near a point of "large" jump.

According to (1.5) the function $\bar{u}_k$ is a good approximation of the solution u in the neighborhood of the point $(x_k, t_0) \in \mathcal{J}(u)$. Hence, given $\eta > 0$, for all h sufficiently small we have (Figure X-3)

$$\frac{1}{h^2} \int_{t_0}^{t_0+h} \int_{x_k-\lambda^\infty h}^{x_k+\lambda^\infty h} \left| u(x,t) - \bar{u}_k(x,t) \right| dx dt \leq \eta, \quad 1 \leq k \leq p,$$

in which we will choose

$$\eta := \frac{\varepsilon^2}{2K\,p^2},$$

K being a uniform Lipschitz constant for all functions $u - \bar{u}_k$, $k = 1, \dots, p$.

Applying Lemma 2.3 with $w(\xi, \tau) := (u - \bar{u}_k)(x_k + \lambda^\infty \xi, t_0 + \tau)$, we deduce that

$$\frac{1}{h} \int_{x_k-\lambda^\infty h}^{x_k+\lambda^\infty h} \left| u(x, t_0 + h) - \bar{u}_k(x, t_0 + h) \right| dx \leq \sqrt{2K\eta} = \frac{\varepsilon}{p} \qquad (2.5)$$

for $1 \leq k \leq p$ and all h sufficiently small. (One can always take $\varepsilon/p < K$ so that the assumption in Lemma 2.1 holds.) Since $v(t_0) = u(t_0)$, we can set $\bar{v}_k := \bar{u}_k$ and the function v satisfies a completely analogous estimate obtained by replacing u and $\bar{u}_k$ by v and $\bar{v}_k$ respectively. Hence, by (2.5) we arrive at

$$\begin{aligned}
&\frac{1}{h} \sum_{k=1}^{p} \int_{x_k-\lambda^\infty h}^{x_k+\lambda^\infty h} \left| u(x, t_0 + h) - v(x, t_0 + h) \right| dx \\
&\leq \frac{1}{h} \sum_{k=1}^{p} \int_{x_k-\lambda^\infty h}^{x_k+\lambda^\infty h} \left(\left| (u - \bar{u}_k)(x, t_0 + h) \right| + \left| (\bar{v}_k - v)(x, t_0 + h) \right| \right) dx \\
&\leq 2\varepsilon
\end{aligned} \qquad (2.6)$$

for all h sufficiently small.

Step 2 : *Using the tame variation property.* Choose $\rho = \rho(\varepsilon) > 0$ such that $2\rho < \min_{k \neq m} |x_k - x_m|$ and for every interval (a, b)

$$TV\big(u(t_0); (a, b)\big) \leq \varepsilon \quad \text{when } b - a < 2\,\rho \text{ and } (a, b) \cap \{x_1, \dots, x_p\} = \emptyset. \qquad (2.7)$$

Next, select points y_l $(l = 1, 2, \dots)$ to obtain a *locally finite* covering of

$$\mathbb{R} \setminus \{x_1, \dots, x_p\}$$

by intervals of the form $(y_l - \rho, y_l + \rho)$. We can always assume that each point of the x-axis belongs to two such intervals, at most.

To describe the domain of determinacy of the interval $I_l(0) = (y_l - \rho, y_l + \rho)$,

$$\Omega_l := \Big\{ (x, t) \,/\, x \in \big(y_l - \rho + \lambda^\infty(t - t_0), \, y_l + \rho - \lambda^\infty(t - t_0) \big), \quad t_0 \le t \le t_0 + h \Big\},$$

we introduce the intervals

$$I_l(h) = \big(y_l - \rho + \lambda^\infty h, \, y_l + \rho - \lambda^\infty h \big).$$

Clearly, (2.7) together with the tame variation property (1.2) imply that the oscillation of u is small in each interval $I_l(h)$, i.e.,

$$\big| u_+(x, t_0 + h) - u_+(y_l, t_0) \big| \le (2\kappa + 1)\,\varepsilon, \quad x \in I_l(h),\ l = 1, 2, \dots \tag{2.8}$$

for h sufficiently small. Of course, the same estimate is satisfied by the function v.

Step 3 : *Estimate in regions of small oscillations.* We define now an approximation adapted to points of approximate continuity of u and to points where the jump in $u(t_0)$ is less than ε. Fix some index $l \in \{1, 2, \dots\}$ and, for simplicity in the notation, set

$$\underline{A} := A\big(u_+(y_l, t_0)\big), \quad \underline{\lambda}_j := \lambda_j\big(u_+(y_l, t_0)\big), \quad \underline{l}_j := l_j\big(u_+(y_l, t_0)\big).$$

Let $\underline{u} = \underline{u}(x, t)$ be the solution of the linear hyperbolic problem

$$\begin{aligned}
\partial_t \underline{u} + \underline{A}\,\partial_x \underline{u} &= 0, \quad t \ge t_0, \\
\underline{u}(t_0) &= u(t_0).
\end{aligned} \tag{2.9}$$

On one hand, we can multiply (2.9) by each $\underline{l}_j$ and obtain N decoupled equations for the characteristic components $\underline{l}_j\,\underline{u}$

$$\partial_t \big(\underline{l}_j\,\underline{u} \big) + \underline{\lambda}_j\,\partial_x \big(\underline{l}_j\,\underline{u} \big) = 0, \quad 1 \le j \le N. \tag{2.10}$$

On the other hand, the solution u of (1.1) satisfies an equation similar to (2.10), but containing a source-term. Namely, since u is a (Φ, ψ)-admissible solution, according to (1.10) it solves the equation

$$\partial_t u + A(u_+)\,\partial_x u = \mu, \tag{2.11}$$

where μ is the measure concentrated on the set $\mathcal{J}(u)$ and given by

$$\mu(B) := \int_{B \cap \mathcal{J}(u)} \big(A(u_+) - \psi(u_-, u_+) \big)\,(u_+ - u_-)\,d\mathcal{H}_1 \tag{2.12}$$

for every Borel set $B \subset \mathbb{R} \times \mathbb{R}_+$. In view of (2.11) we have

$$\partial_t u + \underline{A}\,\partial_x u = \big(\underline{A} - A(u_+) \big)\,\partial_x u + \mu,$$

and so, after multiplication by $\underline{l}_j$ $(1 \le j \le N)$,

$$\partial_t \big(\underline{l}_j\,u \big) + \underline{\lambda}_j\,\partial_x \big(\underline{l}_j\,u \big) = \underline{l}_j\,\big(\underline{A} - A(u_+) \big)\,\partial_x u + \underline{l}_j\,\mu, \tag{2.13}$$

which resembles (2.10). Combining (2.10) and (2.13) we arrive at

$$\partial_t \big(\underline{l}_j\,(u - \underline{u}) \big) + \underline{\lambda}_j\,\partial_x \big(\underline{l}_j\,(u - \underline{u}) \big) = \underline{l}_j\,\big(\underline{A} - A(u_+) \big)\,\partial_x u + \underline{l}_j\,\mu. \tag{2.14}$$

Since the matrix-valued function A depends Lipschitz continuously upon its argument, the coefficient $\underline{A} - A(u_+)$ in the right-hand side of (2.14) satisfies

$$
\begin{aligned}
\left| (\underline{A} - A(u_+))(x, t_0 + h) \right| &\leq C \left| u_+(y_l, t_0) - u_+(x, t_0 + h) \right| \\
&\leq C\,(2\kappa + 1)\,\varepsilon, \quad x \in I_l(h),\, l = 1, 2, \ldots
\end{aligned}
\tag{2.15}
$$

for all small h, where we have used the estimate (2.8).

To estimate the measure μ in the right-hand side of (2.14) we use the consistency property (1.7) together with (2.8). For every region with polygonal boundaries (for simplicity) $B \subset \Omega_l$ we find that

$$
\begin{aligned}
|\mu(B)| &\leq \int_{B \cap \mathcal{J}(u)} \eta(|u_+ - u_-|)\, |u_+ - u_-|\, d\mathcal{H}_1 \\
&\leq 2\eta(\varepsilon)\,(2\kappa + 1)\,\varepsilon \int_{B \cap \mathcal{J}(u)} |u_+ - u_-|\, d\mathcal{H}_1 \\
&\leq C\,\eta(\varepsilon) \int_{t_0}^{t_0 + h} TV\big(u(t); (B)_t\big)\, dt,
\end{aligned}
\tag{2.16}
$$

where $(B)_t := \{ x \,/\, (x, t) \in B \}$. (See (A.11) in the appendix.)

We will now integrate (2.14) on some well-chosen sub-regions of the domain of determinacy Ω_l of the interval $I_l(0)$. For each $j = 1, \ldots, N$ and each x', x'' in $I_l(h)$ we consider (Figure X-4)

$$
\Omega^j_{x', x''} := \Big\{ (x, t) \,/\, x' + (t - t_0 - h)\underline{\lambda}_j \leq x \leq x'' + (t - t_0 - h)\underline{\lambda}_j, \quad t_0 \leq t \leq t_0 + h \Big\},
$$

which is a subset of Ω_l. By using Green's formula for functions with bounded variation, since $\underline{u}(t_0) = u(t_0)$ we obtain

$$
\int_{t_0}^{t_0 + h} \int_{x' + (t - t_0 - h)\underline{\lambda}_j}^{x'' + (t - t_0 - h)\underline{\lambda}_j} \partial_t \big(\underline{l}_j\,(u - \underline{u}) \big) + \underline{\lambda}_j\, \partial_x \big(\underline{l}_j\,(u - \underline{u}) \big) = \int_{x'}^{x''} \underline{l}_j\,(u - \underline{u})(x, t_0 + h)\, dx.
$$

Hence, integrating (2.14) over $\Omega^j_{x', x''}$ we get

$$
\int_{x'}^{x''} \underline{l}_j\,(u - \underline{u})(x, t_0 + h)\, dx = \int_{t_0}^{t_0 + h} \int_{x' + (t - t_0 - h)\underline{\lambda}_j}^{x'' + (t - t_0 - h)\underline{\lambda}_j} \underline{l}_j\,(\underline{A} - A(u_+))\, \partial_x u + \underline{l}_j\, \mu.
$$

Using (2.15) and (2.16) we arrive at the estimate

$$
\begin{aligned}
&\left| \int_{x'}^{x''} \underline{l}_j\,(u - \underline{u})(y, t_0 + h)\, dy \right| \\
&\qquad \leq C\,\eta(\varepsilon) \int_{t_0}^{t_0 + h} TV\big(u(t); (x' + (t - t_0 - h)\underline{\lambda}_j, x'' + (t - t_0 - h)\underline{\lambda}_j)\big)\, dt.
\end{aligned}
\tag{2.17}
$$

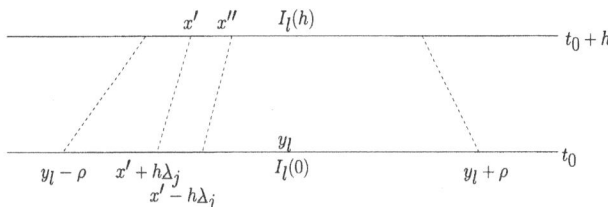

Figure X-4 : In a region with small oscillations.

Next, in view of Lemma 2.4 and summing (2.17) over finitely many intervals we find

$$\int_{I_l(h)} \left| \underline{l}_j \, (u - \underline{u})(x, t_0 + h) \right| dx \leq C \, \eta(\varepsilon) \int_{t_0}^{t_0+h} TV\big(u(t); I_l(t - t_0)\big) dt.$$

The same estimate holds for the solution v as well and, therefore,

$$\int_{I_l(h)} \left| \underline{l}_j \, (u - v)(x, t_0 + h) \right| dx \leq C \eta(\varepsilon) \int_{t_0}^{t_0+h} \Big(TV\big(u(t); I_l(t - t_0)\big) + TV\big(v(t); I_l(t - t_0)\big) \Big) dt$$

for each $j = 1, \ldots, N$. Since $|u - v| \leq C \sum_{j=1}^{N} \left| \underline{l}_j \, (u - v) \right|$ we conclude that

$$\int_{I_l(h)} \left| (u - v)(x, t_0 + h) \right| dx$$
$$\leq C \, \eta(\varepsilon) \int_{t_0}^{t_0+h} \big(TV(u(t); I_l(t - t_0)) + TV(v(t); I_l(t - t_0)) \big) \, dt. \tag{2.18}$$

Step 4 : *Conclusion.* Summing up the estimates (2.18) for all $l = 1, 2, \ldots$ we obtain

$$\sum_{l=1,2,\ldots} \int_{y_l - \rho + \lambda^\infty h}^{y_l + \rho - \lambda^\infty h} \left| (u - v)(x, t_0 + h) \right| dx \leq O(\varepsilon) \int_{t_0}^{t_0+h} \big(TV(u(t)) + TV(v(t)) \big) \, dt$$

$$\leq \big(O(\varepsilon) + \eta(\varepsilon) \big) \, h, \tag{2.19}$$

since two intervals, at most, may overlap and the function $TV(u(.)) + TV(v(.))$ is uniformly bounded. Finally, since the intervals $(y_l - \rho + \lambda^\infty h, y_l + \rho - \lambda^\infty h)$ and the intervals $(x_k - \lambda^\infty h, x_k + \lambda^\infty h)$ form a covering of the real line, we can combine the main two estimates (2.6) and (2.19) and conclude that for each $\varepsilon > 0$ and all sufficiently small h

$$\frac{1}{h} \int_{I\!R} \left| (u - v)(x, t_0 + h) \right| dx \leq O(\varepsilon) + \eta(\varepsilon),$$

hence (2.1) holds. This completes the proof of Theorem 2.1. □

3. Uniqueness theory

We introduce the following notion of semi-group.

DEFINITION 3.1. (General concept of semi-group of entropy solutions.) Consider the hyperbolic system (1.1) together with admissible jumps Φ and speeds ψ satisfying the property (1.7). By definition, a **semi-group of entropy solutions** is a mapping $S : \mathcal{K} \times [0, \infty) \to \mathcal{K}$ defined on a non-empty **subset $\mathcal{K}$ of functions with bounded variation** on $I\!R$ such that the following three properties hold:

- **Semi-group property :** for all $t_1, t_2 \geq 0$ and $u_0 \in \mathcal{K}$, we have $S(t_1)u_0 \in \mathcal{K}$ and $S(t_2) \circ S(t_1)u_0 = S(t_2 + t_1)u_0$.
- **Continuous dependence:** For some fixed constant $K > 0$ and for all $u_0, v_0 \in \mathcal{K}$ and $t \geq 0$,

$$\|S(t)u_0 - S(t)v_0\|_{L^1(I\!R)} \leq K \|u_0 - v_0\|_{L^1(I\!R)}. \tag{3.1}$$

- **Entropy solution:** For each function $u_0 \in \mathcal{K}$ the map $t \mapsto S(t)u_0$ is an entropy solution with tame variation.

□

Of course, since functions of tame variation are Lipschitz continuous in time by Lemma 1.4, a semi-group in the sense of Definition 3.1 satisfies actually

$$\|S(t_2)u_0 - S(t_1)v_0\|_{L^1(I\!R)} \leq K \|u_0 - v_0\|_{L^1(I\!R)} + K |t_2 - t_1| \tag{3.1'}$$

for all $u_0, v_0 \in \mathcal{K}$ and $t_1, t_2 \geq 0$, where $K' := \max\Big(K, 2\lambda^\infty \kappa(\kappa + 1)TV(u(0))\Big)$.

THEOREM 3.2. (Uniqueness of entropy solutions.) *Consider the hyperbolic system (1.1) together with a pair of admissible jumps Φ and speeds ψ satisfying the property (1.7). Assume that there exists a semi-group $S : \mathcal{K} \times [0, \infty) \to \mathcal{K}$ of entropy solutions with tame variation satisfying the following:*
> **Consistency property with single jumps of (Φ, ψ):** *If a function $v = v(x, t)$ is made of a single (admissible) jump discontinuity $(v_-, v_+) \in \Phi$ propagating with the speed $\psi(v_-, v_+)$, then $v(0) \in \mathcal{K}$ and*
> $$v(t) = S(t)v(0), \quad t \geq 0.$$

Then, if u is an entropy solution with tame variation assuming the initial data $u_0 \in \mathcal{K}$ at time $t = 0$, we have
$$u(t) = S(t)u_0, \quad t \geq 0. \tag{3.2}$$
*In particular, there exists a **unique entropy solution with tame variation** of the Cauchy problem associated with (1.1).*

It is clear that the consistency property above is necessary, for otherwise one could find two distinct solutions starting with the same initial data and the conclusion of Theorem 3.2 would obviously fail.

On one hand, for the consistency property to hold, the set Φ must be "sufficiently small", so that any initial data made of a single admissible jump $(u_-, u_+) \in \Phi$ cannot be decomposed (as time evolves) in two (or more) admissible waves. Indeed, suppose there would exist a semi-group of admissible solutions satisfying the consistency property with single admissible jumps and suppose also that

$$(v_-, v_+), (v_-, v_*), (v_*, v_+) \in \Phi,$$

where the three constant states v_-, v_*, and v_+ are distinct and their speeds are ordered:

$$\psi(v_-, v_+) < \psi(v_-, v_*) < \psi(v_*, v_+).$$

Consider the single jump solution

$$v_1(x, t) = \begin{cases} v_-, & x < \psi(v_-, v_+)\, t, \\ v_+, & x > \psi(v_-, v_+)\, t, \end{cases}$$

together with the sequence

$$v_2^\varepsilon(x, t) = \begin{cases} v_-, & x < \psi(v_-, v_*)\, t, \\ v_*, & \psi(v_-, v_*)\, t < x < \varepsilon + \psi(v_*, v_+)\, t, \\ v_+, & x > \varepsilon + \psi(v_*, v_+)\, t. \end{cases}$$

The solution v_2^ε contains two propagating discontinuities and converges in the L^1 norm toward

$$v_2(x, t) = \begin{cases} v_-, & x < \psi(v_-, v_*)\, t, \\ v_*, & \psi(v_-, v_*)\, t < x < \psi(v_*, v_+)\, t, \\ v_+, & x > \psi(v_*, v_+)\, t. \end{cases}$$

According to the consistency property with admissible jumps we have

$$S(t)v_1(0) = v_1(t), \quad S(t)v_2^\varepsilon(0) = v_2^\varepsilon(t).$$

This leads to a contradiction since the semigroup is L^1 continuous, and $v_1(0) = v_2(0)$ but $v_1(t) \neq v_2(t)$ for $t > 0$.

On the other hand, to establish the actual existence of a semi-group of admissible solutions, the set Φ should be "sufficiently large" to allow one to construct the solution of the Riemann problem, at least. As we will check later in Section 4, for several classes of systems and sets of interest, our results in previous chapters of these notes do imply the existence of a semi-group satisfying the properties in Definition 3.1, allowing us to complete the *well-posedness theory*.

The proof of Theorem 3.2 is based on Theorem 2.1 together with the following observation.

LEMMA 3.3. *For every $u_0 \in \mathcal{K}$ and every time-dependent function with bounded variation $u : \mathbb{R} \times \mathbb{R}_+ \to \mathcal{U}$ satisfying the initial condition $u(0) = u_0$ the semi-group of solutions satisfies the estimate*

$$\big\| u(T) - S(T)u_0 \big\|_{L^1(\mathbb{R})} \leq K \int_0^T \liminf_{h \to 0} \frac{1}{h} \big\| u(t + h) - S(h)u(t) \big\|_{L^1(\mathbb{R})}\, dt. \qquad (3.3)$$

PROOF. Consider the (bounded) function

$$L(t) := \liminf_{h \to 0} \frac{1}{h} \big\| u(t + h) - S(h)u(t) \big\|_{L^1(\mathbb{R})}$$

together with the (Lipschitz continuous) functions

$$M(t) := \big\| S(T - t)u(t) - S(T)u(0) \big\|_{L^1(\mathbb{R})}, \quad N(t) := M(t) - K \int_0^t L(s)\, ds.$$

We will show that $N'(t) \leq 0$ for almost every t, which implies the desired inequality

$$N(t) \leq N(0) = 0.$$

By standard regularity theorems there exists a set $\mathcal{Z}$ of zero Lebesgue measure such that for all $t \notin \mathcal{Z}$ the functions M and N are differentiable at the point t, while t is a Lebesgue point of the function L. Hence, we have

$$N'(t) = M'(t) - K\, L(t). \tag{3.4}$$

On the other hand by definition we have

$$
\begin{aligned}
M(t+h) &- M(t) \\
&= \left\| S(T-t-h)u(t+h) - S(T)u(0) \right\|_{L^1(\mathbb{R})} - \left\| S(T-t)u(t) - S(T)u(0) \right\|_{L^1(\mathbb{R})} \\
&\leq \left\| S(T-t-h)u(t+h) - S(T-t)u(t) \right\|_{L^1(\mathbb{R})} \\
&= \left\| S(T-t-h)u(t+h) - S(T-t-h)\, S(h)u(t) \right\|_{L^1(\mathbb{R})} \\
&\leq K \left\| u(t+h) - S(h)u(t) \right\|_{L^1(\mathbb{R})}.
\end{aligned}
$$

Dividing by h and letting $h \to 0$ we find

$$M'(t) \leq K\, L(t). \tag{3.5}$$

The conclusion follows from (3.4) and (3.5).   □

PROOF OF THEOREM 3.2. In view of (3.3) we see that (2.1) precisely implies that the integrand in the right-hand side of (3.3) vanishes almost everywhere, *provided* Theorem 2.1 can be applied.

In fact, to complete the proof we will need a slightly generalized version of Theorem 2.1. We shall say that a point (x,t_0) is a **forward regular point for u** if either it is a (Lebesgue) point of approximate continuity for u in the set $\mathbb{R} \times [t_0, +\infty)$, or else there exist some traces $u_\pm(x,t_0)$ and a speed $\lambda^u(x,t_0)$ such that

$$\lim_{\rho \to 0} \frac{1}{\rho^2} \int_{t_0}^{t_0+\rho} \int_{x-\rho}^{x+\rho} \left| u(y,t) - \overline{u}(y,t) \right| dy\,dt = 0. \tag{3.6}$$

Next, observe that the values $u(t)$ and $v(t)$ with $t \geq t_0$, only, are relevant in the statement of Theorem 2.1. Indeed, consider two functions u and v which are defined and are admissible solutions on the set $\mathbb{R} \times [t_0, +\infty)$, such that for every $x \in \mathbb{R}$ the point (x,t_0) is a forward regular point of both u and v. Given a point of jump (x,t_0), we have $\big(u_-(x,t_0), u_+(x,t_0)\big) \in \Phi$ and $\lambda^u(x,t_0) = \psi\big(u_-(x,t_0), u_+(x,t_0)\big)$. Then, it is clear from the proof of Theorem 2.1 that (2.1) still holds.

Let u be a (Φ, ψ)-admissible solution of (1.1) assuming some initial data u_0 at time $t = 0$. We want to show that $u(t)$ coincides with $w(t) := S(t)u_0$ for all $t \geq 0$. Consider any $t_0 \geq 0$ with $t_0 \notin \mathcal{I}(u)$ which is also is a Lebesgue point of the (bounded) function $t \mapsto TV\big(u(t)\big)$. Define

$$v(t) := S(t-t_0)u(t_0) \quad \text{for } t \geq t_0.$$

We claim that, for every $x \in \mathbb{R}$, the point (x,t_0) is a *forward regular point* of v. Indeed, consider any point of continuity x of the function $u(t_0)$. The tame variation property (1.2) implies easily

$$\lim_{\substack{y \to x,\, t \to t_0 \\ t > t_0}} \left| v(y,t) - u(x,t_0) \right| = 0$$

and that (x,t_0) is a point of approximate continuity for the function v.

Next, consider any point of jump x of the function $u(t_0)$. Since u is an admissible solution the limits $u_\pm := u_\pm(x, t_0)$ determine a pair in the set Φ. Call $\overline{u}$ the jump propagating at the speed $\psi(u_-, u_+)$. According to the *consistency condition* assumed in Theorem 3.2 we have

$$S(h)\overline{u}(t_0) = \overline{u}(t_0 + h) \quad \text{for all } h.$$

Using the Lipschitz continuity of the semi-group, for $\rho = \lambda^\infty h$ we have

$$\frac{1}{h\rho} \int_{t_0}^{t_0+h} \int_{x-\rho+\lambda^\infty(t-t_0)}^{x+\rho-\lambda^\infty(t-t_0)} \left| v(y, t) - \overline{u}(y, t) \right| dy dt$$

$$= \frac{1}{h\rho} \int_{t_0}^{t_0+h} \int_{x-\rho+\lambda^\infty(t-t_0)}^{x+\rho-\lambda^\infty(t-t_0)} \left| S(t-t_0)u(t_0) - S(t-t_0)\overline{u}(t_0) \right| dy dt \qquad (3.7)$$

$$\leq \frac{K}{\rho} \int_{x-\rho}^{x+\rho} \left| u(t_0) - \overline{u}(t_0) \right| dy \to 0,$$

since x is a point of jump of the function $u(t_0)$. Thus (x, t_0) is a point of approximate jump for the function v.

This completes the proof that (x, t_0) is a forward regular point of v for every $x \in I\!R$. In view of the preliminary observation above the conclusion in Theorem 2.1 holds for the two solutions u and v at time t_0:

$$\lim_{\substack{h \to 0 \\ h > 0}} \frac{1}{h} \left\| u(t_0 + h) - S(h)u(t_0) \right\|_{L^1(I\!R)} = \lim_{\substack{h \to 0 \\ h > 0}} \frac{1}{h} \left\| u(t_0 + h) - v(t_0 + h) \right\|_{L^1(I\!R)} = 0.$$

This proves that the integrand on the right-hand side of (3.3) vanishes at almost every t. Thus $u(T) := S(T)u_0$ for every $T > 0$ which completes the proof of Theorem 3.2.

$\square$

4. Applications

This section contains some important consequences to the uniqueness theory presented in Section 3.

For strictly hyperbolic systems of conservation laws

$$\partial_t u + \partial_x f(u) = 0, \quad u = u(x, t) \in \mathcal{U}, \qquad (4.1)$$

it is natural to define the admissible speeds ψ^{RH} from the standard Rankine-Hugoniot relation, that is,

$$-\psi^{RH}(u_-, u_+)(u_+ - u_-) + f(u_+) - f(u_-) = 0, \quad u_-, u_+ \in \mathcal{U}. \qquad (4.2)$$

The second ingredient in Definition 1.2, the set Φ, determines which discontinuities are admissible. Classical solutions are recovered by setting

$$\Phi^c := \big\{ (u_-, u_+) \text{ satisfies Rankine-Hugoniot relations and Liu entropy criterion} \big\}.$$

Another choice is to include, in the set Φ, jumps violating Liu criterion in order to recover nonclassical entropy solutions selected by a kinetic relation (applied to those characteristic fields which are not genuinely nonlinear, only), more precisely:

$$\Phi^{nc} := \big\{ (u_-, u_+) \text{ satisfies Rankine-Hugoniot relations,}$$

$$\text{a single entropy inequality, and a kinetic relation} \big\}.$$

So, to each of the classical or nonclassical solvers constructed in previous chapters we can associate a set of admissible jumps, Φ^c or Φ^{nc}, while defining ψ^{RH} by the Rankine-Hugoniot relation.

Relying on Theorem IX-2.3 for classical entropy solutions and assuming the validity of Theorem IX-4.1 concerning nonclassical entropy solutions, we arrive at the following uniqueness result, which is a corollary of Theorem 3.2.

THEOREM 4.1. (Uniqueness of classical and nonclassical entropy solutions.) *Consider the strictly hyperbolic system of conservation laws* (4.1) *under the following assumptions: The set* $\mathcal{U} \subset \mathbb{R}^N$ *is a ball with sufficiently small radius and each characteristic field of* (4.1) *is either genuinely nonlinear or concave-convex. To each concave-convex field let us associate a kinetic relation as was done in Chapter VI. Let* Φ *be the corresponding family of classical or nonclassical shock waves occurring in the classical or nonclassical Riemann solvers described in Chapter VI, respectively. Let* $\psi = \psi^{RH}$ *be the speed given by the Rankine-Hugoniot relation* (4.2).

- *Then, there exists a unique semi-group of* (Φ, ψ)–*admissible entropy solutions.*
- *Any two* (Φ, ψ)–*admissible entropy solutions with tame variation satisfy the* L^1 *continuous dependence property*

$$\|v(t_2) - u(t_2)\|_{L^1(\mathbb{R})} \le K \, \|v(t_1) - u(t_1)\|_{L^1(\mathbb{R})}, \quad 0 \le t_1 \le t_2, \qquad (4.3)$$

for some fixed constant $K > 0$.

PROOF. Let u_0 and v_0 be some initial data and choose piecewise constant approximations $u^h(0)$ and $v^h(0)$, with uniformly bounded total variation and converging pointwise toward u_0 and v_0, respectively. Consider the approximate solutions u^h and v^h constructed by wave front tracking (Theorems IX-2.3 or IX-4.1) from from the initial data $u^h(0)$ and $v^h(0)$. We rely on the continuous dependence estimate (Chapter IX)

$$\|v^h(t_2) - u^h(t_2)\|_{L^1(\mathbb{R})} \le K \, \|v^h(t_1) - u^h(t_1)\|_{L^1(\mathbb{R})} + o(h), \quad t_1, t_2 \in \mathbb{R}_+, \quad (4.4)$$

where $o(h) \to 0$ when $h \to 0$. Extracting a subsequence if necessary we define

$$S(t) \, u_0 := \lim_{h \to 0} u^h(t), \quad t \ge 0. \qquad (4.5)$$

The notation makes sense since the function $\lim_{h \to 0} u^h$ is independent of the particular discretization of the initial data and the particular subsequence. Indeed, if $u^h(0) \to u_0$ and $v^h(0) \to u_0$ and (for some subsequence) $u^h \to u$ and $v^h \to v$, we have the estimate (4.4) with, say, $s = 0$, and after passing to the limit

$$\|v(t) - u(t)\|_{L^1(\mathbb{R})} \le K \, \|u_0 - u_0\|_{L^1(\mathbb{R})} = 0, \quad t \in \mathbb{R}_+,$$

which implies that $v \equiv u$. By a very similar argument one can check that the formula (4.5) defines a semi-group, that is, the condition 1 in Definition 3.1 holds.

Furthermore, the solutions are known to remain uniformly bounded in the total variation norm. It is not difficult to return to the argument of proof and, applying the same arguments but now *along space-like segments*, to check the tame variation estimate (1.2) for u^h, where the constant κ is independent of h. Fix a time $t \ge 0$ and select piecewise constant approximations $x \mapsto u^h(x,t)$ such that

$$TV(u^h(t); (a,b)) \to TV(u(t); (a,b))$$

for every interval (a, b). Using the same notation as in Definition 1.1 the right-hand side of the inequality

$$TV(u^h; \Gamma) \le \kappa\, TV(u^h; \tilde{\Gamma}) \tag{4.6}$$

converges to $TV(u; \tilde{\Gamma})$ while (by the lower semi-continuity property of the total variation) the limit of the left-hand side is greater than or equal to $TV(u; \Gamma)$. This proves the tame variation estimate (1.2) for the semi-group (4.5).

The first condition in Definition 1.2 is easy since (4.1) is conservative. The second condition was the subject of Sections VII-4 and VIII-4: the approximate solutions converge in a pointwise sense near each discontinuity and the traces of the solution belong the set Φ of admissible jumps and propagate with the speed given by the Rankine-Hugoniot condition. This completes the proof that (4.5) defines a semi-group of admissible solutions in the sense of Definition 3.1. The second property stated in the theorem is immediate from Theorem 3.2. $\qquad\square$

The framework presented in this chapter simplifies if one applies it to genuinely nonlinear systems in conservative form.

DEFINITION 4.2. (Concept of entropy solution of genuinely nonlinear systems.) Consider a system of conservation laws (4.1) whose characteristic fields are genuinely nonlinear and endowed with a strictly convex entropy pair (U, F). Then, a function u with tame variation is called an **entropy solution** of (4.1) if the conservation laws (4.1) and the entropy inequality

$$\partial_t U(u) + \partial_x F(u) \le 0$$

are satisfied in the weak sense. $\qquad\square$

The conditions in Definition 4.1 are equivalent to saying

$$\iint_B \partial_t u + \partial_x f(u) = 0 \quad \text{for every Borel set } B \tag{4.7}$$

and

$$\iint_B \partial_t U(u) + \partial_x F(u) \le 0 \quad \text{for every Borel set } B. \tag{4.8}$$

Under the assumption that (4.1) has only genuinely nonlinear fields, Definition 4.2 is fully equivalent to Definition 1.2. For instance, let us show that a solution in the sense of Definition 4.2 is also a solution in the sense of Definition 1.2. One one hand, from (4.7) we deduce that

$$\iint_B \partial_t u + A(u_+)\, \partial_x u = \iint_B \partial_t u + \partial_x f(u) = 0 \quad \text{for every Borel set } B \subset \mathcal{C}(u).$$

On the other hand, from (4.7) and (4.8) it follows that, at each $(x, t) \in \mathcal{J}(u)$,

$$-\lambda^u(x, t)\left(u_+(x, t) - u_-(x, t)\right) + f(u_+(x, t)) - f(u_-(x, t)) = 0$$

and

$$-\lambda^u(x, t)\left(U(u_+(x, t)) - U(u_-(x, t))\right) + F(u_+(x, t)) - F(u_-(x, t)) \le 0,$$

which, by definition of the families Φ and ψ, is equivalent to (1.9).

Theorem 4.1 is immediately restated as follows.

THEOREM 4.3. (Uniqueness of entropy solutions of genuinely nonlinear systems.) *Consider a system of conservation laws (4.1) whose characteristic fields are genuinely nonlinear and endowed with a strictly convex entropy pair (U, F). Restrict attention to solution taking their values in a ball $\mathcal{U}$ with sufficiently small radius.*

- *Then, there exists a (unique) semi-group of entropy solutions in the sense of Definition 4.2.*
- *Any two entropy solutions with tame variation satisfy the L^1 continuous dependence property (4.3).*

REMARK 4.4. The following example of admissible speeds ψ further illustrates the interest of the general framework proposed in this chapter beyond the class of conservative systems. For simplicity, suppose that $N = 1$ and consider the scalar equation

$$\partial_t u + a(u)\, \partial_x u = 0. \tag{4.9}$$

This equation can be written in conservative form, namely

$$\partial_t u + \partial_x f(u) = 0, \quad f(u) = \int^u a(v)\, dv. \tag{4.10}$$

One may define the speed ψ in agreement with the Rankine-Hugoniot relation associated with the conservative form (4.10), that is,

$$\psi(u_-, u_+) := \frac{f(u_+) - f(u_-)}{u_+ - u_-}. \tag{4.11}$$

However, this choice can be regarded to be somehow arbitrary if no conservative form of (4.9) were specified in the first place.

One could set instead

$$\psi(u_-, u_+) := \frac{h(u_+) - h(u_-)}{g(u_+) - g(u_-)}, \tag{4.12}$$

where the functions $g, h : I\!R \to I\!R$ are chosen so that

$$g'(u) > 0, \quad h'(u) = g'(u)\, a(u), \quad u \in I\!R. \tag{4.13}$$

Both choices (4.11) and (4.12) satisfy the consistency property (1.7). As a matter of fact, the speed (4.12) corresponds also to the standard Rankine-Hugoniot relation, but for *another* conservative form of (4.9), i.e.,

$$\partial_t g(u) + \partial_x h(u) = 0. \tag{4.14}$$

Of course, the admissible speeds *need not* correspond to a conservative form of (4.9). In particular, it need not be a symmetric function in (u_-, u_+). For example, suppose we are given two conservative forms of (4.9), like (4.14), associated with two pairs (g_1, h_1), (g_2, h_2) of conservative variables and flux-functions satisfying (4.13). An admissible speed can be defined by

$$\psi(u_-, u_+) := \begin{cases} \dfrac{h_1(u_+) - h_1(u_-)}{g_1(u_+) - g_1(u_-)}, & u_- < u_+, \\[2ex] \dfrac{h_2(u_+) - h_2(u_-)}{g_2(u_+) - g_2(u_-)}, & u_- > u_+. \end{cases}$$

$\square$

APPENDIX

FUNCTIONS WITH BOUNDED VARIATION

We first introduce some general notations of use throughout these lecture notes. Given an open subset $\Omega \subset \mathbb{R}^n$ and $p \in [1, \infty]$, we denote by $L^p(\Omega)$ the Banach space of all Lebesgue measurable functions whose p-th power is integrable on Ω if $p < \infty$ or which are bounded on Ω if $p = \infty$. The corresponding norm is denoted by $\|.\|_{L^p(\Omega)}$. For each integer $m \in [0, \infty]$, we denote by $\mathcal{C}^m(\Omega)$ the space of all continuous functions whose k-th derivatives ($k \leq p$) exist and are continuous on Ω. The corresponding sup-norm is denoted by $\|.\|_{\mathcal{C}^m(\Omega)}$ whenever it is bounded. The subspace of all functions with compact support is denoted by $\mathcal{C}_c^m(\Omega)$. Similarly, for each real $T \in (0, \infty)$, we can define the space $\mathcal{C}_c^m(\Omega \times [0, T])$ of all functions $v = v(x, t)$, $x \in \Omega$, $t \in [0, T]$, such that for $k \leq m$ all k-th derivatives of v exist and are continuous on $\Omega \times [0, T]$, while v is compactly supported in $\Omega \times [0, T]$. When it is necessary to specify the range of the functions, say $\mathcal{U} \subset \mathbb{R}^N$, we write $L^p(\Omega; \mathcal{U})$, $\mathcal{C}^p(\Omega; \mathcal{U})$, etc. We also set $\mathcal{C}(\Omega) := \mathcal{C}^0(\Omega)$, etc.

Given some bounded or unbounded interval (a, b), a $\mathbb{R}^N$-valued **bounded measure** is a real-valued, bounded linear map μ defined on $\mathcal{C}_c((a, b); \mathbb{R}^N)$. The associated **variation measure** $|\mu|$ is defined by

$$|\mu|\{(a', b')\} := \sup_{\substack{\varphi \in \mathcal{C}_c((a', b'); \mathbb{R}^N) \\ \varphi \neq 0}} \frac{\langle \mu, \varphi \rangle}{\|\varphi\|_{L^\infty((a', b'); \mathbb{R}^N)}}$$

for every $a \leq a' < b' \leq b$. The value $|\mu|\{(a, b)\}$ is called the total mass of the measure μ. Recall the following compactness result.

THEOREM A.1. (Weak-star compactness of bounded measures.) *Given a sequence μ^h of bounded measures whose total mass on the interval (a, b) is uniformly bounded,*

$$|\mu^h|\{(a, b)\} \leq C,$$

there exists a bounded measure μ and a subsequence (still denoted by μ^h) such that

$$\mu^h \rightharpoonup \mu \quad \text{weak-star,}$$

that is,

$$\langle \mu^h, \varphi \rangle \to \langle \mu, \varphi \rangle, \quad \varphi \in \mathcal{C}_c((a, b); \mathbb{R}^N).$$

$\square$

Let (a, b) be a bounded or unbounded interval. A map $u : (a, b) \to \mathbb{R}^N$ defined at *every point* $x \in (a, b)$ is called a **function with bounded variation** in one variable if its **total variation**

$$TV(u;(a,b)) := \sup\left\{ \sum_{k=1}^{q-1} |u(x_{k+1}) - u(x_k)| \, / \, a < x_1 < \ldots < x_q < b \right\} \qquad \text{(A.1)}$$

is finite. When $(a,b) = I\!R$ we use the notation

$$TV(u) := TV(u; I\!R).$$

We denote by $BV((a,b); I\!R^N)$ the Banach space of functions with bounded variation for which (A.1) is finite, endowed for instance with the norm

$$\|u\|_{L^\infty((a,b);I\!R^N)} + TV(u;(a,b)).$$

It is well-known that a function with bounded variation admits countably many points of discontinuity, at most, and at each point of discontinuity, left- and right-limits $u_-(x)$ and $u_+(x)$ respectively. The value $u(x)$ need not coincide with one of these two traces, and it is often convenient to normalize u by selecting, for instance, its **right-continuous representative** u_+ defined at *every point* x by

$$u_+(x) := \begin{cases} u(x) & \text{at points of continuity,} \\ u_+(x) & \text{at points of discontinuity.} \end{cases} \qquad \text{(A.2)}$$

The left-continuous representative u_- could be defined similarly.

An entirely equivalent definition of the notion of bounded variation is given as follows. A function $u : (a,b) \to I\!R^N$, defined almost everywhere for the Lebesgue measure, belongs to $BV((a,b); I\!R^N)$ if its distributional derivative $\partial_x u$ is a bounded measure, the total variation of u being then

$$TV(u;(a,b)) = \sup_{\substack{\varphi \in C_c^1((a,b);I\!R^N) \\ \varphi \neq 0}} \frac{\displaystyle\int_a^b u \cdot \partial_x \varphi \, dx}{\|\varphi\|_{L^\infty((a,b);I\!R^N)}}. \qquad \text{(A.3)}$$

One can check that (A.1) and (A.3) are equivalent, in the following sense: If u is a function defined almost everywhere for which (A.3) is finite, then it admits a representative defined everywhere such that (A.1) is finite and both quantities in (A.1) and (A.3) coincide. Conversely, if u is a function defined everywhere for which (A.1) is finite then the quantity (A.3) is also finite. Furthermore, from (A.3) it follows that when u is smooth

$$TV(u;(a,b)) = \|\partial_x u\|_{L^1((a,b);I\!R^N)}$$

and, for all u with bounded variation

$$\sup_{h>0} \frac{1}{h} \|u(.+h) - u(.)\|_{L^1(a,b)} = TV(u;(a,b)), \qquad \text{(A.4)}$$

provided we extend u by continuity by constants outside the interval (a,b).

The theory of hyperbolic conservation laws uses the following compactness result.

THEOREM A.2. (Helly's compactness theorem.) *Given a sequence of functions with bounded variation $u^h : (a, b) \to \mathbb{R}^N$ (defined for every point x) satisfying, for some constant $C > 0$,*

$$\|u^h\|_{L^\infty((a,b);\mathbb{R}^N)} + TV(u^h; (a,b)) \leq C,$$

there exist a subsequence (still denoted by u^h) and a function with bounded variation, $u : (a, b) \to \mathbb{R}^N$, such that

$$u^h(x) \to u(x) \quad \text{at every } x \in (a, b).$$

Additionally, we have the **lower semi-continuity** *property*

$$TV(u; (a, b)) \leq \liminf_{h \to 0} TV(u^h; (a, b)). \tag{A.5}$$

$\square$

The compactness result in Theorem A.2 extends as follows to time-dependent functions. The regularity assumed here is shared by solutions of hyperbolic conservation laws.

THEOREM A.3. (Time-dependent version of Helly's theorem.) *Given a sequence of Lebesgue measurable functions $u^h : (a, b) \times \mathbb{R}_+ \to \mathbb{R}^N$ satisfying*

$$\begin{aligned}
\|u^h(t)\|_{L^\infty((a,b);\mathbb{R}^N)} + TV(u^h(t); (a,b)) &\leq C, \quad t \in \mathbb{R}_+, \\
\|u^h(t_2) - u^h(t_1)\|_{L^1((a,b);\mathbb{R}^N)} &\leq C\,|t_2 - t_1|, \quad t_1, t_2 \in \mathbb{R}_+,
\end{aligned} \tag{A.6}$$

for some constant $C > 0$, there exists a subsequence (still denoted by u^h) and a function with bounded variation $u : (a, b) \times \mathbb{R}_+ \to \mathbb{R}^N$ such that

$$\begin{aligned}
u^h(x, t) &\to u(x, t) \quad \text{at almost all } (x, t) \\
u^h(t) &\to u(t) \quad \text{in } L^1_{\text{loc}} \text{ for all } t \in \mathbb{R}_+
\end{aligned}$$

and

$$\begin{aligned}
\|u(t)\|_{L^\infty((a,b);\mathbb{R}^N)} + TV(u(t); (a,b)) &\leq C, \quad t \in \mathbb{R}_+, \\
\|u(t_2) - u(t_1)\|_{L^1((a,b);\mathbb{R}^N)} &\leq C\,|t_2 - t_1|, \quad t_1, t_2 \in \mathbb{R}_+.
\end{aligned} \tag{A.7}$$

PROOF. We only sketch the proof. Relying on the first assumptions in (A.6), for each rational point t we can apply Theorem A.1 and extract a subsequence of $u^h(t)$ that converges to some limit denoted by $u(t)$. By considering a diagonal subsequence, we construct a subsequence of u^h such that

$$u^h(t) \to u(t) \quad \text{for all } x \in (a, b) \text{ and all } \textit{rational} \text{ times } t.$$

Then, the second assumption in (A.6) implies that the limiting function $u(t)$ can be extended to irrational times t (in a unique way) and that the desired convergence result holds. $\square$

We now turn to functions with **bounded total variation in two variables**. By definition, $BV(\mathbb{R} \times (0, T); \mathbb{R}^N)$ is the Banach space of all locally integrable functions $u : \mathbb{R} \times (0, T) \to \mathbb{R}^N$ whose first-order distributional derivatives $\partial_t u$ and $\partial_x u$ are vector-valued Radon measures in (x, t) and the mass of variation measures $|\partial_t u|$ and $|\partial_x u|$ are finite in $\mathbb{R} \times (0, T)$. The variation measure $|\partial_x u|$, for instance, is defined by

$$\iint_\Omega |\partial_x u| := \sup_{\substack{\varphi \in C_c^1(\mathbb{R} \times (0,T); \mathbb{R}^N) \\ \varphi \neq 0, \, \text{supp } \varphi \subset \Omega}} \frac{\|u \cdot \partial_x \varphi\|_{L^1(\Omega)}}{\|\varphi\|_{L^\infty(\Omega)}}$$

for every open set $\Omega \subset \mathbb{R} \times (0, T)$. Recall that, by Riesz representation theorem, Radon measures can be regarded as linear functionals on $C_c^0(\mathbb{R} \times (0, T))$.

The key theorem in the theory of such functions is now stated.

THEOREM A.4. (Regularity of functions with bounded variation in two variables.) *Given a function $u \in BV(\mathbb{R} \times (0, T); \mathbb{R}^N)$, there exist a representative of u (which differs from u on a set with zero Lebesgue measure and is still denoted by u) and a decomposition*

$$\mathbb{R} \times (0, T) = \mathcal{C}(u) \cup \mathcal{J}(u) \cup \mathcal{I}(u)$$

such that:

1. *$\mathcal{C}(u)$ is the set of **points of L^1-approximate continuity** (x, t) in the sense that*

$$\lim_{r \to 0} \frac{1}{r^2} \int_{B_r(x,t)} |u(y, s) - u(x, t)| \, dy ds = 0, \qquad (A.8)$$

   where $B_r(x, t) \subset \mathbb{R}^2$ denotes the ball with center (x, t) and radius $r > 0$.

2. *$\mathcal{J}(u)$ is the set of **points of approximate jump discontinuity** (x, t) at which, by definition, there exists a **propagation speed** $\lambda^u(x, t)$ and **left- and right-approximate limits** $u_-(x, t)$ and $u_+(x, t)$, respectively, such that*

$$\lim_{r \to 0} \frac{1}{r^2} \int_{B_r^\pm(x,t)} |u(y, s) - u_\pm(x, t)| \, dy ds = 0, \qquad (A.9)$$

   where

$$B_r^\pm(x, t) = B_r(x, t) \cap \left\{ \pm(y - \lambda^u(x, t) s) \geq 0 \right\}.$$

   Moreover, the set $\mathcal{J}(u)$ is rectifiable in the sense of Federer, i.e., is the union of countably many continuously differentiable arcs in the plane, and

$$\int_{\mathcal{J}(u)} |u_+ - u_-| \, d\mathcal{H}_1 < \infty.$$

3. *Finally, the set of **interaction points** $\mathcal{I}(u)$ has zero one-dimensional Hausdorff measure:*

$$\mathcal{H}_1(\mathcal{I}(u)) = 0.$$

$\square$

In (A.8) and (A.9) we have tacitly extended the function u by zero outside its domain of definition $\mathbb{R} \times (0, T)$. Based on the regularity properties in Theorem A.4 one has, for every Borel set $B \subset \mathbb{R} \times (0, T)$,

$$\iint_B \partial_t u = \iint_{B \cap \mathcal{C}(u)} \partial_t u - \int_{B \cap \mathcal{J}(u)} \lambda^u (u_+ - u_-) \, d\mathcal{H}_1$$

and

$$\iint_B \partial_x u = \iint_{B \cap \mathcal{C}(u)} \partial_x u + \int_{B \cap \mathcal{J}(u)} (u_+ - u_-) \, d\mathcal{H}_1.$$

It makes sense to define the **right-continuous representative** u_+ of a function with bounded variation for $\mathcal{H}_1$-*almost every* (x,t) by

$$u_+(x,t) := \begin{cases} u(x,t), & (x,t) \in \mathcal{C}(u), \\ u_+(x,t), & (x,t) \in \mathcal{J}(u). \end{cases}$$

The left-continuous representative u_- could be defined similarly. If $g : I\!R^N \to I\!R$ is any smooth mapping, the product $g(u_+)\,\partial_x u$ is a vector-valued Radon measure such that

$$\iint_B g(u_+)\,\partial_x u := \iint_{B\cap\mathcal{C}(u)} g(u)\,\partial_x u + \int_{B\cap\mathcal{J}(u)} g(u_+)\,(u_+ - u_-)\,d\mathcal{H}_1. \qquad (A.10)$$

We now restrict attention to functions $u = u(x,t)$ satisfying the conditions (A.7). Since (A.7) implies

$$\|\partial_t u(t)\|_{L^1((a,b);I\!R^N)} \le C, \quad t \in I\!R_+,$$

such a function clearly belongs to $BV(I\!R \times (0,T); I\!R^N)$ for all $T > 0$ and it can be checked that

$$\iint_{(a,b)\times(t_1,t_2)} |\partial_x u| = \int_{t_1}^{t_2} TV(u(t);(a,b))\,dt. \qquad (A.11)$$

Additionally, for all $t_1 < t_2$, $x_1 < x_2$, and $\lambda \in I\!R$, provided the set

$$B := \big\{ (x,t) \,/\, t_1 < t < t_2,\ x_1 + \lambda\,(t - t_1) < x < x_2 + \lambda\,(t - t_1) \big\}$$

is non-empty, the following **Green formulas** hold:

$$\iint_B \partial_t u = \int_{x_1 + \lambda\,(t_2 - t_1)}^{x_2 + \lambda\,(t_2 - t_1)} u(x,t_2)\,dt - \int_{x_1}^{x_2} u(x,t_1)\,dx$$
$$- \int_{t_1}^{t_2} \lambda\,u_-\big(x_2 + \lambda\,(t - t_1), t\big)\,dt + \int_{t_1}^{t_2} \lambda\,u_+\big(x_1 + \lambda\,(t - t_1), t\big)\,dt,$$

$$\iint_B \partial_x u = \int_{t_1}^{t_2} u_-\big(x_2 + \lambda\,(t - t_1), t\big)\,dt - \int_{t_1}^{t_2} u_+\big(x_1 + \lambda\,(t - t_1), t\big)\,dt.$$

Finally, we recall the **chain rule**

$$\partial_x f(u) = \big(Df(u)\big)\widehat{}\,\partial_x u \qquad (A.12)$$

valid for every function with bounded variation $u : I\!R \times (0,T) \to I\!R^N$ and every smooth mapping $f : I\!R^N \to I\!R^N$, where **Volpert's superposition** is defined $\mathcal{H}_1$-almost everywhere by

$$\big(Df(u)\big)\widehat{}(x,t) := \begin{cases} Df(u(x,t)), & (x,t) \in \mathcal{C}(u), \\ \displaystyle\int_0^1 Df\big(\theta\,u_-(x,t) + (1-\theta)\,u_+(x,t)\big)\,d\theta, & (x,t) \in \mathcal{J}(u). \end{cases}$$

In particular, we have $\partial_x f(u) = Df(u)\,\partial_x u$ on $\mathcal{C}(u)$.

BIBLIOGRAPHICAL NOTES

Chapter I. Smooth solutions to strictly hyperbolic systems of conservation laws were studied by many authors; see, for instance, Hughes, Kato, and Marsden (1977), Majda (1984) and the references cited therein. Fundamental notions about systems of conservation laws (Sections I-1 and I-2) were introduced and investigated by Lax (1954, 1957, 1971, 1973). The entropy condition was also studied by Oleinik (1963), Kruzkov (1970), Volpert (1967), Dafermos (1973a), and Liu (1974, 1975, 1976), with many other follow-up works. The existence of strictly convex entropy pairs for systems of two conservation laws was established by Lax (1971) and Dafermos (1987). Friedrichs and Lax (1971), Godunov (see the bibliography in Godunov (1987)), Harten (1983), and Harten, Lax, Levermore, and Morokoff (1998) are good sources for a discussion of the symmetrization of hyperbolic systems via entropy variables. The breakdown of smooth solutions was investigated by Lax (1964, 1973), Liu (1979), John (1974), and Hörmander (1997), as well as, for instance, Chemin (1990ab), Alinhac (1995), Li and Kong (1999), Dias and Figueira (2000), and Jenssen (2000).

The exposition given in Sections I-3 to I-5 follows Hayes and LeFloch (2000) and LeFloch (1993). The kinetic relation for nonclassical shock waves of *strictly hyperbolic* systems (Section I-5) was introduced and discussed by Hayes and LeFloch (1996a, 1997, 1998, 2000) and LeFloch (1999), and further studied in Bedjaoui and LeFloch (2001, 2002ac) and LeFloch and Thanh (2000, 2001a). It represents a generalization of a concept known in material science.

The examples from continuum physics in Section I-4 are taken from Korteweg (1901), Courant and Friedrichs (1948) (a standard textbook on shock waves in fluids), Landau and Lifshitz (1959), Serrin (1979, 1981, 1983), Slemrod (1983ab, 1984ab), Hagan and Serrin (1984), Ericksen (1991), Gurtin (1993ab), and Gavrilyuk and Gouin (1999, 2000).

The modeling of propagating phase boundaries in solid materials undergoing phase transformations has attracted a lot of attention. Various aspects of the capillarity in fluids and solids and the study of a typical *hyperbolic-elliptic* system of two conservation laws (Example I-4.5) are found in Abeyaratne and Knowles (1988, 1990, 1991ab, 1992, 1993), Asakura (1999, 2000), Bedjaoui and LeFloch (2002b), Fan (1992, 1993abc, 1998), Fan and Slemrod (1993), Grinfeld (1989), Hagan and Serrin (1984, 1986), Hagan and Slemrod (1983), Hattori (1986ab, 1998, 2000), Hattori and Mischaikow (1991), Hsiao (1990ab), Hsiao and deMottoni (1990), James (1979, 1980), Keyfitz (1986, 1991), LeFloch (1993, 1998), LeFloch and Thanh (2001b), Mercier and Piccoli (2000), Pego (1987, 1989), Pence (1985, 1986, 1992, 1993), Shearer (1982, 1983, 1986), Shearer and Yang (1995), Slemrod (1983ab, 1984ab, 1987, 1989), and Truskinovsky (1983, 1987, 1993, 1994ab). See also Benzoni (1998, 1999) and Freistuhler (1996, 1998) for stability issues on multi-dimensional problems. Systems of three equations (van der Waals fluids, thermo-elastic solids) were considered by Abeyaratne and Knowles (1994ab), Bedjaoui and LeFloch (2002c), Hoff and Khodja (1993), and LeFloch and Thanh (2002).

Numerical issues related to phase transition dynamics and nonclassical shock waves were discussed by Affouf and Caflisch (1991), Ball et al. (1991), Chalons and LeFloch (2001ab, 2002), Cockburn and Gau (1996), Hayes and LeFloch (1998), Hou, LeFloch, and Rosakis (1999), Hsieh and Wang (1997), Jin (1995), LeFloch (1996, 1998), LeFloch, Mercier, and Rohde (2002), LeFloch and Rohde (2000), Lowengrub et al. (1999), Natalini and Tang (2000), Shu (1992), Slemrod and Flaherty (1986), Vainchtein et al. (1998), and Zhong, Hou, and LeFloch (1996).

Vanishing diffusion-dispersion limits were studied by Schonbek (1982) for scalar conservation laws using compensated compactness arguments (following Murat (1978) and Tartar (1979, 1982, 1983)). Extensions of Schonbek's work were given in Hayes and LeFloch (1997), Correia and LeFloch (1998), LeFloch and Natalini (1999), and Kondo and LeFloch (2002). See also LeFloch and Rohde (2001) for an approach by Dafermos' self-similar method (1973b).

Vanishing dispersion limits are covered by Lax and Levermore's theory; see Lax and Levermore (1983), Goodman and Lax (1988), Hou and Lax (1991), and Lax (1991). Further relevant material on dispersive equations is found in Martel and Merle (2001ab) and the references therein.

Chapter II. The material in Sections II-1 and II-2 concerning the entropy condition and the Riemann problem for one-dimensional conservation laws is standard and goes back to the works by Lax and Oleinik. The Riemann problem with non-convex flux-functions and *single entropy inequality* (Sections II-3 to II-5) was studied by Hayes and LeFloch (1997) (cubic flux-function) and by Baiti, LeFloch, and Piccoli (1999) (general flux-functions). A generalization to hyperbolic *systems of two conservation laws* was given by Hayes and LeFloch (1996, 2000) and LeFloch and Thanh (2001a). For results on Lipschitz continuous mappings (applied here to the function $\varphi^\natural$), see for instance the textbook by Clarke (1990).

Chapter III. Standard textbooks on ordinary differential equations are: Coddington and Levinson (1955), Guckenheimer and Holmes (1983), Hales (1969), and Hartman (1964). *Classical* diffusive and diffusive-dispersive traveling waves for scalar equations and systems were studied by many authors, especially Gilbarg (1951), Foy (1964), Conley and Smoller (1970, 1971, 1972ab), Benjamin, Bona, and Mahoney (1972), Conlon (1980), Smoller and Shapiro (1969), Antman and Liu (1979), Bona and Schonbek (1985), and Antman and Malek-Madani (1988).

Nonclassical diffusive-dispersive traveling waves of conservation laws were discovered by Jacobs, McKinney, and Shearer (1995) for the cubic flux-function (with $b = c_1 = c_2 = 1$). This model is referred to as the *modified Korteweg-de Vries-Burgers* (KdVB) equation. It is remarkable that its nonclassical trajectories can be described by an *explicit formula*. The earlier work by Wu (1991) derived and analyzed the KdVB equation from the full magnetohydrodynamics model. Theorem III-2.3 is a reformulation of Jacobs, McKinney, and Shearer's result (1995) but is based on the concept of a kinetic relation introduced in Hayes and LeFloch (1997).

The effect of the nonlinear diffusion $\varepsilon \left(|u_x| \, u_x \right)_x$ with the cubic flux-function was studied by Hayes and LeFloch (1997). For this model too the nonclassical trajectories are given by an *explicit formula*. As a new feature, the corresponding nonclassical shocks may have arbitrary small strength, that is, the kinetic function *do not coincide* with the classical upper bound $\varphi^\natural$ near the origin. Interestingly enough, this example does enter the existence framework proposed in Section IV-3.

Hayes and Shearer (1999) and Bedjaoui and LeFloch (2001a) established the existence of nonclassical traveling waves for *general* flux-functions. The exposition in Sections III-3 to III-5 follows Bedjaoui and LeFloch (2001a). The behavior of the kinetic function in the large is also derived by Bedjaoui and LeFloch (2001b). Another (fourth-order) regularization arising in driven thin film flows was studied by Bertozzi et al. (1999ab, 2000).

Most of the results in this chapter remain valid for the 2×2 hyperbolic system of elastodynamics (Example I-4.4); see Schulze and Shearer (1999) (cubic flux-functions) and Bedjaoui and LeFloch (2001c) (general flux-functions). Traveling waves of the hyperbolic-elliptic model of phase dynamics (Example I-4.5) were studied by Shearer and Yang (1995) (cubic flux-functions) and Bedjaoui and LeFloch (2001b) (general flux-functions). See also Fan (1992, 1998), Fan and Slemrod (1993), Hagan and Serin (1984, 1986), Hagan and Slemrod (1983), Slemrod (1983ab, 1984ab, 1987, 1989), Truskinovsky (1987, 1993).

Chapter IV. The *explicit formula* in Theorem IV-1.1 is due to Hopf (1950) (Burgers equation) and Lax (1954) (general flux-functions). Many generalizations of the so-called Lax formula are known. See Lions (1985) and the references therein for multi-dimensional Hamilton-Jacobi equations. An explicit formula for the initial and boundary value problem for conservation laws was derived independently by Joseph (1989) and LeFloch (1988b), and, for conservation laws with non-constant coefficients, by LeFloch and Nedelec (1985). The entropy inequality (1.2) was discovered by Oleinik (1963). Interestingly enough, this inequality also holds for approximate solutions constructed by finite difference schemes: Goodman and LeVeque (1986), Brenier and Osher (1988). See also Tadmor (1991) for the derivation of local error estimates. The uniqueness argument in the proof of Theorem IV-1.3 is taken from LeFloch and Xin (1993).

There is an extensive literature on the existence and uniqueness of *classical* entropy solutions, and to review it is out of the scope of these notes. We just mention the fundamental papers by Conway and Smoller (1966), Volpert (1967), Kruzkov (1970), and Crandall (1972). On the other hand, the *wave front tracking scheme* (also called *polygonal approximation method*) for scalar conservation laws (Section IV-2) was introduced by Dafermos (1972). It leads to both a general strategy for proving the existence of discontinuous solutions for scalar conservation laws (as well as for systems of equations, see Chapter VII) and an interesting method of numerical approximation. General flux-functions were considered in Iguchi and LeFloch (2002).

The existence of *nonclassical* entropy solutions (Sections IV-3) was established by Amadori, Baiti, LeFloch, and Piccoli (1999) (cubic flux-function) and Baiti, LeFloch, and Piccoli (1999, 2000) (general flux-function). The concept of minimal backward characteristics used in the proof of Theorem IV-3.2 goes back to the works by Filippov (1960) and Dafermos (1977, 1982).

Theorem IV-4.1 is standard while Theorems IV-4.2 and IV-4.3 are new and due to the author.

Chapter V. The L^1 contraction property for scalar conservation laws (Theorem V-2.2) was originally derived by different methods by Volpert (1967), Kruzkov (1970), Keyfitz (1971), and Crandall (1972). A non-increasing, weighted norm quantifying the rate of decay of L^1 norm (along similar lines as our Theorem V-2.3) was discovered by Liu and Yang (1999a), for piecewise smooth solutions of scalar conservation laws with convex flux. The approach presented here in Sections V-1 and V-2 and based on linear hyperbolic equations was discovered by Hu and LeFloch (2000). The sharp estimate for general solutions with bounded variation was established by Dafermos (2000) using generalized characteristics and, then, by Goatin and LeFloch (2001a) using the technique in Hu and LeFloch (2000). Further generalizations and applications to the framework in this chapter are given in LeFloch (2002).

Chapter VI. Fundamental material on the entropy condition and Riemann problem for strictly hyperbolic systems can be found in Lax (1957, 1970), Liu (1974, 1981), and Dafermos (1978a). The *Riemann problem* described in Sections VI-2 and VI-3 was solved by Lax (1957) and Liu (1974), respectively. Hyperbolic systems under non-convexity assumptions were considered by Oleinik (1957), Ballou (1970), Wendroff (1972ab, 1991), Liu (1974, 1975, 1976, 1981), Dafermos (1984), Menikoff and Plohr (1989), and Zumbrun (1990, 1993).

The concept of a kinetic relation and the generalization of Liu's construction to encompass *nonclassical solutions* (Sections VI-3 and VI-4) is due to Hayes and LeFloch (1997, 2000). See also the notes for Chapter I above for the references in material sciences. Lipschitz continuous mappings are discussed in Clarke (1990), Correia, LeFloch, and Thanh (2002), and Isaacson and Temple (1992). In Hayes and LeFloch (1998), the authors argue that the range of the kinetic functions (enclosed by the extremal choices $\mu_j^\natural$ and $\mu_{j0}^\flat$) may be very narrow in the applications, making particularly delicate the numerical investigation of the dynamics of nonclassical shocks. For numerical works in this direction see Hayes and LeFloch (1998), LeFloch and Rohde (2000), and Chalons and LeFloch (2001ab, 2002).

Important material on the Riemann problem for systems of conservation laws, particularly undercompressive shocks in solutions of non-strictly hyperbolic systems, is also found in Azevedo et al. (1995, 1996, 1999), Canic (1998), Hurley and Plohr (1995), Hsiao (1980), Isaacson et al. (1992), Isaacson, Marchesin and Plohr (1990), Keyfitz (1991, 1995), Keyfitz and Kranzer (1978, 1979), Keyfitz and Mora (2000), Plohr and Zumbrun (1996), Schecter, Marchesin, and Plohr (1996), Schecter and Shearer (1989), and Shearer, Schaeffer, Marchesin, and Paes-Lemme (1987).

Chapter VII. The wave interaction estimates and the general technique to derive uniform total variation bounds go back to Glimm's pioneering work (1965), based on the so-called *random-choice* scheme. A *deterministic* version of this method was obtained by Liu (1977). The *wave front tracking* scheme was initially proposed by Dafermos (1972) for scalar conservation laws, then extended by DiPerna (1973) to systems of two conservation laws, and generalized by Bressan (1992) and Risebro (1993) to systems of N equations. The specific formulation adopted in this chapter is due to Baiti and Jenssen (1998), as far as genuinely nonlinear fields are concerned. Front tracking is also a powerful numerical tool developed by Glimm et al. (1985), Chern et al. (1986), Lucier (1986), Klingenberg and Plohr (1991), and many others. All of the above papers restrict attention to genuinely nonlinear or linearly degenerate fields.

The existence theory for *non-genuinely nonlinear* characteristic fields goes back to the extended work by Liu (1981) (and the references therein), based on Glimm's scheme. The generalization of the wave front tracking scheme to concave-convex characteristic fields (Theorem VII-2.1) is due to the author. The convergence of the wave front tracking scheme for $N \times N$ systems with more general characteristic fields was established by Iguchi and LeFloch (2002). They observed that wave curves of such systems are *only of class* C^1 with second-order bounded derivatives (which is sufficient to apply Glimm's argument in the proof of Theorem VII-1.1).

For a non-convex system of two conservation laws arising in elastodynamics, the existence of solutions with *large total variation* was established by Ancona and Marson (2000). Other interesting developments on Glimm's scheme and its variants (for phase transition dynamics or solutions with a single strong shock, in particular) are found in the following papers: Sablé-Tougeron (1988, 1998), Chern (1989), Temple (1990abc), Schochet (1991ab), LeFloch (1993), Young (1993), Asakura (1994, 1999), Corli and Sablé-Tougeron (1997ab, 2000), Cheverry (1998), and Corli (1999).

The regularity of the solutions of hyperbolic conservation laws (Section VII-4) was investigated by Glimm and Lax (1970), Schaeffer (1973), DiPerna (1976, 1979a), Dafermos (1977, 1982, 1985a), Liu (1981, 2000), and Bressan and LeFloch (1999).

Other approaches to the Cauchy problem for systems of conservation laws were discussed, for instance, in Bereux, Bonnetier, and LeFloch (1996), Chen (1997), Chen and LeFloch (2000, 2002), Chen and Wang (2002), Perthame (1999), Tartar (1979, 1982, 1983), and the many references therein.

Chapter VIII. All of the results in this chapter are based on Baiti, LeFloch, and Piccoli (2002ab).

Chapter IX. The exposition here is based on Hu and LeFloch (2000), which was motivated by the earlier results LeFloch (1990b) and LeFloch and Xin (1993). In this approach, we basically extend Holmgren's method (more precisely, here, the dual formulation due to Haar) to nonlinear systems of conservation laws. Holmgren's method was known to be successful for linear PDE's and, by Oleinik's work (1957), for scalar conservation laws. Finding a suitable generalization to systems was attempted with some success by many authors, including Oleinik (1957), Liu (1976), and LeFloch and Xin (1993), who treated piecewise smooth solutions or special systems, only. Further generalizations and applications to the framework in Section IX-1 were given in Crasta and LeFloch (2002) and LeFloch (2002).

The continuous dependence of solutions for genuinely nonlinear systems was obtained first by Bressan and Colombo (1995ab) (for systems of two conservation laws) and Bressan, Crasta, and Piccoli (2000) (for systems of N equations). These authors developed an *homotopy method* to compare two (suitably constructed, piecewise smooth) approximate solutions and show that the continuous dependence estimate held *exactly* for these approximate solutions. This strategy turned out to be very technical. The method was also applied by Ancona and Marson (2000, 2002) to a non-convex system of two conservation laws of elastodynamics.

Next, Liu and Yang (1999a) discovered a functional (equivalent to the L^1 norm and strictly decreasing in time) for *scalar* conservation laws with convex flux, opening the way to a possible investigation of *systems* of equations. The research on the subject culminated with three papers announced simultaneously in 1998, by Bressan et al. (1999), Hu and LeFloch (2000), and Liu and Yang (1999c). These papers provide now three simple proofs of the L^1 continuous dependence of solutions for systems of conservation laws. A common feature of these proofs is the fact that the continuous dependence estimate is satisfied by the approximate solutions *up to some error term.*

The *sharp L^1 estimate* in Section IX-3 was obtained by Goatin and LeFloch (2001b). The technique of nonconservative product was developed (with different motivations) by Dal Maso, LeFloch, and Murat (1995) and LeFloch and Liu (1994). Lemma IX-3.1 is due to Baiti and Bressan (1997). Theorem IX-3.3 on the convergence of the wave measures was established by Bressan and LeFloch (1999) together with further regularity results on entropy solutions. The L^1 continuous dependence of entropy solutions with *large total variation* for the compressible Euler equations was investigated by Goatin and LeFloch (2002).

Chapter X. The uniqueness of entropy solutions of genuinely nonlinear systems was established by Bressan and LeFloch (1997), who introduced the concept of solutions with *tame variation.* A generalization to solutions with *tame oscillation* was subsequently obtained by Bressan and Goatin (1999). The notion of (Φ, ψ)–admissible entropy solution for *general nonlinear hyperbolic systems* (including conservative systems with non-genuinely nonlinear characteristic fields) was introduced by Baiti, LeFloch, and Piccoli (2001). The earlier work by Bressan (1995) for systems of conservation laws with genuinely nonlinear fields introduced the new concept of semi-group of solutions and established the convergence of the Glimm scheme to a unique limit. See also Colombo and Corli (1999) for a uniqueness result involving phase transitions.

It is an open problem to derive the tame variation property for arbitrary solutions with bounded variation. However, based on Dafermos-Filippov's theory of generalized characteristics, Trivisa (1999) established that the tame variation property is always satisfied by "countably regular" BV solutions of strictly hyperbolic, genuinely nonlinear, 2×2 systems of conservation laws.

Definition X-1.2 covers the concept of weak solutions to *nonconservative systems* in the sense of Dal Maso, LeFloch, and Murat (1990, 1995). See also LeFloch and Tzavaras (1996, 1999). For such systems, the existence of entropy solutions to the Cauchy problem was established by LeFloch (1988a, 1990a, 1991) and LeFloch and Liu (1993).

Among many earlier results on the uniqueness of entropy solutions, we quote the important and pioneering work by DiPerna (1979b) for hyperbolic systems of two equations, extended by LeFloch and Xin (2002) to a class of $N \times N$ systems. DiPerna's method is based on entropy inequalities and covers the case of one arbitrary entropy solution and one piecewise smooth solution. It leads to an estimate in the L^2 norm, to be compared with the L^1 estimate in Theorem X-1.6.

Appendix. For the properties of functions with bounded variation we refer to the textbooks by Evans and Gariepy (1992), Federer (1969), Volpert (1967), and Ziemer (1989).

BIBLIOGRAPHY

Abeyaratne R. and Knowles J.K. (1988), On the dissipative response due to discontinuous strains in bars of unstable elastic materials, Intern. J. Solids Structures 24, 1021–1044.

Abeyaratne R. and Knowles J.K. (1990), On the driving traction acting on a surface of strain discontinuity in a continuum, J. Mech. Phys. Solids 38, 345–360.

Abeyaratne A. and Knowles J.K. (1991a), Kinetic relations and the propagation of phase boundaries in solids, Arch. Rational Mech. Anal. 114, 119–154.

Abeyaratne R. and Knowles J.K. (1991b), Implications of viscosity and strain gradient effects for the kinetics of propagating phase boundaries, SIAM J. Appl. Math. 51, 1205–1221.

Abeyaratne R. and Knowles J.K. (1992), On the propagation of maximally dissipative phase boundaries in solids, Quart. Appl. Math. 50, 149–172.

Abeyaratne R. and Knowles J.K. (1993), Nucleation, kinetics and admissibility criteria for propagating phase boundaries, in "Shock induced transitions and phase structures in general media", IMA Vol. Math. Appl., Vol. 52, Springer Verlag, New York, N.Y., pp. 1–33.

Abeyaratne R. and Knowles J.K. (1994a), Dynamics of propagating phase boundaries: adiabatic theory for thermoelastic solids, Phys. D 79, 269–288.

Abeyaratne R. and Knowles J.K. (1994b), Dynamics of propagating phase boundaries: thermoelastic solids with heat conduction, Arch. Rational Mech. Anal. 126, 203–230.

Affouf M. and Caflisch R. (1991), A numerical study of Riemann problem solutions and stability for a system of viscous conservation laws of mixed type, SIAM J. Appl. Math. 51, 605–634.

Alber H.D. (1985), A local existence theorem for quasilinear wave equations with initial values of bounded variation, Lecture Notes in Math. 1151, Springer Verlag, 9–24.

Alinhac S. (1995), *Blow-up for nonlinear hyperbolic equations*, Progress in Nonlinear Diff. Equa. and Appl. 17, Birkhäuser, Boston.

Amadori D., Baiti P., LeFloch P.G., and Piccoli B. (1999), Nonclassical shocks and the Cauchy problem for non-convex conservation laws, J. Differential Equations 151, 345–372.

Ancona F. and Marson A. (2000), Well-posedness theory for 2×2 systems of conservation laws, Preprint.

Ancona F. and Marson A. (2002), Basic estimates for a front tracking algorithm for general 2 × 2 conservation laws. Math. Models Methods Appl. Sci. 12 (2002), 155–182.

Antman S.S. and Liu T.-P. (1979), Traveling waves in hyperelastic rods, Quart. Appl. Math. 36, 377–399.

Antman S.S. and Malek-Madani R. (1988), Traveling waves in nonlinearly viscoelastic media and shock structure in elastic media, Quart. Appl. Math. 46, 77–93.

Antman S.S. and Seidman T.I. (1996), Quasilinear hyperbolic-parabolic equations of nonlinear viscoelasticity, J. Differential Equations 124, 132–185.

Asakura F. (1993), Decay of solutions for the equations of isothermal gas dynamics, Japan J. Indust. Appl. Math. 10, 133–164.

Asakura F. (1994), Asymptotic stability of solutions with a single strong shock wave for hyperbolic systems of conservation laws, Japan J. Indust. Appl. Math. 11, 225–244.

Asakura F. (1997), Global solutions with a single transonic shock wave for quasilinear hyperbolic systems, Methods Appl. Anal. 4, 33–52.

Asakura F. (1999), Large time stability of the Maxwell states, Methods Appl. Anal. 6, 477–503.

Asakura F. (2000), Kinetic condition and the Gibbs function, Taiwanese J. Math. 4, 105–117.

Azevedo A. and Marchesin D. (1995), Multiple viscous solutions for systems of conservation laws, Trans. Amer. Math. Soc. 347, 3061–3077.

Azevedo A., Marchesin D., Plohr B., and Zumbrun K. (1996), Nonuniqueness of solutions of Riemann problems, Z. Angew. Math. Phys. 47, 977–998.

Azevedo A., Marchesin D., Plohr B., and Zumbrun K. (1999), Bifurcation of nonclassical viscous shock profiles from the constant state, Comm. Math. Phys. 202, 267–290.

Baiti P. and Bressan A. (1997), Lower semi-continuity of weighted path lengths in BV, in "Geometrical Optics and Related Topics", F. Colombini and N. Lerner Eds., Birkhäuser, 1997, pp. 31–58.

Baiti P. and Jenssen H.K. (1998), On the front tracking algorithm, J. Math. Anal. Appl. 217, 395–404.

Baiti P. and Jenssen H.K. (2001), Blowup in L^∞ for a class of genuinely nonlinear hyperbolic systems of conservation laws, Discrete Contin. Dynam. Systems 7, 837–853.

Baiti P., LeFloch P.G., and Piccoli B. (1999), Nonclassical shocks and the Cauchy problem: general conservation laws, in "Nonlinear Partial Differential Equations", Inter. Confer. March 21–24, 1998, Northwestern Univ., G.-Q. Chen and E. DiBenedetto Editors, Contemporary Mathematics, Vol. 238, Amer. Math. Soc., pp. 1–26.

Baiti P., LeFloch P.G., and Piccoli B. (2000), BV Stability via generalized characteristics for nonclassical solutions of conservation laws, in the "Proceedings of Equadiff'99", Berlin, World Scientific Publishing, Singapore. pp. 289–295.

Baiti P., LeFloch P.G., and Piccoli B. (2001), Uniqueness of classical and non-classical solutions for nonlinear hyperbolic systems, J. Differential Equations 172, 59–82.

Baiti P., LeFloch P.G., and Piccoli B. (2002a), Existence theory for nonclassical entropy solutions I. Scalar conservation laws, Preprint.

Baiti P., LeFloch P.G., and Piccoli B. (2002b), Existence theory for nonclassical entropy solutions II. Hyperbolic systems of conservation laws, in preparation.

Ball J.M., Holmes P.J., James R.D., Pego R.L., and Swart P.J. (1991), On the dynamics of fine structure, J. Nonlinear Sci. 1, 17–70.

Ballou D. (1970), Solutions to nonlinear Cauchy problems without convexity conditions, Trans. Amer. Math. Soc. 152, 441–460.

Bedjaoui N. and LeFloch P.G. (2001), Diffusive-dispersive traveling waves and kinetic relations III. An hyperbolic model from nonlinear elastodynamics, Ann. Univ. Ferrara Sc. Mat. 47, 117–144.

Bedjaoui N. and LeFloch P.G. (2002a), Diffusive-dispersive traveling waves and kinetic relations I. Non-convex hyperbolic conservation laws, J. Differential Equations 178, 574–607.

Bedjaoui N. and LeFloch P.G. (2002b), Diffusive-dispersive traveling waves and kinetic relations II. An hyperbolic-elliptic model of phase transitions, Proc. Royal Soc. Edinburgh, to appear.

Bedjaoui N. and LeFloch P.G. (2002c), Diffusive-dispersive traveling waves and kinetic relations IV. Compressible Euler equations, Chinese Annals Math., to appear.

Bénilan P. and Kružkov S. (1996), Conservation laws with continuous flux functions, NoDEA Nonlinear Differential Equations Appl. 3, 395–419.

Benjamin T.B., Bona J.L., and Mahoney J.J. (1972), Model equations for long waves in dispersive systems, Phil. Trans. Royal Soc. London A 272, 47–78.

Benzoni S. (1998), Stability of multi-dimensional phase transitions in a van der Waals fluid, Nonlinear Analysis T.M.A. 31, 243–263.

Benzoni S. (1999), Multi-dimensional stability of propagating phase boundaries, in "Hyperbolic problems: theory, numerics, applications", Vol. I (Zürich, 1998), Internat. Ser. Numer. Math., Vol. 129, Birkhäuser, Basel, pp. 41–45.

Bereux F., Bonnetier E., and LeFloch P.G. (1996), Gas dynamics equations: two special cases, SIAM J. Math. Anal. 28, 499–515.

Bertozzi A.L., Münch A., and Shearer M. (1999a), Undercompressive shocks in thin film flows, Phys. D 134, 431–464.

Bertozzi A.L., Münch A., and Shearer M. (1999b), Undercompressive waves in driven thin film flow: theory, computation, and experiment, in "Trends in mathematical physics" (Knoxville, TN, 1998), AMS/IP Stud. Adv. Math., Vol. 13, Amer. Math. Soc., Providence, RI, pp. 43–68.

Bertozzi A.L. and Shearer M. (2000), Existence of undercompressive traveling waves in thin film equations, SIAM J. Math. Anal. 32, 194–213.

Bona J. and Schonbek M.E. (1985), Traveling-wave solutions to the Korteweg-de Vries-Burgers equation, Proc. Royal Soc. Edinburgh 101A, 207–226.

Brenier Y. and Osher S. (1988), The discrete one-sided Lipschitz condition for convex scalar conservation laws, SIAM J. Numer. Anal. 25, 8–23.

Bressan A. (1992), Global solutions of systems of conservation laws by wave front tracking, J. Math. Anal. Appl. 170, 414–432.

Bressan A. (1995), The unique limit of the Glimm scheme, Arch. Rational Mech. Anal. 130, 205-230.

Bressan A. and Colombo R.M. (1995a), The semigroup generated by 2×2 conservation laws, Arch. Rational Mech. Anal. 133, 1–75.

Bressan A. and Colombo R.M. (1995b), Unique solutions of 2×2 conservation laws with large data, Indiana Univ. Math. J. 44, 677–725.

Bressan A. and Colombo R.M. (1998), Decay of positive waves in nonlinear systems of conservation laws, Ann. Scuola Norm. Sup. Pisa 26, 133–160.

Bressan A., Crasta G., and Piccoli B. (2000), *Well-posedness of the Cauchy problem for $n \times n$ systems of conservation laws*, Mem. Amer. Math. Soc. 694.

Bressan A. and LeFloch P.G. (1997), Uniqueness of entropy solutions for systems of conservation laws, Arch. Rational Mech. Anal. 140, 301–331.

Bressan A. and LeFloch P.G. (1999), Structural stability and regularity of entropy solutions to systems of conservation laws, Indiana Univ. Math. J. 48, 43–84.

Bressan A., Liu T.-P., and Yang T. (1999), L^1 stability estimates for $n \times n$ conservation laws, Arch. Rational Mech. Anal. 149, 1–22.

Bressan A. and Goatin P. (1999), Oleinik-type estimates and uniqueness for $n \times n$ conservation laws, J. Differential Equations 156, 26–49.

Canić S. (1998), On the influence of viscosity on Riemann solutions, J. Dynam. Differential Equations 10, 109–149.

Canić S., Keyfitz B.L, and Lieberman G.M. (2000), A proof of existence of perturbed steady transonic shocks via a free boundary problem, Comm. Pure Appl. Math. 53, 484–511.

Carr J, Gurtin M.E., and Slemrod M. (1984), Structured phase transitions on a finite interval, Arch. Rational Mech. Anal. 86, 317–351.

Casal P. and Gouin H. (1988), Sur les interfaces liquid-vapeur non-isothermes, (in French), J. Mécan. Théor. Appl. 7, 689–718.

Chalons C. and LeFloch P.G. (2001a), High-order entropy conservative schemes and kinetic relations for van der Waals-type fluids, J. Comput. Phys. 167, 1–23.

Chalons C. and LeFloch P.G. (2001b), A fully discrete scheme for diffusive-dispersive conservation laws, Numerische Math. 89, 493–509.

Chalons C. and LeFloch P.G. (2002), Computing undercompressive waves with the random choice scheme: nonclassical shock waves, Preprint.

Chemin J.-Y. (1990a), Remarque sur l'apparition de singularités fortes dans les écoulements compressibles, (in French), Comm. Math. Phys. 133, 323–329.

Chemin J.-Y. (1990b), Dynamique des gaz à masse totale finie, (in French), Asymptotic Anal. 3, 215–220.

Chen G.-Q. (1997), Remarks on spherically symmetric solutions of the compressible Euler equations, Proc. Roy. Soc. Edinburgh A 127, 243–259.

Chen G.-Q. and Kan P.T. (2001), Hyperbolic conservation laws with umbilic degeneracy, Arch. Rational Mech. Anal. 160, 325–354.

Chen G.-Q. and LeFloch P.G. (2000), Compressible Euler equations with general pressure law, Arch. Rational Mech Anal. 153, 221–259.

Chen G.-Q. and LeFloch P.G. (2001), Entropies and entropy-flux splittings for the isentropic Euler equations, Chinese Annal. Math. 22, 1–14.

Chen G.-Q. and LeFloch P.G. (2002), Existence theory for the compressible isentropic Euler equations, Arch. Rational Mech Anal., to appear.

Chen G.-Q.and Wang D. (2002), The Cauchy problem for the Euler equations for compressible fluids, Handbook on Mathematical Fluid Dynamics, Vol. 1, Elsevier Science B.V., pp. 421-543.

Chern I.-L. (1989), Stability theorem and truncation error analysis for the Glimm scheme and for a front tracking method for flows with strong discontinuities, Comm. Pure Appl. Math. 42, 815–844.

Chern I-L., Glimm J., McBryan O., Plohr B., and Yaniv S. (1986), Front tracking for gas dynamics, J. Comput. Phys. 62, 83–110.

Cheverry C. (1998), Systèmes de lois de conservation et stabilité BV, (in French), Mém. Soc. Math. Fr. (N.S.) No. 75.

Clarke F.H. (1990), Optimization and non-smooth analysis, Classics in Applied Mathematics 5, Society for Industrial and Applied Mathematics (SIAM), Philadelphia, PA.

Cockburn B. and Gau H. (1996), A model numerical scheme for the propagation of phase transitions in solids, SIAM J. Sci. Comput. 17, 1092–1121.

Coddington E.A. and Levinson N. (1955), Theory of ordinary and differential equations Mc Graw Hill, New York.

Coleman B.D. and Noll W. (1963), The thermodynamics of elastic materials with heat conduction and viscosity, Arch. Rational Mech. Anal. 13, 167–178.

Colombo R.M. and Corli A. (1999), Continuous dependence in conservation laws with phase transitions, SIAM J. Math. Anal. 31, 34–62.

Colombo R.M. and Corli A. (2000), Global existence and continuous dependence of phase transitions in hyperbolic conservation laws, Trends in applications of mathematics to mechanics (Nice, 1998), Chapman & Hall/CRC Monogr. Surv. Pure Appl. Math., 106, Boca Raton, FL., pp. 161–171.

Conley C. and Smoller J.A. (1970), Viscosity matrices for two-dimensional nonlinear hyperbolic systems, Comm. Pure Appl. Math. 23, 867–884.

Conley C. and Smoller J.A. (1971), Shock waves as limits of progressive wave solutions of high order equations, Comm. Pure Appl. Math. 24, 459–471.

Conley C. and Smoller J.A. (1972a), Shock waves as limits of progressive wave solutions of high order equations, II, Comm. Pure Appl. Math. 25, 131–146.

Conley C. and Smoller J.A. (1972b), Viscosity matrices for two-dimensional nonlinear hyperbolic systems, II, Amer. J. Math. 94, 631–650.

Conlon J. (1980), A theorem in ordinary differential equations with an application to hyperbolic conservation laws, Adv. in Math. 35, 1–18.

Conway E. and Smoller J.A. (1966), Global solutions of the Cauchy problem
    for quasilinear first order equations in several space variables, Comm. Pure
    Appl. Math. 19, 95–105.

Coquel F. and Perthame B. (1998), Relaxation of energy and approximate Rie-
    mann solvers for general pressure laws in fluid dynamics, SIAM J. Numer.
    Anal. 35, 2223–2249.

Corli A. (1997), Asymptotic analysis of contact discontinuities, Ann. Mat. Pura
    Appl. 173, 163–202.

Corli A. (1999), Non-characteristic phase boundaries for general systems of con-
    servation laws, Ital. J. Pure Appl. Math. 6, 43–62.

Corli A. and Sablé-Tougeron M. (1997a), Stability of contact discontinuities under
    perturbations of bounded variation, Rend. Sem. Mat. Univ. Padova 97, 35–
    60.

Corli A. and Sablé-Tougeron M. (1997b), Perturbations of bounded variation of
    a strong shock wave, J. Differential Equations 138, 195–228.

Corli A. and Sablé-Tougeron M. (2000), Kinetic stabilization of a nonlinear sonic
    phase boundary, Arch. Rational Mech. Anal. 152, 1–63.

Correia J.M. and LeFloch P.G. (1998), Nonlinear diffusive-dispersive limits for
    multidimensional conservation laws, in "Advances in nonlinear partial differ-
    ential equations and related areas" (Beijing, 1997), World Sci. Publishing,
    River Edge, NJ, pp. 103–123.

Correia J.M., LeFloch P.G., and Thanh M.D. (2002), Hyperbolic conservation
    laws with Lipschitz continuous flux-functions. The Riemann problem, Bol.
    Soc. Bras. Mat. 32, 271–301.

Courant R. and Friedrichs K.O. (1948), *Supersonic flows and shock waves*, Inter-
    science Publishers Inc., New York, N.Y.

Crasta G. and LeFloch P.G. (2002), A class of non-strictly hyperbolic and non-
    conservative systems, Comm. Pure Appl. Anal., to appear.

Crandall M. (1972), The semi-group approach to first-order quasilinear equations
    in several space dimensions, Israel J. Math. 12, 108–132.

Dafermos C.M. (1969), The mixed initial-boundary value problem for the equa-
    tions of nonlinear one-dimensional visco-elasticity, J. Differential Equations
    6, 71–86.

Dafermos C.M. (1972), Polygonal approximations of solutions of the initial value
    problem for a conservation law, J. Math. Anal. Appl. 38, 33–41.

Dafermos C.M. (1973a), The entropy rate admissible criterion for solutions of
    hyperbolic conservation laws, J. Differential Equations 14, 202–212.

Dafermos C.M. (1973b), Solution for the Riemann problem for a class of hyper-
    bolic systems of conservation laws, Arch. Rational Mech. Anal. 52, 1–9.

Dafermos C.M. (1977a), Generalized characteristics and the structure of solutions
    of hyperbolic conservation laws, Indiana Univ. Math. J. 26, 1097–1119.

Dafermos C.M. (1977b), Generalized characteristics in hyperbolic conservation
    laws: a study of the structure and the asymptotic behavior of solutions,

in "Nonlinear Analysis and Mechanics: Heriot-Watt symposium", ed. R.J. Knops, Pitman, London, Vol. 1, pp. 1–58.

Dafermos C.M. (1982a), Generalized characteristics in hyperbolic conservation laws, Arch. Rational Mech. Anal. 107, 127–155.

Dafermos C.M. (1982b), Global smooth solutions to initial value problems for equations of one-dimensional nonlinear thermo-elasticity, SIAM J. Math. Anal. 13, 397–408.

Dafermos C.M. (1983), Hyperbolic systems of conservation laws, in "Systems of nonlinear partial differential equations" (Oxford, 1982), NATO Adv. Sci. Inst. Ser. C: Math. Phys. Sci., Vol. 111, Reidel, Dordrecht-Boston, Mass., pp. 25–70.

Dafermos C.M. (1984), Conservation laws without convexity, in "Trends and applications of pure mathematics to mechanics" (Palaiseau, 1983), Lecture Notes in Phys., Vol. 195, Springer, Berlin-New York, pp. 20–24.

Dafermos C.M. (1985a), Regularity and large time behavior of solutions of a conservation law without convexity, Proc. Roy. Soc. Edinburgh 99A, 201–239.

Dafermos C.M. (1985b), Dissipation, stabilization and the second law of thermodynamics, in "Thermodynamics and constitutive equations" (Nato, 1982), Lecture Notes in Phys., Vol. 228, Springer, Berlin-New York, pp. 44–88.

Dafermos C.M. (1986), Quasilinear hyperbolic systems with involutions, Arch. Rational Mech. Anal. 94, 373–389.

Dafermos C.M. (1987), Estimates for conservation laws with little viscosity, SIAM J. Math. Anal. 18, 409–421.

Dafermos C.M. (1989), Admissible wave fans in nonlinear hyperbolic systems, Arch. Rational Mech. Anal. 106, 243–260.

Dafermos C.M. (2000), *Hyperbolic conservation laws in continuum physics*, Grundlehren Math. Wissenschaften Series, Vol. 325, Springer Verlag.

Dal Maso G., LeFloch P.G., and Murat F. (1990), Definition and weak stability of nonconservative products, Preprint No. 272, CMAP, Ecole Polytechnique, France.

Dal Maso G., LeFloch P.G., and Murat F. (1995), Definition and weak stability of nonconservative products, J. Math. Pure Appl. 74, 483–548.

Dias J.-P. and Figueira M. (2000), Blow-up and global existence of a weak solution for a sine-Gordon type quasilinear wave equation, Boll. Unione Mat. Ital. Sez. B Artic. Ric. Mat. 8, 739–750.

Dieudonné J. (1960), *Foundations of modern analysis*, Academic Press, New York.

DiPerna R.J. (1973), Existence in the large for nonlinear hyperbolic conservation laws, Arch. Rational Mech. Anal. 52, 244–257.

DiPerna R.J. (1976), Singularities of solutions of nonlinear hyperbolic systems of conservation laws, Arch. Rational Mech. Anal. 60, 75–100.

DiPerna R.J. (1977), Decay of solutions of hyperbolic systems of conservation laws with a convex extension, Arch. Rational Mech. Anal. 64, 1–46.

DiPerna R.J. (1979a), The structure of solutions to hyperbolic conservation laws, in "Nonlinear analysis and mechanics: Heriot-Watt Symposium", Vol. IV, Res. Notes in Math., Vol. 39, Pitman, Boston, Mass.-London, pp. 1–16.

DiPerna R.J. (1979b), Uniqueness of solutions to hyperbolic conservation laws, Indiana Univ. Math. J. 28, 137–188.

DiPerna R.J. (1979c), The structure of solutions to hyperbolic conservation laws, in "Nonlinear analysis and mechanics": Heriot-Watt Symposium, Vol. IV, pp. 1–16, Res. Notes in Math., 39, Pitman, Boston, Mass.-London.

DiPerna R.J. (1985), Measure-valued solutions to conservations laws, Arch. Rational Mech. Anal. 88, 223–270.

DiPerna R.J. (1987), Nonlinear conservative system, in "Differential equations and mathematical physics" (Birmingham, Ala., 1986), Lecture Notes in Math. 1285, Springer, Berlin, pp. 99–109.

Ericksen J.L. (1991), *Introduction to the thermodynamics of solids*, Chapman and Hall.

Evans C. (1990), *Weak convergence methods for nonlinear partial differential equations*, CBMS Regional Conference Ser. in Math., Vol. 74, Amer. Math. Soc.

Evans C. and Gariepy R.F. (1992), *Measure theory and fine properties of functions*, Studies in Advanced Mathematics, CRC Press, Boca Raton, FL.

Fan H.-T. (1992), A limiting "viscosity" approach to the Riemann problem for materials exhibiting a change of phase II, Arch. Rational Mech. Anal. 116, 317–337.

Fan H.-T. (1993a), Global versus local admissibility criteria for dynamic phase boundaries, Proc. Roy. Soc. Edinburgh 123A, 927–944.

Fan H.-T. (1993b), A vanishing viscosity approach on the dynamics of phase transitions in van der Waals fluids, J. Differential Equations 103, 179–204.

Fan H.-T. (1993c), One-phase Riemann problem and wave interactions in systems of conservation laws of mixed type, SIAM J. Math. Anal. 24, 840–865.

Fan H.-T. (1998), Traveling waves, Riemann problems and computations of a model of the dynamics of liquid/vapor phase transitions, J. Differential Equations 150, 385–437.

Fan H.-T. and Slemrod M. (1993), The Riemann problem for systems of conservation laws of mixed type, in "Shock induces transitions and phase structures in general media", R. Fosdick, E. Dunn, and H. Slemrod ed., IMA Vol. Math. Appl. Vol. 52, Springer-Verlag, New York, pp. 61–91.

Federer H. (1969), *Geometric measure theory*, Springer-Verlag, New York.

Filippov A.F. (1960), Differential equations with discontinuous right-hand side, (in Russian), Math USSR Sb. 51, 99–128. English transl. in A.M.S. Transl., Ser. 2, 42, 199–231.

Folland G.B., *Real analysis*, Wiley Interscience New York, N.Y., 1984.

Fonseca I. (1989a), Interfacial energy and the Maxwell rule, Arch. Rational Mech. Anal. 107, 195–223.

Fonseca I. (1989b), Phase transitions of elastic solid materials, Arch. Rational Mech. Anal. 106, 63–95.

Forestier A. and LeFloch P.G. (1992), Multivalued solutions to some nonlinear and non-strictly hyperbolic systems, Japan J. Indus. Appl. Math. 9, 1–23.

Foy L.R. (1964), Steady state solutions of conservation with small viscosity terms, Comm. Pure Appl. Math. 17, 177–188.

Freistuhler H. (1996), Stability of nonclassical shock waves, in "Hyperbolic problems: theory, numerics, applications" (Stony Brook, NY, 1994), World Sci. Publishing, River Edge, NJ, pp. 120–129.

Freistuhler H. (1998), Some results on the stability of nonclassical shock waves, J. Partial Differential Equations 11, 25–38.

Freistuhler H. and Pitman E. (1995), A numerical study of a rotationally degenerate hyperbolic system. II. The Cauchy problem, SIAM J. Numer. Anal. 32, 741–753.

Friedrichs K.O. (1948), Nonlinear hyperbolic differential equations in two independent variables, Amer. J. Math. 70, 555–588.

Friedrichs K.O. and Lax P.D. (1971), Systems of conservation laws with a convex extension, Proc. Nat. Acad. Sci. USA 68, 1686–1688.

Gavrilyuk S. and Gouin H. (1999), A new form of governing equations of fluids arising from Hamilton's principle, Internat. J. Engrg. Sci. 37, 1495–1520.

Gavrilyuk S. and Gouin H. (2000), Symmetric form of governing equations for capillary fluids, in "Trends in applications of mathematics to mechanics" (Nice, 1998), Monogr. Surv. Pure Appl. Math., Vol. 106, Chapman & Hall/CRC, Boca Raton, FL, pp. 306–311.

Gilbarg D. (1951), The existence and limit behavior of shock layers, Amer. J. Math. 73, 256–274.

Glimm J. (1965), Solutions in the large for nonlinear hyperbolic systems of equations, Comm. Pure Appl. Math. 18, 697–715.

Glimm J., Klingenberg C., McBryan, Plohr B., Sharp D., and Yaniv S. (1985), Front tracking and two-dimensional Riemann problems, Adv. Appl. Math. 6, 259–290.

Glimm J. and Lax P.D. (1970), *Decay of solutions to nonlinear hyperbolic conservation laws*, Mem. Amer. Math. Soc. 101.

Goatin P. and LeFloch P.G. (2001a), Sharp L^1 stability estimates for hyperbolic conservation laws, Portugal. Math. 58, 77–120.

Goatin P. and LeFloch P.G. (2001b), Sharp L^1 continuous dependence of solutions of bounded variation for hyperbolic systems of conservation laws, Arch. Rational Mech. Anal. 157, 35–73.

Goatin P. and LeFloch P.G. (2002), L^1 continuous dependence of entropy solutions for the compressible Euler equations, Preprint.

Godunov S. (1987), Lois de conservation et intégrales d'énergie des équations hyperboliques, in "Nonlinear Hyperbolic Problems", Proc. St. Etienne 1986, C. Carasso et al. ed., Lecture Notes in Mathematics 1270, Springer Verlag, Berlin, pp. 135–149.

Goodman J.B. (1986), Nonlinear asymptotic stability of viscous shock profiles for conservation laws, Arch. Rational Mech. Anal. 95, 325–344.

Goodman J.B. and Lax P.D. (1988), On dispersive difference schemes, Comm. Pure Appl. Math. 41, 591–613.

Goodman J.B., Kurganov A., and Rosenau P. (1999), Breakdown in Burgers-type equations with saturating dissipation fluxes, Nonlinearity 12, 247–268.

Goodman J.B. and LeVeque R.J (1988), A geometric approach to high-resolution TVD schemes, SIAM J. Numer. Anal. 25, 268–284.

Goodman J.B. and Majda A. (1985), The validity of the modified equation for nonlinear shock waves, J. Comput. Phys. 58, 336–348.

Gouin H. and Gavrilyuk S. (1999), Hamilton's principle and Rankine-Hugoniot conditions for general motions of mixtures, Meccanica 34, 39–47.

Greenberg J.M. (1971), On the elementary interactions for the quasilinear wave equations, Arch. Rational Mech. Anal. 43, 325–349.

Grinfeld M. (1989), Non-isothermal dynamic phase transitions, Quart. Appl. Math. 47, 71–84.

Guckenheimer J. and Holmes P. (1983), *Nonlinear oscillations, dynamical systems and bifurcations of vector fields*, Applied Math. Sc., Vol. 42, Springer Verlag, New York.

Gurtin M.E. (1984), On a theory of phase transitions with interfacial energy, Arch. Rational Mech. Anal. 87, 187–212.

Gurtin M.E. (1993a), The dynamics of solid-solid interfaces. 1. Coherent interfaces, Arch. Rational Mech. Anal. 123, 305–335.

Gurtin M.E. (1993b), *Thermodynamics of evolving phase boundaries in the plane*, Oxford Univ. Press.

Hagan R. and Serrin J. (1984), One-dimensional shock layers in Korteweg fluids, in "Phase transformations and material instabilities in solids" (Madison, Wis., 1983), Publ. Math. Res. Center Univ. Wisconsin, Vol. 52, Academic Press, Orlando, FL, pp. 113–127.

Hagan R. and Serrin J. (1986), Dynamic changes in phase in a van der Waals fluid, in "New perspectives in thermodynamics", Proceedings, Springer Verlag, Berlin, pp. 241–260.

Hagan R. and Slemrod M. (1983), The viscosity-capillarity admissibility criterion for shocks and phase transitions, Arch. Rational Mech. Anal. 83, 333–361.

Hale J.K. (1969), *Ordinary differential equations*, Wiley-Interscience, New York, N.Y.

Harten A. (1983), On the symmetric form of systems of conservation laws with entropy, J. Comput. Phys. 49, 151–164.

Harten A., Lax P.D., Levermore C.D., and Morokoff W.J. (1998), Convex entropies and hyperbolicity for general Euler equations, SIAM J. Numer. Anal. 35, 2117–2127.

Hartman P. (1964), *Ordinary differential equations*, Wiley, New York, N.Y.

Hattori H. (1986a), The Riemann problem for a van der Waals fluid with the entropy rate admissibility criterion: isothermal case, Arch. Rational Mech. Anal. 92, 246–263.

Hattori H. (1986b), The Riemann problem for a van der Waals fluid with the entropy rate admissibility criterion: non-isothermal case, J. Differential Equations 65, 158–174.

Hattori H. (1998), The Riemann problem and the existence of weak solutions to a system of mixed-type in dynamic phase transition, J. Differential Equations 146, 287–319.

Hattori H. (2000), The entropy rate admissibility criterion and the entropy condition for a phase transition problem: The isothermal case, SIAM J. Math. Anal. 31, 791–820.

Hattori H. and Mischaikow K. (1991), A dynamical system approach to a phase transition problem, J. Differential Equations 94, 340–378.

Hayes B.T. and LeFloch P.G. (1996a), Nonclassical shocks and kinetic relations : strictly hyperbolic systems, Preprint No. 357, Ecole Polytechnique, France.

Hayes B.T. and LeFloch P.G. (1996b), Measure-solutions to a strictly hyperbolic system of conservation laws, Nonlinearity 9, 1547–1563.

Hayes B.T. and LeFloch P.G. (1997), Nonclassical shocks and kinetic relations : scalar conservation laws, Arch. Rational Mech. Anal. 139, 1–56.

Hayes B.T. and LeFloch P.G. (1998), Nonclassical shocks and kinetic relations : finite difference schemes, SIAM J. Numer. Anal. 35, 2169–2194.

Hayes B.T. and LeFloch P.G. (2000), Nonclassical shocks and kinetic relations : strictly hyperbolic systems, SIAM J. Math. Anal. 31, 941–991.

Hayes B.T. and Shearer M. (1999), Undercompressive shocks for scalar conservation laws with non-convex fluxes, Proc. Royal Soc. Edinburgh 129A, 717–732.

Hertzog A. and Mondoloni A. (2002), Existence of a weak solution for a quasilinear wave equation with boundary condition, Comm. Pure Appl. Anal. 1, 191–219.

Hoff D. and Khodja M. (1993), Stability of coexisting phases for compressible van der Waals fluids, SIAM J. Appl. Math. 53, 1–14.

Hoff D. and Smoller J. (1985), Error bounds for Glimm difference approximations for scalar conservation laws, Trans. Amer. Math. Soc. 289, 611–645.

Hopf E. (1950), The partial differential equation $u_t + uu_x = \mu u_{xx}$, Comm. Pure Appl. Math. 3, 201–230.

Hörmander L. (1976), *Linear partial differential operators*, Springer Verlag, New York, Berlin.

Hörmander L. (1997), *Lectures on nonlinear hyperbolic differential equations*, Mathématiques & Applications, No. 26. Springer-Verlag, Berlin.

Hou T.Y. and LeFloch P.G. (1994), Why nonconservative schemes converge to wrong solutions: Error analysis, Math. of Comput. 62, 497–530.

Hou T.Y., LeFloch P.G., and Rosakis P. (1999), A level-set approach to the computation of twinning and phase-transition dynamics, J. Comput. Phys. 150, 302–331.

Hou T.Y. and Lax P.D. (1991), Dispersion approximation in fluid dynamics, Comm. Pure Appl. Math. 44, 1–40.

Hsiao L. (1980), The entropy rate admissibility criterion in gas dynamics, J. Differential Equations 38, 226–238.

Hsiao L. (1990a), Uniqueness of admissible solutions of the Riemann problem for a system of conservation laws of mixed type, J. Differential Equations 86, 197–233.

Hsiao L. (1990b), Admissibility criteria and admissible weak solutions of Riemann problems for conservation laws of mixed type: a summary, in "Nonlinear evolution equations that change type", IMA Vol. Math. Appl., 27, Springer, New York, pp. 85–88.

Hsiao L. and deMottoni P. (1990), Existence and uniqueness of Riemann problem for nonlinear systems of conservation laws of mixed type, Trans. Amer. Math. Soc. 322, 121–158.

Hsieh D.Y. and Wang X.P. (1997), Phase transitions in van der Waals fluid, SIAM J. Appl. Math. 57, 871–892.

Hu J.X. and LeFloch P.G. (2000), L^1 continuous dependence property for systems of conservation laws, Arch. Rational Mech. Anal. 151, 45–93.

Hughes T.J.R., Kato T., and Marsden J.E. (1977), Well-posed quasilinear second-order hyperbolic systems with applications to nonlinear elastodynamics and general relativity, Arch. Rational Mech. Anal. 63, 273–294.

Hurley J.M. and Plohr B.J. (1995), Some effects of viscous terms on Riemann problem solutions, Mat. Contemp. 8, 203–224.

Hwang H.C. (1996), A front tracking method for regularization-sensitive shock waves, Ph. D. Thesis, State Univ. New York, Stony Brook.

Iguchi T. and LeFloch P.G. (2002), Existence theory for hyperbolic systems of conservation laws: general flux-functions, in preparation.

Isaacson E., Marchesin D., Palmeira C.F., and Plohr B. (1992), A global formalism for nonlinear waves in conservation laws, Comm. Math. Phys. 146, 505–552.

Isaacson E., Marchesin D. and Plohr B. (1990), Transitional waves for conservation laws, SIAM J. Math. Anal. 21, 837–866.

Isaacson E. and Temple B. (1992), Nonlinear resonance in systems of conservation laws, SIAM J. Appl. Math. 52, 1260–1278.

Isaacson E. and Temple B. (1995), Convergence of the 2×2 Godunov for a general resonant nonlinear balance law, SIAM J. Appl. Math. 55, 625–640.

Ishii M., *Thermofluid dynamics theory of two-fluid flows*, Eyrolles Press.

Jacobs D., MacKinney W., and Shearer M. (1995), Traveling wave solutions of the modified Korteweg-De-Vries Burgers equation, J. Differential Equations 116, 448–467.

James R.D. (1979), Co-existence of phases in the one-dimensional static theory of elastic bars, Arch. Rational Mech. Anal. 72, 99–140.

James R.D. (1980), The propagation of phase boundaries in elastic bars, Arch. Rational Mech. Anal. 73, 125–158.

Jenssen H.K. (2000), Blow-up for systems of conservation laws, SIAM J. Math. Anal. 31, 894–908.

Jin S. (1995), Numerical integrations of systems of conservation laws of mixed type, SIAM J. Appl. Math. 55, 1536–1551.

Jin S. and Liu J.G. (1996), Oscillations induced by numerical viscosities, Fourth Workshop on Partial Differential Equations (Rio de Janeiro, 1995). Mat. Contemp. 10, 169–180.

John F. (1974), Formation of singularities in one-dimensional nonlinear wave propagation, Comm. Pure Appl. Math. 27, 377–405.

John F. (1975), *Partial differential equations*, Springer Verlag, New York.

Joseph K.T. and LeFloch P.G. (1999), Boundary layers in weak solutions to hyperbolic conservation laws, Arch. Rational Mech Anal. 147, 47–88.

Joseph K.T. and LeFloch P.G. (2002a), Boundary layers in weak solutions of hyperbolic conservation laws II. Self-similar vanishing diffusion limits, Comm. Pure Applied Anal. 1, 51–76.

Joseph K.T. and LeFloch P.G. (2002b), Boundary layers in weak solutions of hyperbolic conservation laws III. Self-similar relaxation limits, Portugaliae Math., to appear.

Kato T. (1975), Quasilinear equations of evolutions with applications to partial differential equations, Lectures Notes Math. 448, Springer Verlag, pp. 25—70.

Keyfitz B.L. (1971), Solutions with shocks: an example of an L^1 contractive semi-group, Comm. pure Appl. Math. 24, 125–132.

Keyfitz B.L. (1986), The Riemann problem for non-monotone stress-strain functions: a "hysteresis" approach, Lect. Appl. Math. 23, 379–395.

Keyfitz B.L. (1991), Admissibility conditions for shocks in conservation laws that change type, SIAM J. Math. Anal. 22, 1284–1292.

Keyfitz B.L. (1995), A geometric theory of conservation laws which change type, Z. Angew. Math. Mech. 75, 571–581.

Keyfitz B.L. and Kranzer H.C. (1978), Existence and uniqueness of entropy solutions to the Riemann problem for hyperbolic systems of two nonlinear conservation laws, J. Differential Equations 27, 444–476.

Keyfitz B.L. and Kranzer H.C. (1979), A system of non-strictly hyperbolic conservation laws arising in elasticity theory, Arch. Rational Mech. Anal. 72, 219–241.

Keyfitz B.L. and Mora C.A. (2000), Prototypes for nonstrict hyperbolicity in conservation laws, in "Nonlinear PDE's, dynamics and continuum physics" (South Hadley, MA, 1998), Contemp. Math., 255, Amer. Math. Soc., Providence, R.I., pp. 125–137.

Klingenberg C. and Plohr B. (1991), An introduction to front tracking, in "Multidimensional hyperbolic problems and computations" (Minneapolis, MN, 1989), IMA Vol. Math. Appl., Vol. 29, Springer Verlag, New York, pp. 203–216.

Knowles J. (1995), Dynamic thermoelastic phase transitions, Internat. J. Solids Structures 32, 2703–2710.

Knowles J. (2002), Impact induced tensile waves in a rubberlike material, SIAM J. Appl. Math., to appear.

Kohn R.V. and Müller S. (1994), Surface energy and microstructure in coherent phase transitions, Comm. Pure Appl. Math. 47, 405–435.

Kondo C. and LeFloch P.G. (2001), Measure-valued solutions and well-posedness of multi-dimensional conservation laws in a bounded domain, Portugaliae Math. 58, 171–194.

Kondo C. and LeFloch P.G. (2002), Zero-diffusion dispersion limits for hyperbolic conservation laws, SIAM J. Math. Anal., to appear.

Korteweg D.J. (1901), Sur la forme que prennent les équations du mouvement des fluides, (in French), Arc. Neerl. Sc. Exactes Nat. 6, 1–24.

Kruzkov S. (1970), First-order quasilinear equations with several space variables, (in Russian), Mat. USSR Sb. 123, 228–255; English Transl. in Math. USSR Sb. 10, 217–243.

Landau L.D. and Lifshitz E.M. (1959), Fluid mechanics, Pergamon Press, New York and Oxford.

Lax P.D. (1954), Weak solutions of nonlinear hyperbolic equations and their numerical computation, Comm. Pure Appl. Math. 7, 159–193.

Lax P.D. (1957), Hyperbolic systems of conservation laws II, Comm. Pure Appl. Math. 10, 537–566.

Lax P.D. (1961), On the stability of difference approximations of dissipative type for hyperbolic equations with variables coefficients, Comm. Pure Appl. Math. 14, 497–520.

Lax P.D. (1964), Development of singularities of solutions of nonlinear hyperbolic partial differential equations, J. Math. Phys. 45, 611–613.

Lax P.D. (1971), Shock waves and entropy, in "Contributions to Functional Analysis", ed. E.A. Zarantonello, Academic Press, New York, pp. 603–634.

Lax P.D. (1972), The formation and decay of shock waves, Amer. Math. Monthly 79, 227–241.

Lax P.D. (1973), Hyperbolic systems of conservation laws and the mathematical theory of shock waves, Regional Conf. Series in Appl. Math. 11, SIAM, Philadelphia.

Lax P.D. (1991), The zero dispersion limit: a deterministic analogue of turbulence, Comm. Pure Appl. Math. 44, 1047–1056.

Lax P.D. and Levermore C.D. (1983), The small dispersion limit of the Korteweg-deVries equation, Comm. Pure Appl. Math. 36, I, 253–290, II, 571–593, III, 809–829.

Lax P.D. and Wendroff B. (1960), Systems of conservation laws, Comm. Pure Appl. Math. 13, 217-237.

Lax P.D. and Wendroff B. (1962) On the stability of difference scheme, Comm. Pure Appl. Math. 15, 363–371.

LeFloch P.G. (1988a), Entropy weak solutions to nonlinear hyperbolic systems in nonconservative form, Comm. Part. Diff. Eqs. 13, 669–727.

LeFloch P.G. (1988b), Explicit formula for scalar conservation laws with boundary condition, Math. Meth. Appl. Sc. 10, 265–287.

LeFloch P.G. (1989a), Entropy weak solutions to nonlinear hyperbolic systems in conservative form, Proc. Intern. Conf. on Hyperbolic problems, ed. J. Ballmann and R. Jeltsch, Note on Num. Fluid Mech., Vol. 24, Viewieg, Braunschweig, pp. 362–373.

LeFloch P.G. (1989b), Shock waves for nonlinear hyperbolic systems in nonconservative form, Institute for Math. and its Appl., Minneapolis, Preprint # 593, 1989.

LeFloch P.G. (1990a), *Sur quelques problèmes hyperboliques nonlinéaires,* (in French), Habilitation à Diriger des Recherches, Université Pierre et Marie Curie, Paris.

LeFloch P.G. (1990b), An existence and uniqueness result for two non-strictly hyperbolic systems, in "Nonlinear evolution equations that change type", ed. B.L. Keyfitz and M. Shearer, IMA Volumes in Math. and its Appl., Vol. 27, Springer Verlag, pp. 126–138.

LeFloch P.G. (1991), Shock waves for nonlinear hyperbolic systems in nonconservative form, Institute for Math. and its Appl., Minneapolis, Preprint No. 593.

LeFloch P.G. (1993), Propagating phase boundaries: formulation of the problem and existence via the Glimm scheme, Arch. Rational Mech. Anal. 123, 153–197.

LeFloch P.G. (1996), Computational Methods for propagating phase boundaries, In "Intergranular and Interphase Boundaries in Materials: iib95", Lisbon, June 1995. Eds. A.C. Ferro, J.P. Conde and M.A. Fortes. Materials Science Forum Vols. 207-209, pp. 509–515.

LeFloch P.G. (1998), Dynamics of solid-solid phase interfaces via a level set approach, Matematica Contemporanea 15, 187–212.

LeFloch P.G. (1999), An introduction to nonclassical shocks of systems of conservation laws, in "An introduction to recent developments in theory and numerics for conservation laws" (Freiburg/Littenweiler, 1997), Lect. Notes Comput. Sci. Eng., Vol. 5, Springer Verlag, Berlin, pp. 28–72.

LeFloch P.G. (2002), in preparation.

LeFloch P.G. and Liu J.-G. (1998), Generalized monotone schemes, extremum paths and discrete entropy conditions, Math. of Comput. 68, 1025–1055.

LeFloch P.G. and Liu T.-P. (1993), Existence theory for nonconservative hyperbolic systems, Forum Math. 5, 261–280.

LeFloch P.G., Mercier J.-M., and Rohde C. (2002), Fully discrete, entropy conservative schemes of arbitrary order, Preprint.

LeFloch P.G. and Mondoloni (2002), in preparation.

LeFloch P.G. and Natalini R. (1999), Conservation laws with vanishing nonlinear diffusion and dispersion, Nonlinear Analysis (Ser. A: Theory Methods) 36, 213–230.

LeFloch P.G. and J.-C. Nedelec (1985), Weighted conservation laws, C. R. Acad. Sc. Paris (Ser. I) 301, 793–796.

LeFloch P.G and Rohde C. (2000), High-order schemes, entropy inequalities, and nonclassical shocks, SIAM J. Numer. Anal. 37, 2023–2060.

LeFloch P.G. and Rohde C. (2001), The zero diffusion-dispersion limit for the Riemann problem, Indiana Univ. Math. J. 50., 1707–1744.

LeFloch P.G. and Thanh M.D. (2000), Nonclassical Riemann solvers and kinetic relations III. A non-convex hyperbolic model for van der Waals fluids, Electron. J. Differential Equations 72, 19 pp.

LeFloch P.G. and Thanh M.D. (2001a), Nonclassical Riemann solvers and kinetic relations. I. An hyperbolic model of phase transitions dynamics, Z. Angew. Math. Phys. 52, 597–619.

LeFloch P.G. and Thanh M.D. (2001b), Nonclassical Riemann solvers and kinetic relations. II. An hyperbolic-elliptic model of phase transition dynamics, Proc. Royal Soc. Edinburgh 131A, 1–39.

LeFloch P.G. and Tzavaras A. (1996), Existence theory for the Riemann problem for nonconservative hyperbolic systems, C.R. Acad. Sc. Paris (Sr. I) 323, 347–352.

LeFloch P.G. and Tzavaras A. (1999), Representation of weak limits and definition of nonconservative products, SIAM J. Math. Anal. 30, 1309–1342.

LeFloch P.G. and Tzavaras A. (2002), Graph-solutions to systems of conservation laws, in preparation.

LeFloch P.G. and Xin Z.-P. (1993a), Uniqueness via the adjoint problems for systems of conservation laws, Comm. Pure Appl. Math. 46, 1499–1533.

LeFloch P.G. and Xin Z.-P. (1993b), Formation of singularities in periodic solutions to gas dynamics equations, Preprint # 287, CMAP, Ecole Polytechnique (France), October 1993.

LeFloch P.G. and Xin Z.-P. (2002), Uniqueness via entropy estimates for systems of conservation laws, in preparation.

LeVeque R.J. (1990), *Numerical methods for conservation laws*, Lectures in Mathematics, ETH Zürich, Birkhäuser Verlag, Basel.

Li T.-T. (1993), *Global classical solutions for quasilinear hyperbolic systems*, Research in Applied Math. Series, Wiley, New York, & Masson, Paris.

Li T.-T. and Kong D.-X. (1999), Explosion des solutions régulières pour les systèmes hyperboliques quasi-linéaires, (in French), C. R. Acad. Sci. Paris (Sr. I) 329, 287–292.

Li T.-T. and Yu W.C. (1985), *Boundary value problems for quasilinear hyperbolic systems*, Duke University, Durham.

Liapidevskii V. (1975), The continuous dependence on the initial conditions of the generalized solutions of the as dynamics systems of equations, USSR Comput. Math. and Math. Phys. 14, 158–167.

Lions P.-L. (1982), *Generalized solutions of Hamilton-Jacobi equations*, Research Notes in Mathematics, No. 69, Pitman (Advanced Publishing Program), Boston, Mass.

Liu T.-P. (1974), The Riemann problem for general 2×2 conservation laws, Trans. Amer. Math. Soc. 199, 89–112.

Liu T.-P. (1975), Existence and uniqueness theorems for Riemann problems, Trans. Amer. Math. Soc. 213, 375–382.

Liu T.-P. (1976), Uniqueness of weak solutions of the Cauchy problem for general 2×2 conservation laws, J. Differential Equations 20, 369–388.

Liu T.-P. (1977), The deterministic version of the Glimm scheme, Comm. Math. Phys. 57, 135-148.

Liu T.-P. (1979), Development of singularities in the nonlinear waves for quasi-linear hyperbolic partial differential equations, J. Differential Equations 33, 92–111.

Liu T.-P. (1981), *Admissible solutions of hyperbolic conservation laws*, Mem. Amer. Math. Soc. 30.

Liu T.-P. (1987), Hyperbolic conservation laws with relaxation, Comm. Math. Phys. 108, 153–175.

Liu T.-P. (1997), Pointwise convergence to shock waves for viscous conservation laws, Comm. Pure Appl. Math. 50, 1113–1182.

Liu T.-P. (2000), *Hyperbolic and viscous conservation laws*, CBMS-NSF Regional Conference Series in Applied Mathematics, Vol. 72. Society for Industrial and Applied Mathematics (SIAM), Philadelphia, PA.

Liu T.-P. and Smoller J. (1980), On the vacuum state for the isentropic gas dynamics equations, Advances Pure Appl. Math. 1, 345–359.

Liu T.-P. and Yang T. (1999a), A new entropy functional for a scalar conservation law, Comm. Pure Appl. Math. 52, 1427–1442.

Liu T.-P. and Yang T. (1999b), L^1 stability of conservation laws with coinciding Hugoniot and characteristic curves, Indiana Univ. Math. J. 48, 237–247.

Liu T.-P. and Yang T. (1999c), Well-posedness theory for hyperbolic conservation laws, Comm. Pure Appl. Math. 52, 1553–1586.

Liu T.-P. and Zumbrun K. (1995a), On nonlinear stability of general undercompressive viscous shock waves, Comm. Math. Phys. 174, 319–345.

Liu T.-P. and Zumbrun K. (1995b), Nonlinear stability of an undercompressive shock for complex Burgers equation, Comm. Math. Phys. 168, 163–186.

Lowengrub J. S., Goodman J., Lee H., Longmire E. K., Shelley M. J., and Truskinovsky L. (1999), Topological transitions in liquid/liquid interfaces, in "Free boundary problems: theory and applications" (Crete, 1997), Res. Notes Math., Vol. 409, Chapman & Hall/CRC, Boca Raton, FL, pp. 221–236.

Lucier B.J. (1985), Error bounds for the methods of Glimm, Godunov and LeVeque, SIAM J. Numer. Anal. 22, 1074–1081.

Lucier B.J. (1986a), A moving mesh numerical method for hyperbolic conservation laws, Math. of Comput. 46, 59–69.

Lucier B.J. (1986b), On non-local monotone difference schemes for scalar conservation laws Math. of Comput. 47, 19–36.

Majda A. (1984), *Compressible fluid flows and systems of conservation laws in several space variables*, Applied Mathematical Sciences, Vol. 53, Springer Verlag, New York, N.Y.

Martel Y. and Merle F. (2001a), Instability of solitons for the critical generalized Korteweg-de Vries equation, Geom. Funct. Anal. 11, 74–123.

Martel Y. and Merle F. (2001b), Asymptotic stability of solitons for subcritical generalized KdV equations, Arch. Rational Mech. Anal. 157, 219–254.

Menikoff R. and Plohr B.J. (1989), The Riemann problem for fluid flow of real materials, Rev. Modern Phys. 61, 75–130.

Mercier J.M. and Piccoli B. (2000), Global continuous Riemann solver for nonlinear elasticity, Arch. Rational Mech. Anal. 156, 89–119.

Métivier G. (2001), Stability of multidimensional shocks, in "Advances in the theory of shock waves", Progr. Nonlinear Differential Equations Appl., Vol. 47, Birkhäuser, Boston, 25–103.

Modica L. (1987), The gradient theory of phase transitions and the minimal interface condition, Arch. Rational Mech. Anal. 98, 123–142.

Moler C. and Smoller J.A. (1970), Elementary interactions in quasi-linear hyperbolic systems, Arch. Rational Mech. Anal. 37, 309–322.

Mondoloni A. (2000), Existence d'une solution faible d'une équation des ondes quasi-linéaire avec conditions aux limites, (in French), Thèse de l'Université de Corse, France.

Morawetz C.S. (1958), A weak solution for a system of equations of elliptic-hyperbolic type, Comm. Pure Appl. Math. 11, 315–331.

Morawetz C.S. (1981), *Lecture on nonlinear waves and shocks*, Tata Institute of Fundamental Research, Bombay.

Murat F. (1978), Compacité par compensation, Ann. Scuola Norm. Sup. Pisa, Sci. Fis. Mat. 5, 489–507.

Murat F. (1981), L'injection du cône positif de H^{-1} dans $W^{-1,q}$ est compacte pour tout $q < 2$, J. Math. Pure Appl. 60, 309-322.

Natalini R. and Tang S.-Q. (2000), Discrete kinetic models for dynamical phase transitions, Comm. Appl. Nonlin. Anal. 7, 1–32.

Nishida T. (1968), Global solution for an initial boundary value problem of a quasilinear hyperbolic system, Proc. Japan Acad. 44, 642–646.

Nishida T. (1978), Nonlinear hyperbolic equations and related topics in fluid dynamics, Publications Mathématiques d'Orsay, No. 78-02, Université de Paris-Sud, Orsay.

Nishida T. and Smoller J.A. (1973), Solutions in the large for some nonlinear hyperbolic conservation laws, Comm. Pure Appl. Math. 26, 183–200.

Nishida T. and Smoller J.A. (1981), Mixed problems for nonlinear conservation laws, J. Differential Equations 71, 93–122.

Ogden R.W. (1984), *Nonlinear elastic deformations*, Ellis Horwood Limited, John Wiley and Sons, Chichester.

Oleinik O. (1957), On the uniqueness of the generalized solution of the Cauchy problem for a nonlinear system of equations occurring in mechanics, (in Russian), Usp. Mat. Nauk (N.S.) 12, 169–176.

Oleinik O. (1963), Discontinuous solutions of nonlinear differential equations, Amer. Math. Soc. Transl. Ser. 26, 95–172.

Pego R.L. (1987), Phase transitions in one-dimensional nonlinear visco-elasticity: Admissibility and stability, Arch. Rational Mech. Anal. 97, 353–394.

Pego R.L. (1989), Very slow phase separation in one dimension, in "PDE's and continuum models of phase transitions" (Nice, 1988), Lecture Notes in Phys., Vol. 344, Springer, Berlin, 216–226.

Pego R.L. and Serre D. (1988), Instability in Glimm's scheme for two systems of mixed type, SIAM Numer. Anal. 25, 965–988.

Pence T.J. (1985), On the asymptotic analysis of traveling shocks and phase boundaries in elastic bars., in "Transactions of the second Army conference on applied mathematics and computing" (Troy, N.Y., 1984), ARO Rep. 85-1, U.S. Army Res. Office, Research Triangle Park, NC, pp. 859–874,

Pence T.J. (1986), On the emergence and propagation of a phase boundary in an elastic bar with a suddenly applied end load, J. Elasticity 16, 3–42.

Pence T.J. (1992), On the mechanical dissipation of solutions to the Riemann problem for impact involving a two-phase elastic material, Arch. Rational Mech. Anal. 117, 1–55.

Pence T.J. (1993), The dissipation topography associated with solutions to a Riemann problem involving elastic materials undergoing phase transitions, in "Shock induced transitions and phase structures in general media", IMA Vol. Math. Appl., Vol. 52, Springer, New York, pp. 169–183.

Perthame B. (1999), An introduction to kinetic schemes for gas dynamics, in "An introduction to recent developments in theory and numerics for conservation laws" (Freiburg/Littenweiler, 1997), Lect. Notes Comput. Sci. Eng., Vol. 5, Springer Verlag, Berlin, pp. 1–27.

Plohr B.J. and Zumbrun K. (1996), Nonuniqueness of solutions of Riemann problems caused by 2-cycles of shock waves, in "Hyperbolic problems: theory, numerics, applications" (Stony Brook, NY, 1994), World Sci. Publishing, River Edge, NJ, pp. 43–51.

Plohr B.J. and Zumbrun K. (1999), Bifurcation of nonclassical viscous shock profiles from the constant state, Comm. Math. Phys. 202, 267–290.

Rauch J. (1986), BV estimates fail for most quasilinear hyperbolic systems in dimensions greater than one, Comm. Math. Phys. 106, 481–484.

Risebro N.H. (1993), A front-tracking alternative to the random choice method, Proc. Amer. Math. Soc. 117, 1125–1139.

Rosakis P. and Knowles J.K. (1997), Unstable kinetic relations and the dynamics of solid-solid phase transitions, J. Mech. Phys. Solids 45, 2055–2081.

Rosakis P. and Knowles J.K. (1999), Continuum models for irregular phase boundary motion in shape-memory tensile bars, Eur. J. Mech. A Solids 18, 1–16.

Rosakis P. (1995), An equal area rule for dissipative kinetics of propagating strain discontinuities, SIAM J. Appl. Math. 55, 100–123.

Rowlinson J.S. and Widom B. (1982), *Molecular theory of capillarity*, Oxford Univ. Press.

Rozdestvenskii and Yanenko N. (1983), *Systems of quasilinear equations and their applications*, A.M.S. Trans. Math. Monograph Series 55, Providence.

Rudin W. (1966), *Real and complex analysis*, Mc Graw-Hill, New York.

Rudin W. (1973), *Functional analysis*, Mc Graw-Hill, New York.

Sablé-Tougeron M. (1998), Stabilité de la structure d'une solution de Riemann à deux grands chocs, (in French), Ann. Univ. Ferrara Sez. 44, 129–172.

Sablé-Tougeron M. (1988), Propagation des singularités faibles en élastodynamique nonlinéaire, (in French), J. Math. Pures Appl. 67, 291–310.

Sablé-Tougeron M. (2000), Non-convexity and hyperbolic phase boundaries, Preprint.

Schaeffer D.G. (1973), A regularity theorem for conservation laws, Adv. in Math. 11, 368–386.

Schatzman M. (1985), Continuous Glimm functionals and uniqueness of solutions of the Riemann problem, Indiana Univ. Math. J. 34, 533–589.

Schatzman M. (1987), Can hyperbolic systems of conservation laws be well-posed in BV?, in "Nonlinear hyperbolic problems" (St. Etienne, 1986), Lecture Notes in Math., Vol. 1270, Springer Verlag, Berlin, pp. 253–264.

Schecter S., Marchesin D., and Plohr B.J. (1996), Structurally stable Riemann solutions, J. Differential Equations 126, 303–354.

Schecter S. and Shearer M. (1989), Riemann problems involving undercompressive shocks, in "PDEs and continuum models of phase transitions" (Nice, 1988), Lecture Notes in Phys., Vol. 344, Springer Verlag, Berlin, pp. 187–200.

Schecter S. and Shearer M. (1990), Undercompressive shocks in systems of conservation laws, in "Nonlinear evolution equations that change type", IMA Vol. Math. Appl., Vol. 27, Springer Verlag, New York, pp. 218–231.

Schecter S. and Shearer M. (1991), Undercompressive shocks for non-strictly hyperbolic conservation laws, J. Dynamics Diff. Equa. 3, 199–271.

Schochet S. (1991a), Glimm's scheme for systems with almost-planar interactions, Comm. Part. Diff. Eqs. 16, 1423–1440.

Schochet S. (1991b), Sufficient conditions for local existence via Glimm's scheme for large BV data, J. Differential Equations 89, 317–354.

Schonbek M.E. (1982), Convergence of solutions to nonlinear dispersive equations, Comm. Partial Differential Equations 7, 959–1000.

Schonbek M.E. and Rajopadhye S.V. (1995), Asymptotic behavior of solutions to the Korteweg-de Vries-Burgers system, Ann. Inst. H. Poincaré Anal. Non Linéaire 12, 425–457.

Schulze S. and Shearer M. (1999), Undercompressive shocks for a system of hyperbolic conservation laws with cubic nonlinearity, J. Math. Anal. Appl. 229, 344–362.

Serrin J. (1979), Conceptual analysis of the classical second law of thermodynamics, Arch. Rational Mech. Anal. 70, 355–371.

Serrin J. (1981), Phase transitions and interfacial layers for van der Waals fluids, in "Recent methods in nonlinear analysis and applications" (Naples, 1980), Liguori, Naples, pp. 169–175.

Serrin J. (1983), The form of interfacial surfaces in Korteweg's theory of phase equilibria, Quart. Appl. Math. 41, 357–364.

Sever M. (1985), Existence in the large for Riemann problems for systems of conservation laws, Trans. Amer. Math. Soc. 292, 375–381.

Sever M. (1988), A class of hyperbolic systems of conservation laws satisfying weaker conditions than genuine nonlinearity, J. Differential Equations 73, 1–29.

Sever M. (1990), The rate of total entropy generation for Riemann problems, J. Differential Equations 87, 115–143.

Sever M. (1992), Separation of solutions of quasilinear hyperbolic systems in the neighborhood of a large discontinuity, Comm. Part. Diff. Eqs. 17, 1165–1184.

Shearer M. (1982), The Riemann problem for a class of conservation laws of mixed type, J. Differential Equations 46, 426–443.

Shearer M. (1983), Admissibility criteria for shock waves solutions of a system of conservation laws of mixed type, Proc. Royal Soc. Edinburgh. 93A, 233–244.

Shearer M. (1986), Non-uniqueness of admissible solutions of the Riemann initial-value problem for a system of conservation laws of mixed type, Arch. Rational Mech. Anal. 93, 45–59.

Shearer M., Schaeffer D., Marchesin D., and Paes-Leme P. (1987), Solution of the Riemann problem for a prototype 2 × 2 system of non-strictly hyperbolic conservation laws, Arch. Rational Mech. Anal. 97, 299–320.

Shearer M. and Yang Y. (1995), The Riemann problem for the p-system of conservation laws of mixed type with a cubic nonlinearity, Proc. Royal Soc. Edinburgh. 125A, 675–699.

Shu C.-W. (1992), A numerical method for systems of conservation laws of mixed type admitting hyperbolic flux splitting, J. Comput. Phys. 100, 424–429.

Slemrod M. (1983a), Admissibility criteria for propagating phase boundaries in a van der Waals fluid, Arch. Rational Mech. Anal. 81, 301–315.

Slemrod M. (1983b), The viscosity-capillarity criterion for shocks and phase transitions, Arch. Rational. Mech. Anal. 83, 333–361.

Slemrod M. (1984a), Dynamic phase transitions in a van der Waals fluid, J. Differential Equations 52, 1–23.

Slemrod M. (1984b), Dynamics of first order phase transitions, in "Phase transformations and material instabilities in solids" (Madison, Wis., 1983), Publ. Math. Res. Center Univ. Wisconsin, 52, Academic Press, Orlando, FL, pp. 163–203.

Slemrod M. (1987), Vanishing viscosity-capillarity approach to the Riemann problem for a van der Waals fluid, in "Non-classical continuum mechanics"

(Durham, 1986), London Math. Soc. Lecture Note Ser., Vol. 122, Cambridge Univ. Press, Cambridge, pp. 325–335.

Slemrod M. (1989), A limiting "viscosity" approach to the Riemann problem for materials exhibiting change of phase, Arch. Rational Mech. Anal. 105, 327–365.

Slemrod M. and Flaherty J.E. (1986), Numerical integration of a Riemann problem for a van der Waals fluid, in "Phase Transformations", E.C. Aifantis and J. Gittus ed., Elsevier Applied Science Publishers, pp. 203–212.

Smoller J.A. (1969), A uniqueness theorem for Riemann problems, Arch. Rational Mech. Anal. 33, 110–115.

Smoller J.A. (1970), Contact discontinuities in quasi-linear hyperbolic systems, Comm. Pure Appl. Math. 23, 791–801.

Smoller J.A. (1983), *Shock waves and reaction diffusion equations*, Springer-Verlag, New York.

Smoller J.A. and Johnson J.L. (1969), Global solutions for an extended class of hyperbolic systems of conservation laws, Arch. Rational Mech. Anal. 32, 169–189.

Smoller J.A. and Shapiro R. (1982), Dispersion and shock-wave structure, J. Differential Equations 44, 281–305.

Smoller J.A., Temple J.B., and Xin Z.-P. (1990), Instability of rarefaction shocks in systems of conservation laws, Arch. Rational Mech. Anal. 112, 63–81.

Swart P.J. and Holmes P. (1992), Energy minimization and the formation of microstructure in dynamic anti-plane shear, Arch. Rational Mech. Anal. 121, 37–85.

Tadmor E. (1984), Numerical viscosity and the entropy condition for conservative difference schemes, Math. of Comput. 43, 369–381.

Tadmor E. (1986), Entropy conservative finite element schemes, in "Numerical methods for Compressible Flows - Finite Difference Element and Volume Techniques", Proc. Winter Annual Meeting, Amer. Soc. Mech. Engin, AMD-Vol. 78, T.E. Tezduyar and T.J.R. Hughes ed., pp. 149–158.

Tadmor E. (1987), The numerical viscosity of entropy stable schemes for systems of conservation laws, Math. of Comput. 49, 91–103.

Tadmor E. (1991), Local error estimates for discontinuous solutions of nonlinear hyperbolic equations, SIAM J. Numer. Anal. 28, 891–906.

Tartar L. (1979), Compensated compactness and applications to partial differential equations, in "Nonlinear analysis and mechanics: Heriot-Watt Symposium", Vol. IV, Res. Notes in Math., Vol. 39, Pitman, Boston, Mass.-London, pp. 136–212.

Tartar L. (1982), Systèmes hyperboliques nonlinéaires, (in French), Goulaouic-Meyer-Schwartz Seminar, 1981-82, No. XVIII, Ecole Polytechnique, France.

Tartar L. (1983), The compensated compactness method applied to systems of conservation laws, in "Systems of Nonlinear Partial Differential Equations", J.M. Ball ed., NATO ASI Series, C. Reidel publishing Col., pp. 263–285.

Temple B. (1982), Global solution of the Cauchy problem for a class of 2×2 non-strictly hyperbolic conservation laws, Adv. in Appl. Math. 3, 335–375.

Temple B. (1985), No L^1-contractive metrics for systems of conservation laws, Trans. Amer. Math. Soc. 288, 471–480.

Temple B. (1987), Degenerate systems of conservation laws, in "Non-strictly hyperbolic conservation laws", B. Keyfitz and H. Kranzer ed., Contemp. Math., Vol. 60, Amer. Math. Soc., Providence, 125–133.

Temple B. (1990a), Weak stability in the global L^p norm for hyperbolic systems of conservation laws, Trans. Amer. Math. Soc. 317, 96–161.

Temple B. (1990b), Sup-norm estimates in Glimm's method, J. Differential Equations 83, 79–84.

Temple B. (1990c), Weak stability in the global L^1-norm for systems of hyperbolic conservation laws, Trans. Amer. Math. Soc. 317, 673–685.

Temple B. and Young R. (1996), The large time stability of sound waves, Comm. Math. Phys. 179, 417–466.

Treloar L.R.G. (1975), *The physics of rubber elasticity*, Oxford Univ. Press.

Trivisa K. (1997), A priori estimate in hyperbolic systems of conservation laws via generalized characteristics, Comm. Part Diff. Equa. 22, 235–268.

Truesdell C. (1991), *A first course in rational continuum mechanics*, Vol. 1, 2nd edition, Academic Press, New York, N.Y.

Truskinovsky L. (1983), Critical nuclei in the van der Waals model, (in Russian), Dokl. Akad. Nauk SSSR 269, 587–592.

Truskinovsky L. (1987), Dynamics of non-equilibrium phase boundaries in a heat conducting nonlinear elastic medium, J. Appl. Math. and Mech. (PMM) 51, 777–784.

Truskinovsky L. (1993), Kinks versus shocks, in "Shock induced transitions and phase structures in general media", R. Fosdick, E. Dunn, and M. Slemrod ed., IMA Vol. Math. Appl., Vol. 52, Springer-Verlag, New York, pp. 185–229.

Truskinovsky L. (1994a), Transition to detonation in dynamic phase changes, Arch. Rational Mech. Anal. 125, 375–397.

Truskinovsky L. (1994b), About the "normal growth" approximation in the dynamical theory of phase transitions, Contin. Mech. Thermodyn. 6, 185–208.

Truskinovsky L. and Zanzotto G. (1995), Finite-scale microstructures and metastability in one-dimensional elasticity, Meccanica 30, 577–589.

Tzavaras A. (1999), Viscosity and relaxation approximations for hyperbolic systems of conservation laws, in "An introduction to recent developments in theory and numerics for conservation laws" (Freiburg/Littenweiler, 1997), Lect. Notes Comput. Sci. Eng., Vol. 5, Springer Verlag, Berlin, pp. 73–122.

Vainchtein A., Healey T., Rosakis P., and Truskinovsky L. (1998), The role of the spinodal region in one-dimensional martensitic phase transitions, Phys. D 115, 29–48.

Volpert A.I. (1967), The space BV and quasilinear equations, Math. USSR Sb. 2, 257–267.

Wendroff B. (1972a), The Riemann problem for materials with non-convex equations of state. I. Isentropic flow, J. Math. Anal. Appl. 38, 454–466.

Wendroff B. (1972b), The Riemann problem for materials with non-convex equations of state. II. General flow, J. Math. Anal. Appl. 38, 640–658.

Wendroff B. (1991), A study of non-uniqueness and instability for convex materials, in "Third International Conference on Hyperbolic Problems", Vol. I & II (Uppsala, 1990), Studentlitteratur, Lund, pp. 957–973.

Whitham G.B. (1974), *Linear and nonlinear waves*, Wiley, New York, N.Y.

Whitham G.B. (1979), *Lecture on wave propagation*, Tata Institute of Fundamental Research, Bombay.

Wu C.C. (1991), New theory of MHD shock waves, in "Viscous profiles and numerical methods for shock waves" (Raleigh, NC, 1990), SIAM, Philadelphia, PA, 1991, pp. 209–236.

Young R. (1993), Sup-norm stability for Glimm's scheme, Comm. Pure Appl. Math. 46, 903–948.

Zhong X., Hou T.Y., and LeFloch P.G. (1996), Computational methods for propagating phase boundaries, J. Comput. Phys. 124, 192–216.

Ziemer W.P. (1989), *Weakly differentiable functions*, Springer Verlag, New York.

Zumbrun K. (1990), Asymptotic behavior of solutions for non-convex conservation laws, Ph.D. Thesis, New York University.

Zumbrun K. (1993), Decay rates for non-convex systems of conservation laws, Comm. Pure Appl. Math. 46, 353–386.